Social-Ecological Transformation

Karl Bruckmeier

Social-Ecological Transformation

Reconnecting Society and Nature

Karl Bruckmeier
National Research University
Higher School of Economics
Moscow, Russia

ISBN 978-1-137-43827-0   ISBN 978-1-137-43828-7 (eBook)
DOI 10.1057/978-1-137-43828-7

Library of Congress Control Number: 2016942407

© The Editor(s) (if applicable) and The Author(s) 2016
The author(s) has/have asserted their right(s) to be identified as the author(s) of this work in accordance with the Copyright, Designs and Patents Act 1988.
This work is subject to copyright. All rights are solely and exclusively licensed by the Publisher, whether the whole or part of the material is concerned, specifically the rights of translation, reprinting, reuse of illustrations, recitation, broadcasting, reproduction on microfilms or in any other physical way, and transmission or information storage and retrieval, electronic adaptation, computer software, or by similar or dissimilar methodology now known or hereafter developed.
The use of general descriptive names, registered names, trademarks, service marks, etc. in this publication does not imply, even in the absence of a specific statement, that such names are exempt from the relevant protective laws and regulations and therefore free for general use.
The publisher, the authors and the editors are safe to assume that the advice and information in this book are believed to be true and accurate at the date of publication. Neither the publisher nor the authors or the editors give a warranty, express or implied, with respect to the material contained herein or for any errors or omissions that may have been made.

Cover illustration: © RooM the Agency / Alamy Stock Photo

Printed on acid-free paper

This Palgrave Macmillan imprint is published by Springer Nature
The registered company is Macmillan Publishers Ltd. London

To the memory of Marina

Preface

The theme of this book is the social-ecological transformation of modern society to a sustainable future society. Difficulties in this process are twofold: complex environmental problems for which technological and engineering solutions are insufficient, and complex processes to be organised in the governance of global change or earth system governance. In sustainable development, as the transformation process is usually and inexactly called, a new democratic world order needs to be built to achieve the transformation to sustainability. Ends and means of global transformation interplay in complicated ways. The lack of success and the distortions of the prior sustainability process can be seen as a consequence of the prevailing policy: the neoliberal "green economy" strategy, aiming more at an ecological modernisation of the global economy than at a transformation into a sustainable economic system. In the 2015 summit of the United Nations a new agenda, "Transforming our World: the 2030 Agenda for Sustainable Development", was adopted. This soft policy document shows still the predominant and incoherent sustainability thinking of the past, in normative terms, without adequate knowledge and governance practices—although the terminology of transformation is now in use.

When the global discourse of sustainable development began, about thirty years ago, the nature of the changes on the way to sustainability was not clear. A series of social, political, economic, and environmental changes paves the way to sustainability. Knowledge practices in the scientific and political discourses of sustainable development and global governance need to be reviewed critically to initiate a transformation: specialised environmental research and governmental policies do not create the knowledge, action capacity, empowerment of actors, and transformative agency necessary to achieve sustainability. To build more coherent strategies and provide more realistic information, social-scientific and ecological knowledge of the changes of modern society and modern ecological systems needs to be synthesised. Meanwhile, sustainable development has been reformulated as another "great transformation", using the term created by Karl Polanyi in his historical analysis of the rise of modern capitalism and its market economy in England. Today the term is used for a new, global transformation of modern society: a rupture of path-dependent development of the modern economic world system that is programmed for self-destructive economic growth and growth of resource use.

The social-ecological transformation is not another phase of modernisation, as discussed in theories of reflexive or ecological modernisation. The development of a collective political subject for global governance that can drive the transformation is a complex social process; it is not achieved with the organisation of cooperation of political actors with different interests in the routines of environmental policies at regional, national, and international levels. Transformative governance, rethought as social-ecological transformation, is higher-order governance for regulating long-term social and ecological change. Such regulation deals less with policy planning or the management and restoration of ecosystems and more with attempts to influence indirectly the autonomous processes of social and ecological development and change that cannot be managed, triggering further changes that result, finally, in the transformation of modern society and its relations with nature. On the way to global sustainability, a process of many decades or even some hundred years, a new mode of production is built, in the terminology of social ecology called a new societal metabolism.

This social-ecological process that touches all spheres of society and nature cannot be foreseen in its course. In the process of transforming society, only the near future is visible. The distant future, approached in subsequent phases of transformation, clarifies gradually with the advancing process.

Karl Bruckmeier
Moscow, Russia
December 2015

Acknowledgements

For this book, a result of the reworking of information from many sources, it is difficult to describe the immediate influences from the communication and collaboration with colleagues, friends, and students. The important personal communication to be mentioned here is that in several conferences and workshops, including human ecological conferences in different European countries, and in my teaching and research at the National Research University—Higher School of Economics at Moscow. From the courses and the discussions with students, I benefited for the work with this book, especially by learning to avoid oversimplification of complex themes. The people to be named more concretely include the ones who worked with the manuscript, reading, reviewing, or editing it. Earlier versions of the chapters were read critically by Inna Deviatko, Iva Miranda Pires, Ana Velasco Arranz, Parto Teherani-Krönner, Imre Kovach, and Pekka Salmi. I am very grateful for their comments that helped to clarify the arguments. I want to thank the editorial team and an anonymous reviewer at Palgrave Macmillan for their editing work and the comments that helped to structure the chapters more clearly. For the book project no financial support was provided, but it would not have been possible without interdisciplinary research projects on the environment and natural resource management in which I participated during the past decade, several of them funded by the European Union. At least three of these projects influenced the ideas of the social-ecological theory drafted in the following chapters: the projects FRAP (Development of a Procedural Framework for Action Plans to reconcile conflicts between large vertebrate conservation and the use of biological resources), CORASON (Rural Sustainable Development in the Knowledge Society), and SECOA (Solution to Contrasts in Coastal Areas). These projects helped to rethink the sustainability process as a conflict-based process, as a knowledge process, and as a transformation process.

Contents

1 Introduction: Developing Social-Ecological
  Concepts and Theories   1

2 Interaction of Society and Nature in Sociology   15

3 Interaction of Nature and Society in Ecology   69

4 Sustainability in Social-Ecological Perspective   125

5 Social-Ecological Systems and Ecosystem Services   183

6 Knowledge Transfer Through Adaptive Management
  and Environmental Governance   235

7 Climate Change and Development of
  Coastal Areas in Social-Ecological Perspective   285

8 Transformation of Industrial Energy Systems   337

9	Conclusion: The Coming Crisis of Global Environmental Governance	385
Index		399

Abbreviations

BRICS	Brazil, Russia, India, China, South Africa
ESI	Environmental sustainability index
ESS	Ecosystem services
EVI	Environmental vulnerability index
GIS	Geographical information system
IAASTD	International Assessment of Agricultural Science and Technology for Development
ICZM	Integrated Coastal Zone Management
IPCC	Intergovernmental Panel on Climate Change
IUCN	International Union for Conservation of Nature
LAMM	Large Areas for Marine Management
NIMBY	"Not in my backyard"
OECD	Organisation for Economic Co-operation and Development
SES	Social-ecological system
UN	United Nations

List of Figures

Fig. 2.1 Social-scientific components of a theory of
society–nature interaction 39
Fig. 3.1 Ecological components of a theory of
nature–society interaction 107

List of Tables

Table 2.1	Social-scientific theories for analysing society–nature interaction	32
Table 3.1	Ecological theories of nature–society interaction	75
Table 4.1	Sustainable development—changing interpretations	131
Table 4.2	Transformation to sustainability—research and scientific discourse	158
Table 5.1	Social-ecological systems (SES) in different theoretical constructions	194
Table 5.2	Typology of social-ecological systems in modern society	200
Table 5.3	Coupling of social and ecological systems	201
Table 5.4	Classification of ecosystem services: Millennium Ecosystem Assessment	213
Table 5.5	Ecosystem services and theories related to that	216
Table 6.1	Success and failure of adaptive management	248
Table 7.1	Constructions of climate change	292

1

Introduction: Developing Social-Ecological Concepts and Theories

This book advances a social-ecological theory aiming to generate knowledge for reconnecting nature and society. Social ecology develops as an interdisciplinary science, using knowledge from the social sciences, especially sociology and economics, and from natural-scientific ecology that connects to further knowledge from biology and physics. This theory described in seven chapters is a theory in progress that deals with the problems of connecting heterogeneous concepts and theories to form a new interdisciplinary theory. Inter- and transdisciplinary knowledge syntheses create new possibilities to understand global environmental change and problems resulting from that. Such knowledge syntheses are used in integrated analyses of global change of climate, land use, and biodiversity, for the purpose of identifying pathways of transformation to a future sustainable society.

The social-ecological theory develops in a situation where environmental research and the understanding of environmental problems change. Problems analysed earlier in environmental research—pollution of air, water and soils from industrial production and urbanisation, deforestation, erosion, and desertification—become in the global change perspective parts of more complex problems that interact in manifold ways.

© The Editor(s) (if applicable) and The Authors(s) 2016
K. Bruckmeier, *Social-Ecological Transformation*,
DOI 10.1057/978-1-137-43828-7_1

The problems can no longer be separated or kept separate for purposes of technical problem solution. The development of an interdisciplinary theory requires accompanying epistemological and methodological reflection. To study such complex problems, the new theory of nature and society needs to deal with various difficulties of knowledge synthesis described in the following seven points:

1. *Disciplinary knowledge is generated with incompatible epistemologies and methodologies that block interdisciplinary integration of social- and natural-scientific knowledge.* Clashes between different disciplinary epistemologies often show in such prejudices as Weisz described. A social-scientific prejudice of a complex society and a less complex natural environment clashes with a prejudice of natural scientists to see natural systems as complex and man as a uniform actor disturbing nature (Weisz 2001: 11, 114). Weisz's discussion of such "specialisation syndromes" shows the difficulty to reflect on these in an interdisciplinary language. The disciplinary cultures in which specialised scientific knowledge is produced seem to exclude each other. Sociological or economic knowledge cannot be reformulated in terms of biological or physical knowledge or vice versa. Knowledge from different disciplines needs to be connected by taking into account the epistemic differences between disciplines. Studies of social and ecological systems in sociology and ecology differ in their conceptualisation, in structures, functions, processes, and problems. The construction of an integrated framework of social-ecological systems requires further epistemological, theoretical, and methodological reflection and specific methodologies for the synthesis of knowledge from sociology, ecology, and other disciplines. Such a methodology of theory construction works with bridging concepts and frameworks that help to connect social- and natural-scientific research and knowledge. Many of the concepts and frameworks discussed in the construction of a social-ecological theory of nature–society integration are of this kind, for example, the interdisciplinary concepts of vulnerability, resilience, and sustainability in ecological research on the interaction between modern industrial society and nature. Bridging concepts that mediate between theoretical, disciplinary, and practical discourses

and hybrid concepts (as "social-ecological systems", "socio-natures", or "technological natures") spread in the research on interactions between nature and society, but their epistemological status remains unclear.

2. *The possibilities of joint languages and terminologies for interdisciplinary communication and synthesis need to be clarified.* When empirical knowledge from different disciplines is synthesised, conceptual structuring is necessary. A joint terminology for several disciplines, as for example in systems theory, does not yet generate joint explanations or interpretations and integration of disciplinary research as the applications of systems theory show. Interdisciplinary communication in discourses about nature and society is not always possible through simplification of theoretical arguments. The terminological problems need to be solved by methodologically clarifying the forms and scopes of social-ecological knowledge synthesis. Neither in empirical social-ecological research nor in the theory of interaction of nature and society, the aim is to replace specialised and disciplinary research. The synthesis makes other use of specialised knowledge for other than the original purposes. In the analysis of problems of global environmental change, it needs to be asked how much knowledge is to be synthesised to understand the phenomena studied. Impossible as the integration of all specialised knowledge is, it is still necessary to achieve broader syntheses to tackle the complex problems, systems, and processes of global change. Inadequate generalisations and explanations are widespread in environmental and ecological research, for example, in ascriptions of environmental problems to human nature, ignoring the significance of differences in cultures, societies, and modes of production. Also seems doubtful is the anthropological "overshoot and collapse" hypothesis of Diamond (2005) which transfers knowledge about the collapse of small-scale societies in human history to a potential collapse of the present global society.

3. *The use of hybrid terminologies in interdisciplinary theories of nature and society is not always sufficiently developed in epistemological and methodological terms.* Adopting a hybrid terminology by blending the terms of society and nature, in sociology, geography, and elsewhere, a hybrid concept of "socio-natures" is constructed (Blok 2010). This concept

evokes the question, which disciplines create adequate knowledge for interdisciplinary and theoretical syntheses on nature and society? Other examples are the concepts of "technonature" (White and Wilbert 2010) and the theory of "technological nature" by Kahn (2011) who discusses in a human-ecological perspective the question of the consequences when actual nature is replaced by technological nature created by humans. Here the question is, whether technology is the variable sufficient to represent society in the theoretical construction of hybrids. Kahn's synthesis builds on selected knowledge from studies of new information technology and from psychological theory to show the importance of nature for human life. In the discussion of a theory of social-ecological transformation, warnings have been articulated: conceptual shortcuts and visionary claims cannot supersede complicated theoretical and empirical analyses of societal dynamics and societal relations to nature (Brand 2015: 12). The broader the synthesis becomes across the boundaries of social and natural sciences, the more difficult is the choice of knowledge and concepts. In a broad theoretical synthesis, as in the theory of nature–society interaction, the relevant forms of knowledge and theories need to be reflected more carefully by working with several types of contrasting concepts, abstract concepts, connected concepts, and bridging concepts. The terms of society and nature should not be given up before research and theoretical analysis of the interaction of society and nature in modern society provides sufficient arguments for that. Ways to better interdisciplinary syntheses and concepts seem to analyse more systematically the complexity of societal systems and interactions of social and ecological systems, to synthesise more knowledge and theories, and to use currently ignored knowledge.

4. *The types and forms of theories that can be connected in a theory of society and nature need to be assessed.* Combination and integration of theories seems possible only under special presuppositions, for example, that they deal with complementary phenomena, or that more specific theories can be integrated into more general ones. A theory of nature and society differs from conventional theories in social and environmental research that are limited through their disciplinary specialisation and as competing theories. Interdisciplinary approaches that support a

theory of society and nature develop in human, cultural, social, and political ecology. The phenomena of global social and environmental change require new knowledge syntheses in a science of complexity that can deal with internally complex system types in society and nature. Such a science develops slowly, in interdisciplinary approaches, with examples as climate research or political and social ecology. Epistemological and methodological problems of analysing and reducing complexity are insufficiently discussed in environmental research. This research, justified through the analysis of problems and search for potential solutions, seems to approach limits of dealing with situations of risk, uncertainty, lack of scientific knowledge, and with contradicting diagnoses whether anthropogenic climate change exists or not. In the ecological discourses in science and politics, limits of knowledge are too quickly stated: they are often consequences of disciplinary specialisation, ignoring or selectively using knowledge from other disciplines but not showing the present limits of knowledge.

5. *An interdisciplinary theory of nature–society interaction requires integration of concepts and knowledge from different theories.* Theories available for that purpose differ in forms and aspirations, levels of abstraction, generality, and explanation, as conceptual frameworks and as explanatory theories. The methods of theory construction and synthesis differ between social and natural sciences (for ecology see Ford 2000, for sociology: Ritsert 1988). The social-ecological theory of nature and society is an interdisciplinary theory connecting various social-scientific and ecological concepts and explanations. The expectation is that this theory helps to analyse and explain the global complexity of the interaction of society and nature in the phenomena of global environmental and social change. This theory is formulated so far only with three thematic components of societal relations with nature, societal metabolism, and colonisation of nature. In this book, further knowledge components to be used in this theory are described and discussed:

   – The ontological components of worldviews and paradigms
   – The epistemological questions of knowledge generation, synthesis, and application with the help of the theory

– The methods of theory construction, especially the differences between a holism ex ante, as in the case of systems theory, and a holism ex post, as in the kind of critical theories of society
  – The normative criteria guiding the construction of the theory and its application, including ethical criteria of a theory connecting humans, society, and nature

6. *The separation between two contrasting knowledge cultures of social and natural sciences weakens with the differentiation and specialisation of knowledge production in both spheres and the many epistemologies within them.* This brings further problems in working with universal concepts as society and nature, or abstract concepts as resources, production, and reproduction. Such concepts cannot easily be replaced in interdisciplinary theory; however, they can be critically reflected in search for an adequate theoretical terminology. Resource as overgeneralised notion for objects of consumption has been criticised for an anthropocentric perspective (Freese 1997: 232). The argument of Freese is: the view that resources serve humans, are consumable, are separate and distinct from the consumers becomes doubtful, when the connections of humans to nature are taken into account. Humans cannot be separated from nature insofar as they cannot survive without it. The soil, the air, the water, minerals, plants, and animals are not only resources which humans consume, but they are also parts of the ecosystems that need to be maintained to allow for the further existence of life and of all species. The concept of resources does not reflect the complexity of the functions required for the existence of life and the survival of living beings. This concept is, in spite of its differentiation and classification of many kinds of resources, a reductionist concept. Freese does not find another term, rather supporting a return to less abstract concepts with his preliminary description of resource functions. For other concepts as production and reproduction, there may emerge other problems, for example, that of incompatibility between the biological, ecological, economic, and sociological versions to connect in a theory of nature–society interaction. In the social-ecological theory, the clarification of theoretical concepts requires a systematic reconstruction of

the interactions between humans, society, and nature at different levels of interacting social and ecological systems.
7. *A general epistemology for interdisciplinary synthesis of knowledge is presently not available and it can be doubted that it is meaningful to aim for such an epistemology.* The epistemological discourses in philosophy and theory of science during the twentieth century resulted in many and competing approaches, but do not help much in the interdisciplinary synthesis of social- and natural-scientific knowledge, not even at lower levels of theoretical knowledge synthesis. Arnason's (2003: 64f) search for an explanatory model in social-scientific research ended in the recognition of differences. Whereas the deductive-nomological model of explanation for the natural sciences remains doubtful, functional analyses are seen as a step forward to interdisciplinary theorising, although functionalist explanations remain unclear in their relations to causal explanations and to social interpretations. Further ideas of explanation remain methodologically underdeveloped as well. Contextual explanations, useful because of the reflexive nature of knowledge production in interactions between social sciences and social practice, limit the possibilities of generalisation and explanation. The figurational model of explanation by Elias has long-term societal processes of change in focus, but its historical orientation shows also that the dynamics of explanation are too complex and dependent from heterogeneous cases to allow for law-like generalisations. Arnason's alternative to idealised rational, causal, or systemic constructions of explanation is to accept the interplay of strategies, constraints, non-intended consequences, and adaptive changes as an explanatory mechanism of higher order. Yet, this kind of explanation of higher order seems preliminary and limited to social-scientific knowledge. The older differentiation between interpretation and explanation has become more complex, but still explanatory models in the social sciences differ from the traditional model in the natural sciences. Much less epistemological debate in the social and natural sciences touches the problems of connecting knowledge from different spheres, and models for interdisciplinary theories and explanations are rare. In the future, it seems necessary to work with combinations of several explanatory approaches in knowledge syntheses.

The above-mentioned seven points describing epistemological and methodological difficulties in interdisciplinary theory and knowledge synthesis are important for the theory discussed in this book. Further questions in the construction of this theory refer to problems in the analysis and transfer of knowledge from science to political and managerial practices. These questions are part of the detailed discussion in the chapters of this book, including that of solving environmental problems through transitions to sustainability. Forms of maladaptive social organisation that undermine the conditions for long-term development and sustainability of interactions of society and nature are known from environmental research. However, it is difficult to find consensus on strategies of societal transformation to sustainability. These strategies require to deal with a variety of contradicting processes: satisfaction of human needs for present and future generations, material and cultural reproduction of society and economy, maintenance of the natural resource base and life-supporting ecosystem functions, social integration of societies and combatting social exclusion and poverty, inequality and discrimination, and strengthening of human and citizenship rights in the perspective of ecological citizenship. It seems impossible to clarify such contrasts and contradictions without theoretical analysis and reflection. Maladaptive change of modern society and risks of "overshoot and collapse" have many reasons and causes, such as dysfunctional political and economic structures and processes as well as dysfunctional scientific knowledge production for environmental policy and resource management, for social-ecological regulation and institutional change. Interdisciplinary research, knowledge synthesis, transfer, and cooperation of scientists and practitioners are underused knowledge practices in environmental policy and natural resource management. Although interdisciplinary cooperation and knowledge synthesis is not a panacea, it supports the search for solutions to global environmental problems and shows reasons and causes for the malfunctioning of regulations and institutions that block societal transition to sustainability.

In the process of strengthening interdisciplinary production, transfer, sharing, and application of knowledge for regulating the interface of society and nature, arise problems of justification and legitimation of the knowledge practices. Should one rely on research more than on political and public debates about solutions? How far are other forms than scientific knowledge and research required? How is the cooperation of scientists, political actors, and decision-making to be organised? How can participatory research and transdisciplinary knowledge production be organised? Who has legitimate roles in the scientific and political processes when they are connected? How can powerful actors and interests and power asymmetries be dealt with in public policies? Such questions regarding knowledge production and application require continuous discussion when complex environmental problems and global change are addressed. The following difficulties are included:

- *Difficulties of adaptive governance:* consequences of climate change that are sometimes seen as causes of future conflicts, wars, and civil wars (Welzer 2008; Dyer 2008) and how to deal with them in multi-scale adaptation strategies.
- *Difficulties of transforming the societal metabolism of combined use of material and energy resources:* including transformation of the industrial energy systems based on limited fossil energy sources (Hall and Klitgaard 2012).
- *Difficulties of international policies and global governance:* major pollution sources shifted, since the turn of the millennium, from the older industrial countries to the newly developing and industrialising countries, with consequences as environmental distribution conflicts (Martinez-Alier 1995).
- *Difficulties of transition to sustainability at local, national and global scale:* long-term transformation of societal systems is discussed since the Brundtland report from 1987, but in the international policy processes strategies of transformation remain controversial.
- *Difficulties to reduce exponential growth of the globalised market economy:* in the past decades, there has been a discussion of natural limits to growth (Meadows and Meadows), social limits to growth (Hirsch), zero-growth in ecological economics (Daly 1997), and self-destruction of the growth society (Zinn 1980); however, a broader scientific and public debate about degrowth has developed only since the last decade.
- *Difficulties to identify and form new social subjects for socio-ecological transformation of society:* these include the crisis of new social move-

ments, their lacking knowledge and capacities to initiate societal transformation, and the limits of cooperation between actors with different interests.

These difficulties, addressed in the chapters of the book in the perspective of theoretical knowledge synthesis, require an opening of the social and natural sciences towards inter- and transdisciplinary environmental research ('new knowledge production': Gibbons et al. 1994; Nowotny et al. 2001) and methodological improvements of such approaches. The development of a social-ecological theory is part of the broader processes of inter- and transdisciplinary knowledge generation. The formulation of this theory is carried out in the sequence of chapters as follows:

Chapters 2 and 3 outline a suggestion of an interdisciplinary theory of interaction between society and nature starting from different knowledge sources in social and ecological research, supplementing the components of social-ecological theory developed by the German and Austrian authors of social ecology, that of societal relations of nature, societal metabolism, and colonisation of nature (see Bruckmeier 2013). Chapter 2 describes social-scientific concepts and components from the discourse of theory of society in sociology. Chapter 3 gives a parallel description of ecological components of the theory and follows with the question how does one connect vulnerability, resilience, and sustainability analyses with each other in social-ecological research?

In Chaps. 4 and 5, complementary operational concepts are described and critically discussed to connect the social-ecological theory of society and nature with empirical research and to integrate further knowledge from present environmental research in the theory. Problems of matching empirical research and theoretical reflection in the analysis of coupled social-ecological systems include (in Chap. 4) a critical analysis of the discourse of sustainability and sustainable development. It is asked as to how the debate can be transferred into one about the transformation of global society and economy. Chapter 5 discusses two operational concepts that social ecology inherits from the ecological discourse: the framing concept of social-ecological systems, and the operational concept of ecological and socio-cultural ecosystem services. Both concepts require further epistemological and methodological elaboration for their use in a theory of nature–society interaction. The heuristic notion of social-ecological systems is unclear with regard to its theoretical assumptions and implications about the constitution and interaction of social and ecological systems. The concept of ecosystem services is unclear and controversial with regard to its assumptions about the functions and capacities of ecosystems and social systems. Although the concepts emerged separately, the first in scientific, the second in policy related discourses, they are interconnecting through their complementary use in a broader theory of nature and society.

Chapter 6 discusses problems of knowledge transfer from science to the practices of natural resource management and environmental policy, using the debate about adaptive management and environmental governance as paradigmatic cases. In social-ecological knowledge practices, the processes of research and policy interact and need to be reflected in their interconnections. Specific forms of knowledge integration, connected to the theory nature and society, have been discussed mainly in the discourse of critical theory, and less in the ecological discourse. From critical theory, the social-ecological discourse inherits the question, how does one connect theoretical knowledge and societal practice to initiate transformations of societal systems?

In Chaps. 7 and 8, two important themes of environmental research for the purpose of system transformation are discussed in an exemplary way. Chapter 7 follows the research on the problems of climate change adaptation in coastal areas in the contexts of resilience and sustainability. Chapter 8 takes up the discussion of energy problems and the transformation of industrial energy regimes by way of using renewable energy sources. The basic ideas of developing sustainable energy systems through the use of renewable energy sources are clear: energy cannot be reused—it is lost through the dissipation from higher to lower quality according to thermodynamic physics. Fossil energy resources are limited and cause environmental pollution and climate change. To develop energy systems with renewable sources of energy is a way to deal with the problems by transforming industrial energy systems. In practice, the development and use of renewable energy sources have shown that energy conversion technologies are of a complicated nature, and that renewable energy sources can create unwanted environmental problems and socially unwanted consequences of land use.

Following this thematic outline, the book provides a new look into several approaches and theories in sociology and ecology. A variety of theoretical concepts and approaches are used to build an interdisciplinary theory of nature–society interaction that is connected with empirical research. However, not all knowledge from the disciplines studying nature and society can be synthesised. Therefore, this synthetic theory uses an open model of theory construction; the theory can be connected for specific purposes with further knowledge and theories; and the knowledge integrated can be modified and recombined when the construction of the theory and the synthesis advance. The methodology of comparing, broadening, and combining concepts, frameworks and theories is reflected with regard to the limits of the emerging theory. Empirical research is used to illustrate the requirements of interdisciplinary boundary crossing and theoretical reflection, working towards knowledge synthesis at several levels of research, of concept formulation, and theory development.

References

Arnason, J. P. (2003). *Civilizations in dispute: Historical questions and theoretical traditions*. Leiden: Brill.

Blok, A. (2010). Divided socio-natures: Essays on the co-construction of science, society, and the global environment. PhD thesis, Department of Sociology, University of Copenhagen.

Brand, U. (2015, June 10–12). *How to get out of the multiple crisis? Contours of a critical theory of social-ecological transformation*. Paper presented at the conference The Theory of Regulation in Times of Crises, Paris. Accessed November 10, 2015, from https://www.eiseverywhere.com/retrieveupload.php?

Bruckmeier, K. (2013). *Natural resource use and global change: New interdisciplinary perspectives in social ecology*. Houndmills, Basingstoke: Palgrave Macmillan.

Daly, H. E. (1997). *Beyond growth: The economics of sustainable development*. Boston: Beacon Press.

1 Introduction: Developing Social-Ecological Concepts

Diamond, J. (2005). *Collapse: How societies choose to fail or survive.* London: Penguin Books.

Dyer, G. (2008) *Climate Wars.* Random House Canada.

Ford, E. D. (2000). *Scientific method for ecological research.* Cambridge: Cambridge University Press.

Freese, L. (1997). Environmental connections. *Advances in Human Ecology,* Supplement 1, Part B.

Gibbons, M., Limoges, C., Nowotny, H., Schwatzman, S., Scott, P., & Trow, M. (1994). *The new production of knowledge. The dynamics of science and research in contemporary societies.* London: SAGE.

Hall, C. A. S., & Klitgaard, K. A. (2012). *Energy and the wealth of nations: Understanding the biophysical economy.* New York: Springer.

Kahn, P. (2011). *Technological nature: Adaptation and the future of human life.* Cambridge, MA: MIT Press.

Martinez-Alier, J. (1995) 'Political Ecology, Distributional Conflicts and Economic Incommensurability', *New Left Review,* 211: 70-88.

Nowotny, H., Scott, P., & Gibbons, M. (2001). *Re-thinking science: Knowledge and the public in an age of uncertainty.* Cambridge: Polity Press.

Ritsert, J. (1988). *Gesellschaft: Einführung in den Grundbegriff der Soziologie.* Frankfurt am Main: Campus.

Weisz, H. (2001). Gesellschaft-Natur Koevolution: Bedingungen der Möglichkeit nachhaltiger Entwicklung. Kulturwissenschaftliches Seminar (Dissertation), Humboldt Universität, Berlin.

Welzer, H. (2008). *Klimakriege: Wofür im 21. Jahrhundert getötet wird.* Frankfurt am Main: Fischer.

White, D. F., & Wilbert, C. (Eds.). (2010). *Technonatures: Environments, technologies, spaces and places in the twenty-first century.* Waterloo, ON: Wilfried Laurier University Press.

Zinn, G. (1980). *Die Selbstzerstörung der Wachstumsgesellschaft.* Reinbek: Rowohlt.

2

Interaction of Society and Nature in Sociology

This chapter discusses sociological theories of society. The following chapter covers ecological theories. Both chapters use the same guiding questions in seeking constituents of an interdisciplinary theory of interaction of society and nature:

- How far is the theoretical analysis of *interaction, reproduction, and change of social and ecological systems* developed in the theoretical discourses of sociology and ecology?
- How far is the integration of knowledge from social-scientific and natural-scientific knowledge advanced in these disciplinary discourses towards an *interdisciplinary, social-ecological theory of nature and society*?

The forms and the limits of sociological analyses of nature and society are discussed in the following section. Then follow reflections about possibilities to dissolve the limits and a description of interdisciplinary theories that deliver further knowledge for a social-ecological theory of nature and society. Four influential theories of society are reviewed in their analysis of nature and society in the Appendix.

Sociological Theories and Their Reflection of Society–Nature Interaction

In the sociological discourse on modern society, a series of theories were developed that compete with regard to the explanation of modern society (Ritsert 1988, 2009; Jain 2006; Delanty 2006; Bruckmeier 2015a, 2015b). The following theories of society provide knowledge for the analysis of relations between nature and society:

1. *Classical theories* of society by Marx, Durkheim, and Weber still influence the theoretical debates today, although that influence has lessened. They gave theoretical analyses of the capitalist mode of production and the socio-culturally shaped human and societal relations with nature.
2. *Grand theories*, including the system theories of Parsons and Luhmann as variants of conventional sociological theories of modernity and the theory of communicative action of Habermas and world system theory of Wallerstein as variants of critical theory of modern capitalism, show how reflections on nature and society changed in sociological theory during the twentieth century.
3. *Theories of social transformation* indicate a rupture with the "grand theory" tradition towards the end of the twentieth century. These transformation theories include theories of the post-industrial society (Bell), risk society (Beck), reflexive modernisation (Giddens), post-modern society (Bauman), ecological modernisation and sociology of environmental flows (Mol, Spaargaren), actor-network theory (Callon, Latour, Law), and network society (Castells) in which relations between nature and society are reflected with regard to specific environmental problems.

The three groups include theories of society in their historical sequence. These theories are explicitly formulated as theories of society. More social-scientific theories work with implicit assumptions and ideas about the constitution of modern society, however, hardly in systematically elaborated forms. In the discourse on modern society, it is neither consensus about the importance of societal theory, nor about the development of modern society. The analysis of relations between nature and society is a marginal theme in most sociological theories. In the discourse of critical theory, referring to all theories that connect in one way or the other to the theory of Marx, the theorising of nature was stronger (Biro 2011). In the recent sociological discourse, the influence of critical theories weakened and the boundaries between critical and traditional theories became unclear. Yet, efforts can be observed to renew critical theory, a term adopted also by Beck in the debate of cosmopolitanism, and in attempts to reformulate Luhmann's systems theory as a critical theory of modern capitalism, focusing on the changing normative orders in society (Amstutz and Fischer-Lescano 2013).

Outside the analysis of modern society as a system there developed in sociology specialised analyses of nature and society in human ecology. In the 1920s and 1930s, human ecology had a strong influence on the development of sociology in the USA, namely in the Chicago school of sociology (Park, Burgess, McKenzie), which brought the perspective and the terminology of the newly emerging ecology in the analyses of urban development. This interdisciplinary approach was later outcompeted by the new sociological approaches of symbolic interactionism and theories of social action and systems. Human ecology survived as a heterodox and marginal approach. It was renewed in many countries during the 1970s, with the broadening of environmental research in the social and natural sciences, in the newly emerging environmental sociology, or as a separate interdisciplinary discourse. Human ecology has, in contrast to the new social ecology discussed in this book, not advanced to a systematic theory of interaction between nature and society. The syntheses towards the end of the twentieth century (Freese 1997) structured a mass of empirical knowledge about the biological evolution of humans and environmental problems in human history. This knowledge was too broad to be reproduced in a coherent theory of nature–society interaction. A synthesis of knowledge in human ecology as it adopted the classical critique of political economy from Marxist theory (Schnaiberg 1980) was exceptional. In environmental sociology, the actor-network theory gained some influence (Law and Hassard 1999; Latour 1999; Blok 2010a, b; see also Chap. 7). This theory did not synthesise knowledge about modern society in a theory of society and nature, but remained part of specialised research;

with the term "actants", similar capacities were attributed to ecosystems as described in sociology as action.

Sociological Theory of Society in the Twentieth Century

Sociological theory in the twentieth century was strongly influenced by different forms of grand theory. The important examples are that of Parsons, Luhmann, and Habermas where each attempts a synthesis of prior theories. Grand theory comprises examples from traditional theory (sociological systems theory) and from critical theory (referring to Marx and political economy). In these theories, the analysis of nature–society interaction was a neglected theme as shown in the examples of Luhmann and Habermas (see the Appendix). The theories of social transformation developing towards the end of twentieth century analyse the relations between nature and society more in the form of empirical research on environmental problems, for example, in the theory of risk society by Beck (see the Appendix).

In the development of sociological theory, five trends show the increasing difficulties of analysing nature–society interaction in modern sociology:

1. *Competing interpretations of modern society:* In sociology, there existed some theories influential at certain times, but never a generally accepted theory as in biology's synthetic theory of evolution. The theoretical concepts and analyses of modern society differ strongly in interpreting the development of this society and its relations to nature. In critical theory, the inequalities in modern society are seen as resulting from a global system, the capitalist economic world system that develops with different cultural and political orders at national levels. For the large-scale socio-cultural systems of Western, East Asian, and Indian cultures, the term of civilisation (Arnason 2003) can be used to complement that of society. The sociological systems theories by Parsons and Luhmann as variants of traditional theory ended with the thin concept of world society by Luhmann (see the Appendix). For these theories the analysis of inequality, of the modes of production

and reproduction of economic systems, was less important than for critical theories.

Theoretical concepts for the many facets of modern society include that of Western society, industrial society, bourgeois society, capitalist mode of production, modern world system, world society, civil society, and the more recent terms of post-industrial and post-modern society, risk society, reflexive modernisation, knowledge or information society, and network society. These terms highlight different social, cultural, economic, and political facets rather than providing alternative theoretical conceptions of modern society. The terms become relevant at certain stages of elaborating a theory of society and its relations with nature. Whether modern society is one global system (capitalist mode of production, world system, world society) or a plurality of culturally, socially, or politically different orders and national societies, is still disputed today. The controversy dissolves with the description and explanation of the multi-scale organisation of a global societal system with different cultural and political orders. This modern capitalist world system was built as a global system during the long processes of colonisation and globalisation, with varying political orders and the building of modern nation states and national societies continuing until today.

2. *Dissolution of the contrasts between traditional and critical theory of society:* Critical theory has throughout its history kept the theme of societal relations with nature. With the reinterpretation of critical theory in recent theoretical debates the contours and differences between theories of modern society dissolve. In the earlier discourse of critical theory of the Frankfurt School, "societal relations with nature" was a formula inherited from the early theory of Marx to analyse and reflect the relations between nature and society in philosophical and epistemological terms of historical materialism. This remained a specificity of the Frankfurt School. The political-economic analysis of modern capitalism by Marx remained the influential systems analysis of modern capitalist society in critical theory. With the late theory of communicative action by Habermas and Honneth's theory of recognition, the political-economic systems analysis was given up. In the recent development of critical theory, the discussion of Arnason (1976: 7)

concerning how nature could be integrated into the discourse of critical theory of society was ignored. Arnason describes a cognitive programme in three steps: an anthropological analysis of man–nature relations, a theory of society, and the justice-related normative reflections of social emancipation from relations of domination. This programme follows to a large degree the intentions from older critical theory, but was subsequently given up by Arnason (2003) in his civilisational analysis. In other variants of critical theory, Bourdieu's theory of practice, Touraine's theory of social movements, and the power-centred theoretical reflections of Foucault, nature remained a neglected theme. Also the reflections of the post-socialist condition by Fraser (1997) do not comprise nature among the three main points: the lack of an alternative vision to the neoliberal order, a weakening of social politics and its decoupling from cultural politics, and a decentering of equality claims with rising material inequalities in a marketised society.

Critical theory dissolves in its late forms in a series of theories that develop through varying forms of combining elements from critical and traditional theory. This is done in the theory of risk society by Beck, the theory of reflexive modernisation by Giddens, and the recent critical systems theory (Amstutz and Fischer-Lescano 2013). With these modifications, the discourse of critical theory lost direction and the compass of a guiding idea for the critique of modern society. Although motives of critical analyses of modern society and its systemic conflicts are valid for all of these theories, they do not evolve towards new forms of systems analysis of modern society and its interaction with nature. In other critical discourses nature is a main theme in ecofeminism (Merchant 1980; Mellor 1997; Warren 1997), not in feminism generally (Wallace 1989), and not in theories of postcolonialism (Gandhi 1998), theories of inequality in capitalist societies (Rehbein and Souza 2014), or theories of new normative orders (Forst and Günther 2011).

3. *The decline of grand theory in sociology:* With the theories of Luhmann and Habermas ended the discourse of grand theory in sociology. This end of grand theory coincided with critical debates about the possibility of universal theories and about the necessity of a theory of society in

sociology. Much of the critique of grand theory was formulated with epistemological arguments, as in the discourses of post-structuralism and post-modernism that paved the way for the relativisation and fragmentation of theories. The post-modernist critique of sociological theory argues that grand theory followed doubtful self-descriptions of modern society in the grand narratives of modernity and Western culture. This modernity ended in the twentieth century in a series of wars and system collapses. Sociological theory of society is gradually replaced by a series of specialised and sub-disciplinary theories as the theories of social transformation mentioned above (the last section). The variants of sociological theory that developed simultaneously with post-modernism (Welsch 2003), but in contrast to that, are theories of post-industrial society, risk society, and reflexive modernisation. These theories address the theme of nature and society with sociological knowledge, in different forms and perspectives, but do not aim at a critical and systematic theory of nature–society interaction (see the Appendix).

The lack of success in renewing the theory of modern society and the discourse of critical theory in debates since the 1990s (Miller and Soeffner 1996) supported the trend to devalue society as core theme of sociology. Additionally, social class analyses are devalued in sociology (Clark 1991), which is visible in the theory of communicative action of Habermas and other sociological theories. Urry (2000) suggested giving up the concept of society as core concept in sociology, whereas Jain (2002, 2006) criticised this suggestion. The restructuring of the discourse can be interpreted in different ways:

- As reducing theoretical aspirations in reaction to the complexity of societal relations that cannot be sufficiently explained with knowledge from sociological research
- As assuming that older theories failed or are no longer adequate for present development of modern society
- As an implicit or explicit critique of ideology of the narratives of modernisation that failed

New sociological theories tend towards "theories with descriptive terms" as information society or network society. Beetz (2010) gives an example of a new theory of society with simplified analyses of social organisation and public life. This theory illustrates the new forms of sociological theory without systematic analysis of historically specific societal systems, modes of production and reproduction, and multiple cultural modernities. In theories of modern economy (e.g. in economic sociology), the analysis of society is reduced to partial theories of socio-economic change (Beckert 2009; Stehr 2007). Castells (2000: 10), in the theory of the network society, does not give much theoretical explanation of systemic properties of the new globally networked economy created through globalisation.

4. *Loss of criteria for a theory of modern society:* In the development of the sociological discourse, theoretically complex concepts and precise criteria for the analysis of modern society dissolve, reducing the reflection of societal relations to nature in specialised research fields and in environmental sociology. The guiding interest to explain the system of modern society and its mode of production and reproduction was replaced by simpler analyses describing society in terms of differentiation and integration, social and system integration, and system and lifeworld in sociological theories under the influence of systems theory. Sociological systems theory worked with concepts transferred from the natural sciences, especially from general systems theory in biology. The general theory of action by Parsons, aiming at reintegration of disciplines, did not develop beyond the integration of social-scientific knowledge (Loubser et al. 1976). Luhmann's theory exhibits the rupture with the sociological tradition in his concept transfer from biology and ecology. He adopted the biological theory of autopoietic systems of Maturana and Varela, the terminology of biological theory of evolution as variation and selection, and the term reproduction in its biological sense for species or living systems. With this biological terminology, the synthesis is limited to concept transfer, thus creating another language for the theory of modern society. The use of biological concepts as that of autopoietic or living systems continues in metaphoric and analogical thinking.

5. *New social-scientific research on global social change:* This research includes various thematic fields that develop through specialisation by going away from general theories of modern society. From none of the

research areas mentioned below developed so far attempts to a broader analysis or synthetic theory of nature and society. The main difficulties in the elaboration of encompassing theories are the methodological problems of interdisciplinary knowledge synthesis and the complexity of social relations and problems. The following aspects of this specialised research may influence the interaction between society and nature, but are not yet brought together in integrated theoretical analyses:

- Globalisation or global social-economic change resulting in the networking of national societies and economies supporting growth and intensification of natural resource use
- Technical change with new technologies as the Internet-based development of modern society as information society and new technologies directly transforming nature (genetic modification of living nature, geo-engineering)
- Political change, internationalisation of states, and global governance resulting in a new world order with the emergence of global environmental policies and governance
- Cultural change, especially from research on non-Western societies and post-colonial development, showing the varying influence of modernisation on the countries and cultures of the Global North and South, where the appropriation of nature and use of natural resources do not appear only in manifold cultural views, but in social inequalities of access to resources
- Scientific change as reflected in the discourses of inter- and transdisciplinarity that is seen by proponents of that discourse as "updating" the sociological theory of modernity (Nowotny); nature is not theorised, but interdisciplinary perspectives develop that can lead to new analyses of nature–society interaction
- Global environmental change that affects societal development in manifold ways, especially through climate change and land-use change

The reasons for refraining from attempts to formulate new systematic theories of society are manifold, but not always reflected in this research. The critique of universal and grand theories, of their eurocentrism and neglect of societal development in the Global South, is among these reasons. Furthermore, epistemological reflections tend to see the complexity of global systems and global change processes as making coherent theories, causal explanations, and systematic analyses impossible.

The five trends to describe the development of sociological theory of society indicate some of the difficulties in approaching the analysis of nature–society interaction. In sociological theorising nature appeared as "the opposite of society", excluded from the theory of modern society. Also in the newer theories of post-modernism, if they argue sociologically (Bauman 1992), the interaction of society and nature is not the main theme; these theories reflect phenomena of a changing modernity, in some aspects similar to the theories of Giddens and Beck. Theories relevant for nature–society interaction include the cultural-anthropological theory of Diamond or cultural-ecological theories (Steward, Harris, Vayda, Rapoport), but these focus on historical, local, and non-Western societies. The deficits of sociological theories in theorising nature are inverse to these anthropological and ecological theories. Anthropological analyses of the interaction of humans with nature in the twentieth century included, furthermore, the German tradition of philosophical anthropology (Scheler, Gehlen, Plessner) and the French tradition of critical anthropology (Morin, Moscovici; see further: Bruckmeier 2013). These theories did not reconstruct the transformation of human interaction with nature in the modern economic world system; they followed the tradition in cultural anthropology to explain societies through their cultural specificities. The explanations are specified for local societies and cultures or for large systems in terms of civilisations, but not for modern capitalist society as a global system.

Interdisciplinary analyses of socio-spatial and socio-temporal relations in modern societies do not advance to the integration of social-scientific and ecological knowledge. Such analyses exist in fragmented forms, unfolding in analyses of domestication of space and time in modern capitalism and the mobility of people and resources. The first process relates to the human appropriation and transformation of nature and space and is analysed, for example, in the critical sociological theory of everyday life by Lefèbvre (1991) and in sociological analyses of space (Löw 2001) and time (Bergmann 1992). The processes of social structuring of time refer to various forms of social interpretation of time, for example, the long historical process of societal development (Giddens: "*longue durée*") and the economic rationalisation of time in capitalist industrial production, which forms, according to Marx, the innermost secret of capitalist economy. Domestication of time and space in society associates with the political-economic analyses of the processes of work and production, which include the "work of nature", that is, energy (Luhmann 1999: 22) and the "work of society", that is, socially structured human labour. Both forms of work are connected in economic production processes where nature is transformed and appropriated by humans. The political-economic analysis of human labour in the transformation of nature is neglected in most newer theories in sociology, economics, and anthropology. This neglect is not always based on explicit theoretical arguments as in the critical theory of Marcuse who sees human labour as successively replaced by knowledge and technology that do the work of humans.

Limits of Sociological Analyses of Society–Nature Interaction

The analysis of the development of the sociological discourse above resulted in the diagnosis that the analysis of the systemic nature of modern society as capitalist world system which structures societal relations with nature is vanishing in sociological theories. The following four points describe possibilities to dissolve the deficits of sociological theories that impede the further elaboration of a theory of nature and society.

1. *After the decay of grand theory in the sociological discourse, the theory of modern society fell apart in specific theories.* These take up parts of the older theories in critical reflection of modernisation, but do not longer develop a new encompassing concept of modern society. With that the theoretical analysis of societal relations with nature dissolves in specialised research. Newer theories, including the third group of theories described at the beginning as theories of social transformation, are formulated for the recent processes of change in late modernity, updating earlier theories. In this pluralisation of sociological theories,

nature became a theme of specialised theories of risk society, reflexive modernisation, ecological modernisation, or actor-network theory. None of these theories is interdisciplinary in the sense of integrating natural-scientific knowledge in the theory; they refer selectively or for purposes of illustration to empirical knowledge from non-sociological and environmental research.

The analysis of capitalism in political economy where societal interaction with nature was part of the analysis of the modes of production lost significance. In the recent discourse of critical theory, the focus is on analysis and critique of normative orders and recognition processes (Honneth), showing a cultural turn in critical theory. Analyses of modern capitalism that maintain a memory of political-economic systems analysis are the examples of Wolfs' (1982) anthropological theory and Wallerstein's (2000) world system theory. World system theory developed from Marxist sources as a historically specified theory of modern capitalism as economic world system that is based on the separation between core and periphery countries. This separation shows different forms of the inclusion of countries and economies in the capitalist economy, as poor, extractive economies in mainly agricultural societies, or as rich, productive, and processing economies in industrial countries. The changing constellation of countries and national economies in the core or periphery of the world system in different phases of modernity is analysed in this theory, as well as the global inequality in the North–South division of the global capitalism. Wolf (1982) connected and discussed the macroscopic analysis of the economic world system with the microscopic analyses of local and cultural orders in cultural-anthropological research. He showed that modern capitalism developed throughout its history with the continuous economic (not cultural) inclusion and integration of other modes of production, societies, cultures, and civilisations. Cultural orders become in the capitalist process of modernisation and globalisation a secondary codification of societies and their economic structures that are continually integrated in the global order of modern capitalism in its historical change, without dissolving the normative orders in a unified global culture. World system theory interpreted in this sense seems to bridge the gap between sociological theories of modern society and cultural anthropological theories (Moscovici, Morin); thus it comes also closest to a social-scientific theory of nature–society interaction.

2. *For a theory of the interaction of society and nature, it remains necessary to analyse the production and reproduction of society and economy.* The analyses of the modes of production and reproduction reveal the system-specific forms of interaction between nature and society for the modern society. Both forms of analyses show

- how society connects to nature through human labour and its technologies, and
- how society is dependent from nature in the use of natural resources and the societal metabolism.

None of the newer sociological theories mentioned above systematically analyses the "systemic distortions" of reproduction through marketisation and commodification of nature, which result in social, economic, and ecological reproduction crises. Such combined reproduction crises include a disturbance of

- *economic reproduction* of the global market economy where financial assets, capital, and commodities are devalued in dimensions beyond the magnitude of the global economic crash in 1929;
- *symbolic reproduction* of society in the culturally shaped processes of education, socialisation, and social integration that came into crisis through privatisation, deregulation, and deconstruction of the institutions of the welfare state;
- *biological and ecological reproduction,* including demographic processes, at various levels in the interacting social and ecological systems (risks through overuse of natural resources and manipulation of natural material cycles).

The economic and social crises include the global oil price crisis in 1974, global financial crisis in 1987, the continuous financial crises of the developing countries and their subjection to austerity programmes, the economic collapse of socialist countries in Eastern Europe in 1990, and the global financial crises since 2007. These economic events show in varying degrees the features of combined crises of reproduction of societal and ecological systems that are so far insufficiently analysed in sociological, economic, and ecological theories.

3. *The neglect of the theme of society and nature in sociological theories is self-critically discussed in environmental sociology, without significant advances towards theoretically reflected analyses of nature–society interaction and knowledge syntheses.* Environmental sociology fragmented the analysis of society–nature relations. The specialised research works with elements from human ecology, political economy, and new themes as environmental movements, environmental consciousness, and environmental politics, without searching for theoretical connections of these themes. Further social-scientific reflection requires a controversial theme in ecological research: the construction of ecosystem services, attempting to show the dependence of humans and society from nature, without theoretically developing the idea through analyses of interaction between social and ecological systems. Other possibilities to overcome the disciplinary limits of sociological and ecology, broadening interdisciplinary knowledge integration, developed late. In the social-ecological discourse, an interdisciplinary theory of society and nature is in progress where the difficulties of recombining theories and interdisciplinary knowledge synthesis become visible. The theory reflects the interaction of modern society and nature beyond simple assumptions of the dependence of society and humans from nature or the coupling of social and ecological systems. The interdisciplinary subjects of human, social, cultural, and political ecology take up the theme of nature–society interaction in different forms. These subjects received new significance only with the broader environmental discourse towards the end of the twentieth century, and still later with attempts to connect interdisciplinary research with a theory of modern society.

4. *Which knowledge from sociology, ecology, and other disciplines is necessary for the construction of a social-ecological theory of nature and society?* This question cannot be answered with the choice of a specific theory of society nor from a single disciplinary perspective. Not all theoretical knowledge from sociology and theory of society is necessary for social-ecological theory, nor is all knowledge from specialised research and from other disciplines. In the elaboration of this interdisciplinary

theory it needs to be determined, which knowledge can be used for the different parts of the theory. In relation to the knowledge from specialised environmental research, an interdisciplinary theory of nature and society needs to be constructed with several levels of abstraction and forms of knowledge integration, thus keeping the core theory limited to the analysis of systemic reproduction and interaction between nature and society. Not all empirical knowledge about cultural variations of interactions between humans and nature needs to be integrated into such a theory. The sociological and economic knowledge relevant for the social-ecological theory includes analyses of capitalist modes of production and reproduction and analyses of the modern world system:

– Analyses of *capitalist modes of production and reproduction* can be specified in different capitalist accumulation regimes, in successive historical phases, for the present phase taking up the analyses of Fordist and post-Fordist accumulation regimes.
– Analyses of *the modern world system* describe the historically long process of globalisation since the beginning of Western modernity, with changing constellations of countries and national economies in the core, semiperiphery, and periphery. These positions show the global division of labour in the modern world system, with different functions regarding natural resource use (extracting economies and industrial economies) and the historical changes in that societal division of labour.

Connected to the core of the social-ecological theory, but not part of it, are the specifics of empirical, cultural, and historical studies of interactions between humans and nature:

– Analyses of *culturally structured societal orders* (multiple cultural modernities, civilisational analysis, including views and constructions of the world and nature) show the interaction of nature and society beyond economic structuring in secondary forms of social structuring of natural resource use as cultural differentiation of the unified global economic system.
– *Theories of late modernity* (risk society, reflexive modernisation, ecological modernisation, post-industrial and post-modern societies)

are complementary to the social-ecological theory in the sense that they describe various aspects of societal change in the history of modern society. In addition, the spatial and temporal structuring of societal processes are of secondary importance to the theory as they do not show the systemic constituents of connected societal and ecological systems.

To some degree the limitations of sociological theories described in the four points above can be seen as consequences of the disciplinary specialisation of sociology. With this specialisation appear new cognitive problems

- the difficulties in the theoretical analysis of societal systems and their interaction with ecological systems;
- the lacking epistemological and methodological reflection of interdisciplinary knowledge synthesis.

Beyond the use of theoretical knowledge from sociological theories, the further development of the social-ecological theory requires an interdisciplinary broadening of the analyses of modes of production and reproduction to include the ecological components (societal metabolism, social-ecological regimes) and theoretical typologies of coupled social and ecological systems. This starts with the following analysis of *interdisciplinary theories of society–nature interaction emerging from the social sciences.*

Interdisciplinary Broadening: Social-Scientific Analyses of Society and Nature

Since the 1970s environmental research has grown rapidly in many countries. Among the interconnecting reasons are

- the growing public awareness of environmental disruption in modern industrial society and resource use crises connected to these;
- the strengthening of environmental policies, nationally and internationally;
- the emergence of new social and environmental movements that made environmental problems to their themes;

- the emergence of environmental sociology in the 1970s;
- the renewal of the earlier interdisciplinary human-ecological discourse on the interaction of man, society, and nature.

With the rapidly developing environmental research, a theory of the relations between modern society and nature and the changes of these relations advances albeit slowly. This theory requires interdisciplinary knowledge syntheses, social- and natural-scientific knowledge. Without such knowledge it is impossible to discuss competing hypotheses whether modern society is controlling nature, detaching from it, fusing with it, or still dependent and embedded in nature.

In interdisciplinary analyses of society and nature, Moscovici (1982) played a pioneering role. For Moscovici theoretical reflection of nature–society interaction needs to be historically specific. His approach reflects the connections between nature and society with new interdisciplinary knowledge synthesis. This attempt remained eclectic: not always based on in-depth analysis, and not clarifying the epistemological and methodological problems of interdisciplinary syntheses including social- and natural-scientific knowledge. Moscovici gave up the critical term of capitalism for the less theoretically elaborate of industrial society. This seems to disconnect his analysis from the discourse of critical theory of society, but it helped to focus social-ecological analysis on industrial components of the mode of production, environmental consequences, and the social and physical limits of resource use. Moscovici's view that with regard to nature a society is neither capitalist nor socialist but industrial or agricultural does not necessarily mean to reject a system analysis of modern capitalism as mode of production, but rather to complete it. The deficits and the limited discussion of Moscovici's theory in the discourse of theory of society do not reduce its significance for a social-ecological theory of nature and society.

The main social-scientific components and variants for the development of a critical social-ecological theory of society and nature are summarised in Table 2.1. These concepts and theories are used for the analysis of

- *economic processes* (economic reproduction, globalisation, global capital, and resource flows);

Table 2.1 Social-scientific theories for analysing society–nature interaction

Theme	Theories and authors	Main ecological assumptions and theorems	Assessment
(1) Modes of production and reproduction: Economic modes of production Economic/material reproduction and accumulation/accumulation regimes Reproduction of society in cultural and symbolic processes	Political-economic theories: interaction of society and nature through human labour and technology Political economy of the environment: O'Connor (1998), Schnaiberg (1980), Schnaiberg and Gould (1994) (human ecology), Altvater (1991) Critical theory of society (symbolic and material reproduction); the interaction society/ nature in theories of the Frankfurt School (Biro 2011)	The historical factors that specify society/nature relations are the producing branches of agriculture, handicraft, industry: in modern society industry is "the real historical relation of nature and therefore of natural science to the human being" (Marx and Engels 1932: 122) Second contradiction of capitalism—impossibility to solve environmental problems (O'Connor) Environmental problems are inherent in industrial production: material wellbeing requires more and more use of natural resources; environmental policy is insufficient: required is the change and reduction of resource use and consumption (Schnaiberg and Gould)	Theoretical core of classical political economy and critical theory—material and symbolic reproduction; developed further and integrated partially in social-ecological theories of societal relations with nature and societal metabolism Intensive discussion of nature/society interaction in the Frankfurt School (Biro 2011), remaining in the limits of philosophical and sociological reflection

(2) Global resource flows, physical economy	Physical economy (Tennenbaum 2015), environmental sociology, (Rice 2007), social ecology (Fischer Kowalski and Haberl 2007)	Material and energy flow accounting (MFA, MEFA): Measuring global resource flows in modern society in physical and monetary terms	Interdisciplinary synthesis of thinking about economy and nature (physical economy); operational parts of the social-ecological theory of societal metabolism; connecting with the theories of unequal ecological exchange (Rice) and ecological distribution conflicts (Martinez-Alier)
(3) Interaction of society and nature: – Theory of societal metabolism (social ecology) – Theory of unequal ecological exchange (environmental sociology, ecological economics, cultural anthropology) – Theory of ecological distribution conflicts (environmental economics, political ecology)	Fischer Kowalski and Haberl (2007) Rice (2007), Escobar (2006) Martinez-Alier (1995, 2004)	Cross-national unequal exchange: "The socioeconomic metabolism of the world-system consists of the interdependent flows of energy, natural resources, and waste products between countries as it shapes the differential processes of production-consumption-accumulation at different positions in the global economy." (Rice 2007: 46)	Developing in the tradition of critical theory of society, broadening the scope of the theory through interdisciplinary knowledge and analyses

(continued)

Table 2.1 (continued)

Theme	Theories and authors	Main ecological assumptions and theorems	Assessment
(4) Civilisational-cultural analysis	Arnason (1988, 2003; 2006)	Interdisciplinary analysis of civilisations and their development; comparative analysis of civilisations and cultures in their social and historical contexts.	Somewhat vague and under-theorised concept of society; interaction between nature and culture as potential theme of inter-civilisational comparison of cultures (Arnason 2003: 62f, 304)
(5) Economic globalisation, global social change: – globalisation theories – world system theory – world ecology analyses (interaction of societal and ecological systems at global levels)	Globalisation theories: Featherstone et al. (1995), Scholte (2000), Held and McGrew (2007), Walby (2009) Critical theories of – capitalism (Piketty 2014) – world system (Wallerstein 2000) – world ecology (Hornborg and Crumley 2006)	Description of globalisation in its manifold forms In critical theories (Wallerstein 2000; Hornborg and Crumley 2006; Walby 2009) analysis of multi-scale systems with complex inequalities	Complementary to sociological theories of society and the social-ecological theory, overlapping theme of interaction of global social and ecological systems Components of the unfolding perspective of interdisciplinary and social-ecological world ecology analysis

(6) New social movements and civil society action	Environmental movements (Roth and Ructh 2008) Knowledge practices of environmental movements (Jamison 2001, 2010) Greening of social theory (Barry 1999; Jamison 2001)	Problems of the environment and global environmental change need to be taken up by environmental movements who think, speak and act for all citizens and the whole society although they remain minoritarian ("alarm bell" function)	Complementary part of the theory of society and its relations with nature: theories of social or collective action, related to lifeworld and civil society, reflecting knowledge practices in environmental movements
(7) Governance and regulation – Political governance, global governance, earth system governance, environmental governance – Regulating nature–society interaction	Policy-related theories (Biermann 2011) Interdisciplinary theories of global governance, climate change adaption: (Brunnengräber 2009; Brunnengräber et al. 2008; Brand 2015a, b)	Interaction of society and nature is to be dealt with in management, public policy, governance, regulation (as far as it can be dealt with in social/collective action)	Governance theories connect to the social-ecological theory, but not its core components; in relation to this theory they are "transfer theories" to generate knowledge for policy and governance processes

(continued)

Table 2.1 (continued)

Theme	Theories and authors	Main ecological assumptions and theorems	Assessment
(8) Other sociological and social-scientific theories – Post-modernism – Ecofeminism – Post-colonialism – Actor-network theory	Bauman (1992) Mellor (1997) Gandhi (1998) Law and Hassard (1999), Latour (1999)	Nature–society interaction analysed in specific perspectives, as the names of the theories indicate	Although nature is part of the themes of these theories, they are not converging to a systematic theory of modern society and its interaction with nature; they have either specific and limited theoretical perspectives (actor-network theory), or are part of broader critical, social and philosophical discourses (Post-modernism, Ecofeminism, Post-colonialism)

Sources: Quoted in the table

- *political processes* (global and environmental governance, regulation of nature–society interaction);
- *complex social processes of change and transformation* (societal transformation to sustainability).

The examples show a selective discussion of nature–society interaction, limited through the thematic scope of sociological analyses, where social inequality and class structures, political structures and power relations, cultural and value systems, gender relations, forms of social consciousness, and social practices in the lifeworld or lifestyles are core themes. These analyses remain parts of specialised sociological theories that can inform the social-ecological theory for the analysis of societal metabolism and its historical changes, power relations, collective action, and conditions to transform societal systems in the practices of environmental policy and governance. The sociological theories of modern society discussed at the beginning of the chapter are not included in Table 2.1 although parts of their analyses can be used in interdisciplinary theorising.

The theories providing knowledge for an interdisciplinary social-ecological theory differ in their analyses of the society–nature dynamics:

1. *The theoretical core of a social-ecological theory of society and nature* develops from three groups of theoretical analyses in theories of economic and societal production and reproduction, of global resource flows, and of societal metabolism (themes 1–3 in Table 2.1). Production and reproduction mean, in the final analysis, production and reproduction of human life, but cannot be described in one discipline or theory. In human history, with the development of culture and society, production and reproduction processes differentiated in various forms that require analysis of production as economic and ecological process, of reproduction as biological reproduction, economic reproduction, and societal reproduction in symbolic forms.
2. *The civilisational-cultural theory of Arnason (theme 4 in Table 2.1) and the recently developing theories of governance and regulation (theme 7 in Table 2.1) connect in different ways to the core themes of the social-ecological theory.* Civilisational analysis allows for comparative global analyses of the cultural framing of nature–society interaction, its secondary codification

beyond economic structuring. These thematically specific theories do not cover fully the complexity of societal transformation of socio-metabolic regimes: civilisational analyses focus on cultural and symbolic processes, governance theories focus on policy, and power-based processes.
3. *Theories of social system dynamics and social movements* include economic globalisation, global social change in cultural, political, and technological forms, and new environmental movements (themes 5 and 6 in Table 2.1). These theories can be seen as variants of "middle range theories" with limited importance for the formulation of a social-ecological theory. They analyse parts of the interaction of social and ecological systems that connect to overarching theories of societal reproduction with empirical research on changing quantity and quality of resource use. Other theories (theme 8 in Table 2.1) do not analyse the relations between humans and nature in form of a systematic theory of modern capitalist society and nature.
4. *Further theories not mentioned in Table 2.1* but in the discussion of sociological theories above (theories of social space and time, of lifeworld and lifestyles, consumption cultures, mobility of people and resources) are thematically limited or insufficiently developed as theories of societal interaction with nature. As far as they analyse the spatio-temporal processes of mobility and global flows of people and resources in modern society, they describe the "communication surface" of modern capitalism with trade, exchange, global communication, and action over distance. They do not connect to the theoretical analysis of modes of production and reproduction of economy and society (or only in exceptional forms: Lefèbvre 1991, connecting analysis of lifeworld processes to Marxist theory).

For an interdisciplinary theory of modern capitalist society and nature, several components from social-scientific theories can be used (see Fig. 2.1). Two important components of critical theories are

– the analysis of societal modes of production and reproduction specified in the social-ecological theory of metabolic regimes of societal metabolism, and

- the analysis of theory–practice relationship to specify forms of agency for the transformation towards sustainability.

The ways of societal transformation are finally found in the social practices of transformation: not in the theory and through research, but through application of the knowledge reflected in the theory in environmental governance and transformative action. In traditional theories of society, this knowledge transfer is reduced to the conventional forms of knowledge transfer from research to technologies and consulting of decision-makers, resource managers, and social movements. In these traditional forms, the dominance and superior quality of scientific knowledge are claimed, ignoring the social-emancipatory forms of agency and the necessities of transformation of societal systems.

The social-scientific components, theories, and themes of a theory of nature and society summarised in Fig. 2.1 need to be completed through the ecological components discussed in Chap. 3 before the contours of the social-ecological theory of nature become visible. Ecological research covers part of the missing analyses in social research: analyses of vulnerability, resilience, and sustainability of the coupling of social and ecosystems,

Fig. 2.1 Social-scientific components of a theory of society–nature interaction. *Sources*: Own compilation (theories discussed in this chapter)

of the global material and energy flows, and of the planetary boundaries of resource use. The further development of the theory and the problems to be addressed are discussed in the following section.

Social-Ecological Theory of Society and Nature

With the notion of global social change, a broadened theoretical perspective of economic, political, and cultural globalisation emerges. The social processes of globalisation are interwoven with ecological processes through natural resource use. The social-ecological theory integrates knowledge from social research on globalisation and ecological research on global climate change and land-use change. This integration is concretised successively in the following chapters with research on resilience and sustainability (Chaps. 3 and 4), on the interaction and governance of social and ecological systems (Chaps. 5 and 6), and on climate change adaptation and transformation of energy regimes (Chaps. 7 and 8). Three general questions need to be answered for all these themes in the further elaboration of the social-ecological theory:

1. *How can the historically changing relations between society and nature be described in terms of alienation, detachment, hybridisation, or integration?* The various answers are still disputed. Sociological theories tend more towards detachment hypotheses, and ecological theories tend more to integration hypotheses. The main variants of these hypotheses are the following, implying contradicting diagnoses of society–nature interaction. The hypotheses require theoretical contextualisation to describe their relative validity.

    – In modern society, the boundaries between society and nature dissolve: *hypothesis of fusion and hybridisation,* for example, in sociological actor-network theory and in the ecological theory of technological natures.
    – Modern society is progressively detaching from the forces and laws of nature: *hypothesis of autonomy or emancipation of nature*

from society, in sociology, for example, in Castells' theory of network society and in the mainstream thinking in sociology.
- Modern capitalist society transforms nature with new forms of industrialised production: *hypothesis of co-production of nature and society through human labour and natural resource use that transforms nature and society*, in the historical materialism of Marx or in newer theories of historical-geographical materialism. In this critical theory develops simultaneously the *hypothesis of human alienation from nature in modern capitalism*.
- Modern society is embedded in nature; attempts of disembedding based on modern science and technology result in environmental disruption: *hypothesis of society as part of nature*, in various forms of the ecological discourse (e.g. the "new ecological paradigm" in human ecology and environmental sociology, ecological economics, ecological anthropology).
- Humans in modern society are manipulating and overwhelming the great forces of nature thus creating new global environmental risks and dangers: *hypotheses of humans as destroyers of nature, in various forms*. "Death of nature" is attributed to the mechanistic view of modern natural sciences (Merchant 1980). Humans are, because of climate change and other modifications of global nature, seen as a geological force since the industrial society (or the anthropocene).
- Modern societies are colonising and modifying nature in different forms, from extensive to intensive forms of land use and production (agriculture, forestry, urbanisation, and industry) to modification of nature (genetic modification of animals and plants, commodification of life-supporting ecological processes): *social-ecological hypothesis of "colonisation of nature"* that happens in all historical forms of society, but in specific forms in modern society (Fischer-Kowalski et al).

The contradictions between the diagnoses of the situation in modern society in these hypotheses cannot be dissolved by proving or rejecting each of them with limited empirical knowledge. All these hypotheses refer to different theories of society. The selective knowledge use and theory connection need to be specified to show the relative validity of the hypotheses for specific societies and processes of nature–society interaction. The variety of these hypotheses corresponds to the partial and fragmented state of theories of nature and society in different variants in political economy, critical theory, in human, social, cultural, and political ecology.

2. *How does the social-ecological theory as a historically specific theory explain societal transformations since the beginning of human history?* Past transformations are already studied in the social-ecological theory with regard to the changing social-metabolic profiles and societal transformation in modes of production (Fischer-Kowalski et al. 1997). As historical and interdisciplinary theory, the social-ecological theory needs to deal with questions of "biological embodiedness and ecological embeddedness of human beings and human society" (Barry 1999: 204) that emerge from natural-scientific research. The forms of embeddedness are more complicated in modern society, whereas in earlier societies the societal relations with nature were more transparent, could be "read off" from the natural resource use practices. A historically informed theory of possible paths of transformation to a future sustainable society cannot be derived from historical transformations. The changing spatial and temporal dynamics, different forms of coupling of social and ecosystems, and complexity of modes of production and reproduction need to be analysed to explain societal transformations. Assumptions about determination, embeddedness, and co-evolution need to be specified in concrete historical descriptions of relations between nature and society. Nature and society have only a limited number of common trajectories of development that vary for every historical epoch and form of society; to identify such trajectories or their change is a core theme of social ecology.

3. *How does the social-ecological theory describe possibilities and forms of global transformation to sustainability?* This seems the most important question for the present environmental discourse that suffers from policy failure, resignation, and regression to visions and simpler ideas. Sometimes, for example, in the analysis of Benson and Craig (2014), it is suggested to give up the idea of sustainable development by referring to the many failures in the debate and the failure of the "Rio+20"

conference. The authors want to restrict the debate about the future to that of resilience. Yet, it remains doubtful whether arguments of complexity, uncertainty, or necessity of growth are sufficient to give up the idea of sustainability. The term sustainability is not well elaborated in theoretical regards; it is seen as an "essentially contested concept" which allows for many different and contradicting interpretations. Although sustainability may not be a lasting term, it cannot be easily replaced by other terms. What implies "beyond" modern and industrial society in terms of societal system transformation towards sustainability remains unclear. The situation of insecurity and ignorance is evident: as for the contradicting hypotheses above, supporting knowledge can be found for contradicting interpretations of sustainability. One of the few things known about this transformation is that it requires reduction of natural resource use and the acceptance of planetary boundaries of growth. All earlier transformations had vast and unused space and nature to develop through intensification of resource use. The ignorance about such transitions resembles that in earlier societal transformations in the history of human societies. It was never possible to plan great transformations of societies and modes of production and to foresee where they might end.

The assumption that modern scientific knowledge has created the possibility to dominate nature and foresee the future, a Promethean view, seems to focus on advances of modern science and technology, whereas the consequences of progress and development of society are ignored. Such a view ignores the environmental damages and burdens, the limits of natural resources, and the limits of scientific knowledge. The results of environmental research provide evidence that new environmental and ecological technologies may solve prior problems, but these technologies may create new problems and risks unknown at the time of their initiation. With all increase of scientific knowledge, the "veil of ignorance" about the long-term future of modern society remains. This does not mean that science, policy, and society are blind in "navigating" the ways towards sustainability. Scientific knowledge can be used for transition management and environmental governance, but not in the forms applied hitherto in planning and management of natural resource use that have only short-term perspectives. New forms of environmental governance and regulation require epistemological reflections and revisions of conventional assumptions about the growth of scientific knowledge.

For the elaboration of an interdisciplinary social-ecological theory, two theoretical models exist that can be useful for formulating strategies for transformation to sustainability. The first dates back to the Marxist theory of historical materialism reformulated by Swyngedouw et al. The second is the theory of societal metabolism, which was formulated in the new social ecology by Fischer-Kowalski et al. described in Table 2.1.

Swyngedouw (2010) describes in the footsteps of Marx's dialectical concept of historical materialism the basic idea of how social and ecological knowledge about the interaction of society and nature can be integrated in the perspective of "historical-geographical materialism". Living organisms need to transform nature in metabolic processes where humans and nature are changed simultaneously. The metabolic transformation of nature is not only physical, chemical, or biological, but also a social and historical process of specific relations of production. Nature and humans are social and historical from the beginning of human history. The social appropriation and cultural transformation of nature produces historically specific social and physical natures that are interwoven with social power relations. Thus, nature can be seen as socially produced, and this process is embedded in a series of social, political, cultural, and economic relations at different geographical scales (Swyngedouw 2010: 11).

These ideas, used to account for social transformation of nature, seem too abstract and general in their theoretical formulations. The approach relies on older reflections of societal relations to nature from historical materialism. Although used in empirical studies, for example, on urban metabolism and development, there seems to lack theoretical analysis of the societal metabolism that can mediate the general reflections about hybridisation with empirical research on the use of natural resources. The empirically specified profiles of social metabolism from social ecology seem more useful for the analysis and comparison of socio-metabolic regimes in different forms of human society.

Conclusion: Conceptualising the Relations Between Society and Nature

From the social-scientific theories described in Table 2.1 the interdisciplinary theories of political economy, of civilisational analysis, of global governance, and social-ecological theories take up more systematically the analysis of nature–society interaction than sociological theories. The social-scientific components of a theory of nature–society interaction are summarised in Fig. 2.1, but not discussed further here with regard to a systematic theory of nature and society. Such a theory does not exist in sociology and cannot be constructed with sociological knowledge. The social-ecological theory discussed in this book is not a sociological theory, but an interdisciplinary theory in open, discursive form, with several thematic components discussed successively in the following chapters.

In the elaboration of a social-ecological theory of the relations between nature and society, several epistemological questions need to be answered that are not answered in sociology:

1. *What are the theoretical concepts to construct a theory of nature and society?* Abstract interdisciplinary terms from systems theory are not enough. Systems theory in sociology has opened the theory of society for interdisciplinary communication, but not for systematic knowledge exchange between sociology and ecology, which is required to understand the interaction between society and nature, social and ecological systems, and social and environmental change. The theories of Parsons, Luhmann, and Habermas remained specialised theories of society. Also with the use of a natural-scientific terminology, the distinction between society and nature was more taken for granted than theoretically analysed. For an interdisciplinary theory of nature and society, much more natural- and social-scientific knowledge is required.

    The first epistemological problem is how to deal with concepts transferred across disciplinary boundaries between natural sand social sciences. How are the concepts transformed from metaphors into theoretical and explanatory concepts and theorems in a new knowledge domain? Natural-scientific metaphors used in the sociological analysis of modern society are confronted with that problem: the older term of societal metabolism, the term of autopoietic systems, the recent terms of vulnerability, resilience, and sustainability. In the interdisciplinary social-ecological theory, it becomes evident, that most of the concepts transferred into social-scientific discourses have not been sufficiently developed as theoretical concepts. Some concepts advanced to the status of new theoretical concepts simply by attaching the label social, as, for example, in the notions of social and social-ecological resilience. The core concept of the social-ecological theory under development—societal metabolism—gives an example of how a natural scientific metaphor is transformed through theoretical reflection and connection to knowledge from several disciplines into a new theoretical and an interdisciplinary concept. Also in an interdisciplinary theory theoretical concepts can be used and combined that maintain their original disciplinary meaning; this is the case with most of the sociological concepts discussed in this chapter when they are used in the new theory. A third form, between disciplinary and interdisciplinary concepts, are the ones with double codification, in social-scientific and natural-scientific versions.

2. *How can knowledge from different disciplines and specialised research in the social and natural sciences be integrated?* Beyond the simple interdisciplinary knowledge exchange in the form of concept transfer this is the more important and less answered question of knowledge synthesis and theory construction. Further questions to deal with as part of the construction of the social-ecological theory include therefore epistemological and methodological reflections of the construction of interdisciplinary theory and knowledge synthesis. The theory of nature–society interaction deviates from theories formulated in disciplinary perspectives through its synthetic, interdisciplinary, and intertheoretical approach connecting knowledge components from heterogeneous theories and disciplines. Such an interdisciplinary theory can be described as a "discursive theory with various components and open boundaries". A super-theory in the sense of Luhmann's attempt to formulate a general theory of all social systems and processes is of a different nature. The theory discussed here develops as a

historically situated and changing theory, through a discourse; it does not become a closed theoretical system.
3. *How can the boundaries between society and nature be identified and assessed?* The existence of boundaries between society and nature is more accepted in sociology than in ecology, but not sufficiently clarified in either subject areas. In social ecology, the argument develops that there is no strict boundary between society and nature; the two spheres are interacting and interwoven as they always have been in the history of human societies—only in changing forms. Similarities to the epistemology of historical materialism are evident and need to be reflected. Ecologists tend to see the connection between nature and society trivial as evidenced by the sufficiently proven ecological research and in the lifeworld through everyday experience (e.g. in local ecological knowledge about natural resource use, agriculture, fishery, and forestry). Such views underestimate the problems of a theoretical conceptualisation of nature–society relations. The boundaries between nature and society may be changing or dissolving in social practices of production and resource use. Yet it is necessary for explaining interactions and couplings between social and ecological systems to maintain theoretical awareness of the origin of the processes analysed and conceptualised. Not all processes of social or environmental change can be simply described as hybrid processes.
4. *The unclear distinction between nature and society and social and ecosystems is reflected in the debate about autonomy or dependence of society from nature.* The conventional social-scientific view of society is that in modernity and through the manifold processes of technical, economic, political, and cultural modernisation, the sphere of social life has become more and more detached from nature—disembedded or independent from natural processes. Society has become an autonomous sphere, independent in its cultural and material reproduction from the forces of nature that set tight limits of development in earlier phases of human history. This hypothesis, repeated by Castells in the theory of the network society, confronts the understanding of the relationship between nature and society in human, cultural, and social ecology. In these subject areas, there is no assumption of a linear change or continually reducing dependence of society from nature.

The qualitative changes observed in the interaction of society and nature cannot consequently be understood as independence of society from the laws of nature. Also with significant modifications and manipulations of natural processes in modern society the impression that humanity has "overwhelmed" the great forces of nature in the epoch of the anthropocene still seems insufficient to catch the complex interactions and relations between society and nature. Without theoretically reflected concepts, the interpretations move erratically between the extremes of autonomy and dependence.

5. *Global environmental change as described in ecological research can be understood as a consequence of global social change through industrialisation and as anthropogenous change.* Beyond man-made change there are also forms of environmental change that cannot be attributed to the influence of humans. To account for social and environmental change, trough theoretical reconstruction and possibilities to distinguish between them, is more difficult than reflected in most theories. The climate change discourse (see Chap. 7) gives examples for continuously unsolved questions with the controversy as to how far climate change is man-made and how far it is part of natural processes, and how man-made and natural climate change relate to each other. For the social-ecological theory under construction, it seems meaningful to keep explanations open and to accept controversies as indicators of knowledge problems that cannot be ignored in theoretical interpretations and explanations.

6. *The normative assumptions in the reconstruction of nature–society relations need to be critically reflected.* Assuming that normative assumptions influence all theoretical reflections, it is still necessary to distinguish "simple normativity" as prevailing in large parts of the ecological discourse from "reflected or critical normativity", where normative assumptions are crosschecked with the data, the knowledge, the theories, and their explanations. A renewal of the critical theory of societal relations with nature through social-ecological research follows other normative assumptions and emancipatory perspectives than does the utopia of a society free of domination and exploitation appearing in the older forms of critical theory. The social-emancipatory perspective of a critical theory of societal relations to nature is one to

combat progressing valorisation and commodification of nature, to defetishise commodity production, and to find criteria for human use of natural resources that is fair and just in terms of social and environmental justice. This implies complicated readjustments of economic institutions of reciprocity, redistribution, exchange, and household to be balanced with each other. Such ideas were developed in the critical institutional economy of Polanyi (1944, 1979) and later analyses of the self-destructive mechanisms of the economy of growth (Zinn 1980; Girardet 2007), or in the discourse of degrowth.

Appendix: Relations Between Society and Nature According to Main Sociological Theories

(1) Traditional Theory: Luhmann's Theory of Social Systems

Luhmann's sociological theory of modern society is a theory of a functionally differentiated society that develops from Parsonian theory. Luhmann modifies this theory, maintaining the theorem of functional differentiation of societal subsystems as state, economy, science, culture, and others. His theory results in a more abstract description of modern society, abstracting from territorial segmentation with the concepts of functional differentiation and world society. This happens in a complicated theoretical language, in constructivist thinking, difficult to formulate in other concepts and theoretical languages.

Modern society as world society is built through global communication: the operation to produce and reproduce society is communication [Luhmann (1997: 3); for further discussion of the idea of world society see Stichweh (2000)]. The theoretical term of society is simplified with the term communication. World is not understood in a geographical meaning as the physical world or the global space, although such "background meaning" is always present when Luhmann writes about society. The world of world society is an abstract functional concept in the sense of a horizon for all meaningful communication. More than in spatial dimensions and as contrast to national societies or territories is the world society understood in a temporal logic as simultaneity through global communication without physical presence.

The explanation of the systemic constitution and reproduction of the modern economic world system is in this theory reduced to loose forms of system integration and interaction within and between functionally differentiated subsystems of society. Functionally differentiated societal subsystems develop their own internal codes and forms of communication. Communication that cannot be structured in the binary logics of money for the economy, power for politics, and truth for science does not succeed in modern societies. The system-theoretical analysis of society is guided by the idea of complexity reduction through the building of social systems. The aspiration of explanation is reduced to description, although in complicated constructivist terms. Theories of society produce theoretically concentrated variants of the self-description of modern society. Self-description is understood by Luhmann (1997: 867) as communication within society about society.

Analyses of societal relations with nature, natural resource use, and environmental problems are marginal themes in Luhmann's final theory. There they appear in theoretically distorted forms as historical variants of self-description of society as nature (Luhmann 1997: 989ff), otherwise as forms of communication of "ecological problems" in modern society (Luhmann 1986). In the theory of power and money as generalised media of societal communication, it is described how political and economic systems communicate about nature. In the reflections on "ecological communication", Luhmann argues why the ecological discourse cannot succeed in a functionally differentiated modern society; for nature there exists no subsystem in society and no effective generalised mechanism of communication. Environmental problems cannot be articulated in the specialised system languages and codes of economy, politics, and science. Each of these societal subsystems has a specific function and does not care for the functions of other subsystems nor for society at large or global problems. Nature and environmental problems do not find a place in the order and discourse of modern society. A solution for environmental problems seems impossible in the functionally specialised modern society, with the systemic mechanisms of industrial society where the codes of conduct for the use of money and power dictate the standards for environmental action. However, a transformation of society to another mode of production or socio-metabolic regime appears also as impossible. According to Luhmann, modern society can only continue to evolve on the path of specialisation and differentiation.

The theoretical construction of modern society with the systemic communication media of power and money helps to get rid of complicated explanations in earlier theories. There production and reproduction, relations of power, domination, exploitation of humans and nature, capital accumulation, and appropriation of nature appeared more as problems created through modern society and problems to explain with the help of theory. Reproduction is in Luhmann's theory a concept copied from the biological forms of reproduction of species or living systems, without further theoretical significance for the theory. The biological term remains in the theory of society a metaphor exceeding its cognitive limit when applied to society at large or when social systems are understood as living systems. The theoretically upgraded term of communication shows the systemic distortion of societal communication and as a consequence the inefficient solution of environmental and other problems. Luhmann's theory describes the modern world society as an incomplete society, with imperfect societal synthesis, as a truncated system with the global communication and interaction functional for the world market.

(2) Critical Theory: Habermas' Theory of Communicative Action

In the early works of Marx, where critical theory takes shape through the critique of Hegelian philosophy, a theoretical programme for the analysis of relations between society and nature was drafted. These relations cannot be directly read off from relations of individual humans to nature or their consciousness about these relations. Relations between humans and nature and their alienation from nature need to be decoded in the analysis of social relations between humans, from the societal nature of humans and their cooperation within the modes of production in human history.

The transformation of nature happens through human work which is itself transformed through the increasing use of scientific knowledge in production and resource use. The anthropological argument of critical theory to explain human society and societal relations to nature unfolds in an intersubjective theory of the human condition and the societal relations between humans (Anacker 1974: 149).

The theoretical analysis of the relations between nature, humans, and society was specified in the political-economic systems analysis of modern society explaining the capitalist mode of production and economic reproduction. Part of this theory is the analysis of social and class structures that developed in modern capitalist society. Other components included the theory of symbolic and material reproduction of modern societies and of social consciousness, developing as theory of class consciousness, ideology, and manipulated consciousness. The guiding idea of a normative order in critical theory is that of social emancipation from relations of domination and exploitation.

The turn of critical theory towards a negative anthropology began with the "Dialectic of Enlightenment" by Horkheimer and Adorno (1971), where nature appeared as an object of human domination. Nature appears in a series of philosophical concepts, showing the disturbed relations between humans and nature in modern society: nature as outside society, "the other" that humans try to dominate; second nature as reified social relations in society; nature as the unknown and unknowable; human thinking about nature; human nature which is analysed in an anthropological theory. In the second generation of authors from the Frankfurt School, Habermas has in his writings on the philosophical discourse of modernity broken with the enlightenment analysis of Horkheimer and Adorno. He saw enlightenment as an "unfinished project" that requires a renewal of its emancipatory perspectives. This implies a critique of the naïve forms of enlightenment thinking that inaugurated human domination of nature (Vietta 1995: 169ff). In his sociological work, Habermas did not renew the analysis of societal relations with nature.

Habermas (1968) brought in his early theoretical reflections a long debate about the critique of society to the epistemological answer that describes a specific property of critical theory: critique of society requires a theory of society that includes a critique of the forms and interests of cognition in society. A critical diagnosis of society needs to take the form of theory before it can become practical knowledge for the transformation of society. The theory of communicative action (Habermas 1981) brought significant changes in the explanation of society, with a synthesis of sociological thinking about society, including traditional and critical theory. The influence of Marxian theory weakened, and with it class analysis with the argument that it becomes irrelevant when class consciousness is fading away. The concept of communicative action has three components described by Habermas as that of reconstructing communicative rationality, reconstructing a concept of society with the components of system and lifeworld, and explaining the paradoxes of modernity as dominance of systems over the lifeworld. Communicative action becomes the core concept of the theory to explain the reproduction of society replacing the political-economic analysis of productive forces and relations of production. The functionalist theories of Parsons and Luhmann are used in the theoretical systems analysis of the core systems of economy and politics in modern society. A critical intention of the theory is to correct the distorted system descriptions of traditional functionalism and the instrumental rationality in the use of power and money through the analysis of society as lifeworld. Communicative rationality unfolds in the lifeworld as the capacity to achieve social equality, justice, and social emancipation. The concept of communicative action is used to analyse "floating inequalities", which appear no longer as class-bound, are part of a colonialisation of the lifeworld through political and economic systems. According to this analysis of distorted social communication, the communicative mechanisms of power and money require enclosures and need to be controlled to prevent that they damage and destroy the spheres of power-free and non-alienated communication and the autonomy of lifeworld. The argument can be criticised as showing a "sociological naivety" in the theoretical analysis of problems and conflicts of modern society. With that reasoning Habermas does not achieve a critical analysis of the paradoxes of modern society and of the interaction between society and nature. The reasoning can also be criticised to imply a misleading separation of the spheres of social systems and social lifeworld a "fallacy of misplaced concreteness"

(Parsons, Whitehead) where an analytical distinction of system and lifeworld appears as social separation of contrasting social spheres.

The neglect of environmental problems was a main critique of this theory (Scheunemann 2009: 8; Eckersley 1992), answered by Habermas in an astonishing way: he does not feel competent to reflect about ecological themes (Scheunemann 2009: 17). Nature and the environment appear in the theory of communicative action as the material basis of the lifeworld: the physical and biological nature and the nature transformed by humans, all outside of society—although connected to society through specific forms of social action and communication. In instrumental action human labour and its modification of nature can be included, and in communicative action the symbolic and cultural communication about nature. A critical analysis of nature–society relations could be developed in the logic of communicative action as part of the critique of the colonialisation of lifeworld through imperatives of the political and economic systems of modern society. At the interfaces of system and lifeworld appear the manifold conflicts in modern society that are a consequence of this colonialisation. The new social movements become the social actors that articulate and communicate the critique of colonialisation in civil society action.

After Habermas, in the work of Honneth, the theme of nature–society interaction is still more disconnected from sociological theory. In the analyses of Honneth critical theory develops back to a social-philosophical discourse about normative integration and problems of recognition in modern societies. A critical theory of contemporary capitalism is reduced to the themes of fights and conflicts about recognition, equal rights, and social emancipation in the vague sense of normative orders of a just society. These themes are disconnected from earlier variants of critical theory through a doubtful revitalising of sociology as moral science. Honneth, arguing in the tracks of Parsons' theory, eliminates from sociology components that try to connect theory of modern capitalism with non-normative, material analyses of relations of production and reproduction. Other sociological variants of critical theory, for example, the analysis of social space by Lefèbvre (1991), or approaches as radical geography (Brenner 1997), maintain important components of a theory of nature–society relations. At the end of the twentieth century, critical theory in sociology shows a tendency to dissolve into different approaches. These include Beck's ideas of a theory of cosmopolitan risk society, and attempts to formulate a critical theory of another kind from Luhmann's system theory (Amstutz and Fischer-Lescano 2013), and the theory of kaleidoscopic dialectic (Rehbein and Souza 2014). Critical theory falls apart in contrasting variants that can only be kept under this label by assuming that different theories are necessary to develop a renewed critical analysis of modern capitalism.

(3) Societal Transformation in Sociological Views: Post-industrial Society

Bell (1973) saw the coming post-industrial society as one where the service sector became the main sector replacing industrial production that becomes in its new branches science-based production. As a consequence of these projected changes, class structure was also expected to change with the rise of new technical elites. The basic idea of post-industrial society remained simple, following prior observations of Fourastié who saw development and modernisation as a shift from the first economic sector (dominance of primary production in agriculture) to the second (industry), and, finally, to the third sector (services). The future post-industrial society is unknown in its systemic forms; its description as a service-based economy is not sufficient to describe the complexity of societal change, and not to describe relevant changes regarding the environment and use of natural resources. With the present changes of economic systems through information technologies, rather a further stage of industrial society is reached than a post-industrial society with a changed mode of production: a society where industrial production is relocated to the Global South and newly industrialising countries. "Services", including many heterogeneous services produced by public and private organisations, do not show a clear idea and direction of societal change. According to the early ideas of post-industrial society, a country like the USA could already be seen as post-industrial when the industrial sector in the national economy and in the national gross product became smaller than that of the expanding third sector. The assumption, dynamics of economic development are dynamics of national economies and nationally organised societies, is not sufficient to understand the historical development of industrial capitalism that required as its counterpart the non-industrialised economies from which large parts of the resources came. The assumption is theoretically devalued further with the take-off of economic globalisation since the 1970s and the late industrialisation in the Global South. These changes show that industrialisation is a de-synchronous process, part of the global economy, and changes its quality mainly through technological innovations. The second industrial revolution towards the end of nineteenth century was seen as resulting in mass production and the Fordist accumulation regime. The present third industrial revolution (Rifkin) is seen as the digitalisation of manufacturing.

The qualitative changes of industrial production are not analysed in detail in Bells theory; his description of the change as science-based production remained rather vague, not much differing from the broader discourse of the knowledge-based economy or information society. Post-industrial society may even melt with description of post-Fordism as a new capitalist accumulation regime. The question of nature–society interaction is irrelevant for the theory of post-industrialism. It comes up in critical analyses of industrial capitalism, in the context of critique of a growth- and technology-based economy and its environmental consequences by Illich (1973), and as part of the then emerging debate on limits to growth.

In critical theory where the negative consequences of a technocratic society were discussed, the idea of post-industrial society was not theoretically elaborated further. The late diagnosis of Marcuse (1964) was that modern capitalist and socialist societies converge to similar forms of industrial societies, reducing the system differences between capitalism and socialism. This diagnosis where older forms of analyses of capitalism are blended with the assumption of technological determinism in contradicting forms was criticised by other critical theorists (Habermas 1978, see especially Offe and Bergmann, ibid., 73ff, 89ff). In the ecological discourse to which Marcuse contributed in his late writings, his utopian ideas of a liberated society were taken up as critique of industrial society. The idea of post-industrial society as critique and transformation of industrial society was connected with ecological visions of an environmentally sustainable future society. This future society should be achieved through the restructuring of large societal systems to a multiplicity of locally organised societies and economic systems where the accumulation of money and power, of capital and repressive power are reduced through degrowth. But the transformation towards a global post-industrial society in this sense is not yet on the way. Forms of de-industrialisation in some Western countries since the late twentieth century are not showing the development of a future post-industrial society.

The relocation of industrial production from Western countries to newly industrialising countries that become territorially (but not in terms of property and control) the industrial producers for the world is not identical with a transformation of industrial society. It shows new forms of international division of labour in industrial production. What happened in fact with the de-industrialisation in European countries was the relocation of polluting industries to other countries, supported by globalisation and deregulation of national economies. Environmental pollution has not reduced in the past decades, the pollution increases where the production goes. Industrialisation is expected to continue for decades before its long-term consequences in forms of pollution, climate change, exhaustion of resources come to block its further spreading.

The term of post-industrial society seems already to say with this name that one does not know the form of such a future society in terms of its systemic structures, modes of production and reproduction, relations to nature and societal metabolism. The parallel processes of industrialisation and urbanisation continue. The competing interpretations, ideas, and visions of a future society as technology-based information society, as globalised hyper-capitalism, or as ecologically sustainable society in networked local systems make it further complicated to interpret the term of post-industrial society. This term comprises heterogonous and contradicting forms and interpretations of economic change. Ecologically seen industrial society cannot exist forever; natural resource use and environmental pollution are beyond the global limits before industrialisation reaches a large part of the global population. In the early ecological discourse, the idea of the post-industrial society has been called a utopian idea that comes decades too early (Bühl 1983). This is again the conclusion from more advanced, theoretically guided analyses of societal metabolism in recent social-ecological research.

Theoretical terms as post-industrial society, risk society, information society, reflexive modernisation, or network society do not show a great transformation of modern, industrial, or capitalist society. Most of these theories do not analyse systematically the symbolic and material reproduction of modern society in their global forms. In some variants of post-modern theorising, doubtful assumptions are found; for example, the hypothesis that in late capitalism societal reproduction changed from economic to cultural modes (Welsch 2003). Globally seen, the industrial society has not ended: in many parts of the world it has hardly started. The attempts of late industrialisation and modernisation of national economies in the Global South happen at a time when ecological research and theories can already show that it is impossible for all humans to live in industrial socio-metabolic regimes because of the limits of natural resources.

(4) Social Transformation in Sociological Views: Reflexive Modernisation and Cosmopolitan Risk Society (Beck and Giddens)

The theories of risk society (Beck), reflexive modernisation (Giddens), and ecological modernisation (Mol) describe processes of social change, but not structures of a future society. These theories leave the directions of transformation of society open, describing modernisation as taking other forms and as a slow structural change, sometimes described as slow transformation. Descriptions of early and late, first and second modernity, simple and reflexive, conventional and ecological modernisation give only names to historical phases of modern society. The concept of modernity is used once more to describe the development of society in the forms of a never-ending modernisation. This was already seen as a paradox of "the future cannot begin" by Luhmann (1976). He analysed temporal structures of modern societies, based on a distinction of the immediate past and future (which are always reference points of social action) and distant past and future. These reflections remain at the level of abstract philosophical reasoning about the time of social systems and "nontemporal extensions" of time, with a differentiation between sequential or action-related forms of time and structural or system-related forms. System-related forms of time indicate differences between system and environment in complex forms of multiple temporality. This analysis does not result in a sociological conceptualisation of a new society, but ends in the diagnosis that the complexity of time regimes in modern society makes a transformation of this society an unrealistic view of the future.

The theories of Beck and Giddens can be understood as part of a theory of globalisation, describing some aspects of social change: dissolution of class structures or individualisation (Beck), a modernity determined by systemic risks, and reflexive modernisation as attempts to deal with such risks in the industrial market economy in attempts of internalisation of external effects. The idea of reflexive modernisation aims at dealing with problems and risks, and does not fully take into account the "misadventures of capitalist nature" (O'Connor 1993) and the countertrends of societal "de-modernisation" or dissolution of social organisation. These are insufficiently explored in sociology, discussed as a "new barbarism" with the exploding intra- and international violence in regional conflicts, wars, and civil wars since the end of the East–West confrontation and the collapse of socialism (Miller and Soeffner 1996).

Beck's hypothesis of individualisation is a variant of the sociological reflections of dissolving class structures in modern society without dissolution of the capitalist mode of production. Individualisation is not a sociological proof for the transformation of modern societies. It can also be doubted that individualisation is a form of de-structuration, rather it is part of multiple and contradicting forms of changing social and class structures. Modern capitalism changes significantly with the processes of economic globalisation and the technological innovations of the past decades. But this can more convincingly be explained as part of the global spreading of the capitalist mode of production than as its transformation into another form of economy and society. Sociological and economic assumptions that see the beginning of globalisation in the recent past, since the 1970s, when the deregulation of international capital flows and the expanding significance of financial capital began, seem to ignore what is known from older critical and political-economic theories:

- The theorem of the critique of political economy by Marx sees the modern capitalist mode of production as constantly revolving productive forces, technologies, and means of production. This is shown in the "long waves", or Kondratiev cycles, of capitalist development since the beginning of the industrial revolution.
- The historically more specific theory of the modern world system does not adopt the notion of globalisation in its sense of recent neoliberal deregulation of the markets. The beginning of international spreading of modern capitalism and its continued economic and technological modernisation began with the building of the modern world system or world market in the early sixteenth century, after the opening of the seaways for global trade, with the European colonisation of the Global South.

Beck has outlined a cosmopolitan vision of world risk society (1996, 2002, 2009) where he tries to deal with the multi-scale phenomena of societal organisation in terms of local, national, and global structuring. His account of globalisation as de-territorialisation of production and cosmopolitanism illustrates more the confusion about globalisation (for the continuing controversy about national and cosmopolitan orders see Frödin 2013), playing with a variety of ideas and terms: globalisation through information technology, through changing scopes of action and decision-making frames, through de-territorialisation of economic production, transnationalisation of the world market, globalisation of trade and global intra-firm trade. For all these phenomena, the new term of cosmopolitanism does not find explanations—it only adds observations to the idea of a qualitative rupture of national and international processes. In the Internet-based economy, the internal space of national and local ways of business and trade is transformed through new options and decision-making under the influence of possibilities of global communication and action. Beck interprets this as a paradigm shift from territorial production for local or national markets to de-territorialised forms of production for several national markets or the world market. The statistics based on national economies and international exchange lose in value as information. International trade is transformed to intra-firm trade without buying and selling, only pushing back and forth goods within firms operating transnationally. At the turn of the century, estimates were that 40–60 % of international trade are "intra-firm non-trade" (Beck 2002: 32). The meaning of transformation is reduced to that of changes in international trade- and market-based processes of which the consequences seem dramatic, but are not analysed further in their consequences.

Giddens elaborated a more critical analysis of modern society in his early theoretical critique of historical materialism and his critique of the conventional forms of systems and action theory. He tries to integrate system and action theories in a theory of structuration and modernisation that takes up insights from critical theory (Held and Thompson 1990). In its further development his theory, with the core theory of reflexive modernisation, ended in similar diagnoses as Beck's theory of risk society: as modernisation that cannot end, can only improve its forms through becoming self-reflexive to maintain growth and development. Reflexive modernisation can be seen as an overarching term to describe processes of change that have also been reflected in the discourse of post-industrial society: development and modernisation require in late modernity a reorganisation of the economy to deal with the systemic risks. The modernisation that characterises risk society turns out to be reorganisation and adaptation to global social and environmental change; it can be seen as similar to processes of adaptation described in the ecological discourse of resilience. Social and economic systems can acquire capacities of dealing with risks, disturbances, and shocks and can even remain in states of "sustained crisis" as the long crisis of financial capital from the last decade shows. But no transformation of the societal or economic system is initiated. Similar as in the theory of ecological modernisation of Mol, the new social and environmental movements are seen as articulating the requirements of reflexive modernisation by trying to adapt to the industrial system in attempts to change it "from within", in contradicting ideas of sustainable development without system transformation.

For the new theories of post-industrial society, risk society, and reflexive modernisation, the analysis of modes of production and reproductive mechanisms of society are no longer required. These theories make use of supporting arguments from sociological analyses of globalisation and cultural and political change. In the new theories, the common denominator is an idea that survived since the early evolutionary theory of Spencer in the nineteenth century, although does no longer take the form of naïve evolutionism or the Dominant Western Worldview in industrial society (Dunlap and Dunlap 1992/1993). This idea is, that modernisation and industrialisation, or, in a more valuing term, progress can go on forever. Today progress or development is no longer seen as linear and irreversible, but as threatened through risks, ruptures, and catastrophes. Modernisation, seen as reflexive modernisation, implies adaptation to changing contexts and conditions of development.

The sociological theories of Beck (cosmopolitan risk society), of Giddens (reflexive modernisation), and of Mol (ecological modernisation, environmental flows: Mol and Spaargaren 2006) took up the nature–society theme in fragmented form. The authors do not aim at grand theory and big knowledge syntheses. All three theories develop from the conventional sociological term and perspective of modernisation assuming that the modernisation process is continuing in changing forms. The theories show that modern industrial societies cannot escape their systemic constraints in their relations with nature. Economic growth and industrial production require more use of natural resources and cause environmental problems. But in the forms of ecological or reflexive modernisation industrial societies appear as blocked in their transformation towards sustainability. Efforts to deal with environmental problems by way of internalisation of pollution costs in the economic process, through cost assessment and pricing of nature, show limited success in regulating societal interaction with nature (see Chap. 5).

References

Altvater, E. (1991). *Die Zukunft des Marktes. Ein Essay über die Regulation von Geld und Natur nach dem Scheitern des „real existierenden Sozialismus".* Münster: Westfälisches Dampfboot.

Amstutz, M., & Fischer-Lescano, A. (Eds.). (2013). *Kritische Systemtheorie: Zur Evolution einer normativen Theorie.* Bielefeld: Transcript Verlag.

Anacker, U. (1974). *Natur und Intersubjektivität: Elemente zu einer Theorie der Aufklärung*. Frankfurt: Suhrkamp.
Arnason, J. P. (1976). *Zwischen Natur und Gesellschaft. Studien zu einer kritischen Theorie des Subjektes*. Frankfurt a. M.: Europäische Verlagsanstalt.
Arnason, J. P. (1988). Social theory and the concept of civilisation. *Thesis Eleven, 20*(1), 87–105.
Arnason, J. P. (2003). *Civilizations in dispute: Historical questions and theoretical traditions*. Leiden: Brill.
Arnason, J. P. (2006). Civilizational analysis, social theory and comparative history. In G. Delanty (Ed.), *Handbook of contemporary european social theory*. London: Routledge.
Barry, J. (1999). *Environment and social theory*. London: Routledge.
Bauman, Z. (1992). *Intimations of postmodernity*. London: Routledge.
Beck, U. (1996). World risk society as cosmopolitan society? Ecological questions in a framework of manufactured uncertainties. *Theory, Culture & Society, 13*, 1–32.
Beck, U. (2002). The cosmopolitan society and its enemies. *Theory, Culture & Society, 19*(1–2), 17–44.
Beck, U. (2009). Critical theory of world risk society: A cosmopolitan vision. *Constellations, 16*(1), 3–22.
Beckert, J. (2009). Wirtschaftssoziologie als Gesellschaftstheorie. *Zeitschrift für Soziologie, 38*(3), 182–197.
Beetz, M. (2010). *Gesellschaftstheorie zwischen Autologie und Ontologie: Reflexionen über Ort und Gegenstand der Soziologie*. Bielefeld: Transcript Verlag.
Bell, D. (1973). *The coming of post-industrial society: A venture in social forecasting*. New York: Basic Books.
Benson, M. H., & Craig, R. K. (2014). The end of sustainability. *Society and Natural Resources, 27*(7), 777–782.
Bergmann, W. (1992). The problem of time in sociology. *Time & Society, 1*(1), 81–134.
Biermann, F. (2011). *Reforming global environmental governance: The case for a United Nations Environment Organisation (UNEO)*. Amsterdam: Earth System Governance Project and VU University Amsterdam.
Biro, A. (Ed.). (2011). *Critical ecologies: The Frankfurt School and contemporary environmental crises*. Toronto: University of Toronto Press.
Blok, A. (2010a). Topologies of climate change: Actor-network theory, relational-scalar analytics, and carbon-market overflows. *Environment and Planning D: Society and Space, 28*, 896–912.

Blok, A. (2010b). Divided socio-natures: Essays on the co-construction of science, society, and the global environment. PhD thesis, Department of Sociology, University of Copenhagen.

Brand, U. (2015a, June 10–12). *How to get out of the multiple crisis? Contours of a critical theory of social-ecological transformation.* Paper presented at the conference The Theory of Regulation in Times of Crises, Paris. Accessed November 10, 2015, from https://www.eiseverywhere.com/retrieveupload.php?

Brand, U. (2015b). Sozial-ökologische Transformation als Horizont praktischer Kritik: Befreiung in Zeiten sich vertiefender imperialer Lebensweise. In D. Martin, S. Martin, & J. Wissel (Eds.), *Perspektiven und Konstellationen kritischer Theorie*. Münster: Westfälisches Dampfboot.

Brenner, N. (1997). Global, fragmented, hierarchical: Henri Lefebvre's geographies of globalization. *Public Culture, 10*(1), 137–169.

Bruckmeier, K. (2013). *Natural resource use and global change: New interdisciplinary perspectives in social ecology*. Houndmills, Basingstoke: Palgrave Macmillan.

Bruckmeier, K. (2015a). *Soziologitcheskjie teorii obshestvo (Sociological theories of society)*. Report, Department of Sociology, National Research University – Higher School of Economics, Moscow.

Bruckmeier, K. (2015b). Systemtheorie in der Humanökologie. In K.-H. Simon & F. Tretter (Eds.), *Systemtheorie und Humanökologie: Positionsbestimmungen in Theorie und Praxis*. Edition Humanökologie Band 9. München: Ökom-Verlag.

Brunnengräber, A. (2009). Die politische Ökonomie des Klimawandels Ergebnisse sozial-ökologischer Forschung, Band 11. München: Ökom-Verlag.

Brunnengräber, A., Dietz, K., Hirschl, B., Walk, H., & Weber, M. (2008). *Eine sozial-ökologische Perspektive auf die lokale, nationale und internationale Klimapolitik*. Münster: Westfälisches Dampfboot.

Bühl, W. (1983). Die 'Postindustrielle Gesellschaft': eine verfrühte Utopie? *Kölner Zeitschrift für Soziologie und Sozialpsychologie, 35*(4), 771–780.

Clark, T. N. (1991). Are social classes dying? *International Sociology, 6*(4), 397–410.

Castells, M. (2010) *The Rise of the Network Society*. Chichester, UK: Wiley-Blackwell (2nd ed.).

Delanty, G. (Ed.). (2006). *Handbook of contemporary European social theory*. London: Routledge.

Dunlap, R., & Catton, W. (1992/1993). Toward an ecological sociology: The development, current status, and probable future of environmental sociology. *The Annals of the International Institute of Sociology, 3,* 263–284.
Eckersley, R. (1992). *Environmentalism and political theory: Toward an ecocentric approach.* Albany, NY: State University of New York Press.
Escobar, A. (2006). Difference and conflict in the struggle over natural resources: A political ecology framework. *Development, 49*(3), 6–13.
Featherstone, M., Lash, S., & Roberston, R. (1995). *Global modernities.* London: Sage.
Fischer Kowalski, M., & Haberl, H. (Eds.). (2007). *Socioecological transitions and global change. Trajectories of social metabolism and land use.* Cheltenham: Edward Elgar.
Fischer-Kowalski, M., Haberl, H., Hüttler, W., Payer, H., Schandl, H., Winiwarter, V., et al. (1997). *Gesellschaftlicher Stoffwechsel und Kolonisierung von Natur: Ein Versuch in Sozialer Ökologie.* Amsterdam: G+B Verlag Fakultas.
Forst, R., & Klaus, G. (Eds.). (2011). Die Herausbildung normativer Ordnungen: Interdisziplinäre Perspektiven. Normative Orders Bd. 1. Frankfurt am Main: Campus.
Fraser, N. (1997). *Justice interruptus.* New York: Routledge.
Freese, L. (1997). Environmental connections. *Advances in Human Ecology,* Supplement 1, Part B.
Frödin, O. (2013). Political order and the challenge of governance: Moving beyond the nationalism-cosmopolitanism debate. *Distinktion, 14*(1), 65–79.
Gandhi, L. (1998). *Postcolonial theory: A critical introduction.* New York: Columbia University Press.
Girardet, H. (Ed.). (2007). *Surviving the century.* London: Earthscan.
Habermas, J. (1968) *Erkenntnis und Interesse.* Frankfurt am Main: Suhrkamp.
Habermas, J. (Ed.). (1978). *Antworten auf Herbert Marcuse.* Frankfurt am Main: Suhrkamp.
Habermas, J. (Ed.). (1981). *Theorie des kommunikativen Handelns, 2 Bände.* Frankfurt am Main: Suhrkamp.
Held, D., & McGrew, A. (2007). *Globalization theories: Approaches and controversies.* Cambridge: Polity Press.
Held, D., & Thompson, J. B. (Eds.). (1990). *Social theory of modern societies: Anthony Giddens and his critics.* Cambridge: Cambridge University Press.
Horkheimer, M., & Adorno, T. W. (1971 [1947]). *Dialectic of enlightenment.* New York: Herder and Herder.

Hornborg, A., & Crumley, C. L. (Eds.). (2006). *The World System and The Earth System: Global socioenvironmental change and sustainability since the neolithic.* Walnut Creek, CA: Left Coast Press.
Illich, I. (1973). *Tools for conviviality.* New York: Harper and Row.
Jain, A. K. (2006, April 4). *Jenseits der Gesellschaft? Soziologische Konzepte für das neue Jahrtausend.* http://www.power-xs.net/jain/pub/jenseitsdergeselslchaft.pdf
Jamison, A. (2001). *The making of green knowledge: Environmental politics and cultural transformation.* Cambridge: Cambridge University Press.
Jamison, A. (2010). Climate change knowledge and social movement theory. *Wiley Interdisciplinary Reviews: Climate Change, 1*(1), 811–823.
Latour, S. (1999). On recalling ANT. In K. Law & J. Hassard (Eds.), *Actor network theory and after* (pp. 15–25). Oxford: Blackwell.
Law, J., & Hassard, J. (Eds.). (1999). *Actor network theory and after.* Oxford/Keele: Blackwell/Sociological Review.
Lefebvre, H. (1991 [1974]). *The production of space.* Oxford: Basil Blackwell.
Loubser, J., et al. (Eds.). (1976). *Explorations in general theory in social science: Essays in honor of Talcott Parsons, 2 Vols.* New York: Free Press.
Löw, M. (2001). *Raumsoziologie.* Frankfurt am Main: Suhrkamp.
Luhmann, N. (1976). The future cannot begin: Temporal structures in modern society. *Social Research, 43,* 130–152.
Luhmann, N. (1986). *Ökologische Kommunikation: Kann die moderne Gesellschaft sich auf ökologische Gefährdungen einstellen?* Opladen: Westdeutscher Verlag.
Luhmann, N. (1997). Die Gesellschaft der Gesellschaft, 2 Bände. Frankfurt am Main: Suhrkamp.
Luhmann, N. (1999). *Gesellschaftsstruktur und Semantik.* Frankfurt am Main: Suhrkamp.
Marcuse, H. (1964). *One dimensional man.* Boston: Beacon.
Martinez-Alier, J. (1995). Political ecology, distributional conflicts and economic incommensurability. *New Left Review, 211,* 70–88.
Martinez-Alier, J. (2004). Ecological distribution conflicts and indicators of sustainability. *International Journal of Political Economy, 34*(1), 13–30.
Marx, K., & Engels, F. (1932). Ökonomisch-philosophische Manuskripte aus dem Jahre 1844. In *Marx Engels Gesamtausgabe, Abteilung I, Band 3* (pp. 29–172). Berlin: Marx-Engels-Verlag.
Mellor, M. (1997). *Feminism and ecology.* Cambridge: Polity Press.
Merchant, C. (1980). *The death of nature: Women, ecology, and the scientific revolution.* San Francisco: Harper and Row.

Miller, M., & Soeffner, H.-G. (Eds.). (1996). *Modernität und Barbarei: Soziologische Zeitdiagnose am Ende des 20. Jahrhunderts*. Frankfurt am Main: Suhrkamp.

Mol, A., & Spaargaren, G. (2006). Towards a sociology of environmental flows: A new agenda for 21st century environmental sociology. In G. Spaargaren, A. P. J. Mol, & F. H. Buttel (Eds.), *Governing environmental flows: Global challenges to social theory* (pp. 39–82). Cambridge, MA: MIT Press.

Moscovici, S. (1982 [1977]). *Versuch über die menschliche Geschichte der Natur*. Frankfurt am Main: Suhrkamp.

O'Connor, M. (1993). On the misadventures of capitalist nature. *Capitalism, Nature, Socialism, 4*(3), 7–40.

O'Connor, J. R. (1998). *Natural causes: Essays in ecological Marxism*. New York: Guilford Press.

Piketty, T. (2013). *Le capital au 21e siècle*. Paris: Editions du Seuil.

Polanyi, K. (1944). *The great transformation*. New York: Farrar & Rinehart.

Polanyi, K. (1979). *Ökonomie und Gesellschaft*. Frankfurt am Main: Suhrkamp.

Rehbein, B., & Souza, J. (2014). *Ungleicheit in kapitalistischen Gesellschaften*. Weinheim: Beltz Juventa.

Rice, J. (2007). Ecological unequal exchange: Consumption, equity, and unsustainable structural relationships within the global economy. *International Journal of Comparative Sociology, 48*(1), 43–72.

Ritsert, J. (1988). *Gesellschaft: Einführung in den Grundbegriff der Soziologie*. Frankfurt am Main: Campus.

Ritsert, J. (2009). *Schlüsselprobleme der Gesellschaftstheorie: Individuum und Gesellschaft – Soziale Ungleichheit – Modernisierung*. New York: Springer.

Roth, R., & Rucht, D. (Eds.). (2008). *Die sozialen Bewegungen in Deutschland seit 1945: Ein Handbuch*. Frankfurt: Campus.

Scheunemann, E. (2009 [2008]). *Vom Denken der Natur: Natur und Gesellschaft bei Habermas*. Hamburg-Norderstedt: Books on Demand.

Schnaiberg, A. (1980). *The environment: From surplus to scarcity*. New York: Oxford University Press.

Schnaiberg, A., & Gould, K. A. (1994). *Environment and society: The enduring conflict*. New York: St. Martin's Press.

Scholte, J. A. (2000). *Globalization: A critical introduction*. New York: MacMillan.

Stehr, N. (2007). *Moralisierung der Märkte: eine Gesellschaftstheorie*. Frankfurt am Main: Suhrkamp.

Stichweh, R. (2000). *Die Weltgesellschaft*. Frankfurt am Main: Suhrkamp.

Swyngedouw, E. (2010). Impossible/undesirable sustainability and the post-political condition (reprint). In M. Carreta, G. Concillo, & V. Monno (Eds.), *Strategic spatial planning* (Urban and landscape perspectives). Heidelberg: Springer.

Tennenbaum, J. (2015). *The physical economy of national development*. Berlin. Accessed February 27, 2016, from www.physicaleconomy.com

Urry, J. (2000). *Sociology beyond Societies: Mobilities for the twenty-first century*. London: Routledge.

Vietta, S. (1995). *Die unvollendete Speculation führt zur Natur zurück*. Leipzig: Reclam.

Walby, S. (2009). *Globalization and inequalities: Complexity and contested modernities*. London: Sage.

Wallace, R. (Ed.). (1989). *Feminism and sociological theory*. London: Sage.

Wallerstein, I. (2000). Globalization or the age of transition? A long-term view of the trajectory of the world-system. *International Sociology, 15*(2), 249–265.

Warren, K. J. (Ed.). (1997). *Ecofeminism: Women, culture, nature*. Bloomington, IN: Indiana University Press.

Welsch, W. (2003 [1990]). *Ästhetisches Denken*. Stuttgart: Reclam.

Wolf, E. (1982). *Europe and the people without history*. Berkeley, CA: University of California Press.

Zinn, G. (1980). *Die Selbstzerstörung der Wachstumsgesellschaft*. Reinbek: Rowohlt.

3

Interaction of Nature and Society in Ecology

This chapter discusses ecological research on nature–society interaction in the perspective of an interdisciplinary theory. Ecology developed as a science of the relations between living systems and their environment where the study of human–environment relations has brought interdisciplinary approaches in human, cultural, social, and political ecology. Attempts to connect specific theories in ecology in a unifying and general theory are from recent time (Scheiner and Willig 2011). The theories of ecosystem development include succession theory as core and model of the general theory, the theories of ecosystems ecology and of global change, and other, more specific theories and models. The discussion shows that many of the ecological theories are mathematical or statistical models that have yet to be formulated and developed in broader theories. The development of interdisciplinary theories in human, cultural, social, and political ecology has not been included in this attempt to integrate ecological theories. The interaction between human society and nature, relevant for social-ecological theory, comes into view in global change theory. This theory is discussed and developed in this chapter with the new ecological research on vulnerability, resilience, and sustainability.

The following section discusses how interdisciplinary knowledge synthesis is developed in ecology. Ecological concepts and theories are assessed with regard to their significance for an encompassing social-ecological theory of nature and society. Initially, these concepts and theories are used to assess the broader theories and their limits, and thereafter the recent ecological research on global environmental change in the perspectives of vulnerability, resilience, and sustainability.

Ecology as an Interdisciplinary Science of Humans in Ecosystems

The themes studied in ecology include interactions between the non-living physical nature, the living nature of fauna and flora, and humans as biological species and as social actors with culture, consciousness, and capacities of reflection. The biological and social properties of humans can only be divided analytically into a social and a biological lifeworld. Humans are integrated in ecosystems as members of a biological species and in social systems as members of society. But there is no simple and unchangeable connection between society and nature. In the analysis of the varying forms of interaction and coupling of social and ecological systems, it needs to be shown how the integration processes in social and ecosystems connect to each other. With the paradigm shift in ecology described by Scoones (1999), the cognitive problems with different forms of abstraction in ecological and social concepts and theories have become evident. The new ecological thinking is described by Scoones with two components:

- *Interdisciplinarity:* in ecological studies the focus is on non-equilibrium, dynamics, and spatial and temporal variation of ecosystems where complexity and uncertainty prevail.
- *Espistemic changes:* spatial and temporal dynamics are reconstructed in situated analyses of people in places; the environment is understood as both the product of and the setting for human interactions; complexity and uncertainty in social-ecological systems (SES) show that prediction, management, and control are unlikely, if not impossible.

SES is used to frame the analysis of nature–society interaction, but is not yet developed as a theoretical concept. The other concepts in the description of Scoones come from systems theory. A system-theoretical language does not yet show how knowledge from the different disciplines can be integrated: this requires specific methodologies for knowledge synthesis. Two points made by Scoones need to be discussed further in the social-ecological theory: the interaction and coupling of social and ecological systems; and the argument that complexity and uncertainty in the dynamics of coupled social and ecological systems make management and control of the systems unlikely or impossible. The ecological analyses of vulnerability, resilience, and sustainability discussed in this chapter provide further knowledge about the systemic processes in SES, including managed and non-managed change.

The construction of SES-created epistemological and methodological problems is described in the following three points:

1. *Early attempts of conceptualising the interaction between social and ecological systems in ecology show a lack of criteria, differentiated concepts, and frameworks for the analysis of interfaces between social and ecological systems.* Dykes (1988) discusses the evolutionary dynamics of complex systems by connecting biology to philosophy to reflect on the questions of explanation, determination, teleology, reductionism, and hierarchy. How biological concepts are interwoven with social system concepts that are part of complex adaptive and interacting systems is not yet shown in this epistemological discussion. Pickett et al. (2005) study biocomplexity in interacting social and ecological systems in a simplified framework of spatial, organisational, and temporal dimensions with limited use of social-scientific knowledge. Complexity is seen as increasing when connections develop to configuration and functional interdependence of the system elements. Complex interaction processes like biological and social reproduction, societal metabolism, or societal transformation cannot be adequately analysed in this framework that develops from formal and quantitative analyses of intersystemic relations inexactly described as "human-natural systems". Turchin and Hall (2003) attempt to connect the sociological world system theory with ecosystem concepts and theories. They see

this attempt as speculative and limit the analysis to possibilities of synchronisation of processes in connected systems. Two points from these abstract reflections seem important for the analysis of SES:

- Spatial and temporal connections in interacting systems that include various forms of synchronisation, and
- the requirement of multi-scale analyses in the dynamics of SES.

Scoones (1999) describes the ecological analysis of intersystem dynamics more systematically, but a social-ecological theory is not yet developing with that.

2. *In interdisciplinary social-ecological analyses two forms of knowledge need to be combined:* ecological knowledge from ecosystem analyses and social-scientific knowledge about human action in social systems. Social action in resource-use practices is shaped through the normative ideas and aspirations of collective actors, their varying interests and objectives. Beyond the mapping of interests through empirical research as given facts, the interests are influenced from the social structures of society, the historically specific forms of social division of labour, the systemic structures of modes of production, reproduction, and systems transformation. These connections between actors and systems are analysed in the theory of modern society and its relations with nature (discussed in Chap. 2). How to communicate the social and cultural structuring of action in ecology, in analyses that work with functionalist terms, is methodologically difficult. Incompatible concepts, epistemologies, and ontologies block the interdisciplinary communication. Abstract terms such as "socionatures" are not sufficient to describe the differentiated structures, processes, and relations that mediate between social and ecological systems (see Chap. 5). To support the communication the social-ecological theory develops more specific interdisciplinary concepts and forms of analysis.

3. *The interdisciplinary theory of society and nature includes interactions between the physical, biological, and social nature of humans.* The interactions between the ecological and social systems as core theme of the theory show many forms of exchange of material, energy, and information, and symbolic processes of communication. All these forms are included in the abstract term of system dynamics that is too inexact to

describe SES theoretically. Knowledge and terminology from social-scientific theories about the dynamics of human societies, their social structures, their historically specific modes of production and reproduction, development and transformation are, so far, hardly used in ecological research.

The three kinds of epistemological and methodological difficulties are not yet solved. This makes the interdisciplinary knowledge exchange and synthesis of social and ecological knowledge selective, insufficiently reflected, and structured in epistemological and methodological terms. Ecological concepts and theories of nature–society interaction need to be connected with social-scientific concepts and with social knowledge practices in the processes of resource use and management—in methodological forms that need to be elaborated in the development of the interdisciplinary social-ecological theory.

Ecological Theories of Nature–Society Interaction

In the ecological discourse, the term of the anthropocene (Ehlers and Krafft 2006) is more and more used to describe a new historical epoch by connecting the concepts for the geological epochs in earth history with the concepts for the description of modern society. The anthropocene is described in conventional social-scientific terms as industrial revolution and industrial society. The cognitive status of the term anthropocene—as concept, model, framework, theorem, theory, or hypothesis—is disputed. The main argument for the use of this term is formulated in the hypothesis that humankind has become a geological force in modernity through its strong modification of the ecosystems on earth. The discussion shows the different understandings of theory in interdisciplinary discourses. For ecology, as discussed by Scheiner and Willig (2011), theory means often mathematical and statistical models or less formalised conceptual models and frameworks that do not develop into systematic theories and explanations as found in the social sciences. The ecological community assembly theory (Weiher et al. 2011) that deals with niche-based interaction processes between species shows the highly specialised theories and models used in ecology. These models can only be used for limited and specific forms of interdisciplinary communication and synthesis (in this example between ecology and biogeography). Few systematic theories exist in biology and ecology, with the synthetic theory of evolution (Mayr and Provine 1980) being the widely accepted one. From the practice of theorising in ecology, it can be concluded that

- general theories of the explanatory kind are hardly formulated;
- the theory concept remains epistemologically vague;
- models and conceptual frameworks with the function to guide empirical research prevail in the practice of ecological research.

The concept of the anthropocene came into use in interdisciplinary theorising about humans and nature. It is of interest for a social-ecological theory of nature and society not for the hypothesis of mankind as a geological force, but for the reason that this hypothesis indicates the necessity of much more differentiated empirical and theoretical reconstructions of the interaction between humans, society and nature. A coming theory of the anthropocene can be a part of the broader theory of nature–society interaction. In its present undeveloped state it is one of the few attempts to start to formulate from natural-scientific knowledge a theory of modern society that requires further steps of knowledge synthesis. Such synthesis is in an exemplary way developing with the social-ecological theory.

The core processes to be analysed and explained in the social-ecological theory include that of production and reproduction in the systems in nature and society. A series of ecological theories provide knowledge for this theory (Table 3.1). In the following section, the theme is the relevance of these theories for an interdisciplinary theory. Thereafter the reasons for the slow and difficult development of such a theory are discussed.

The unifying theory of ecology (Scheiner and Willig 2011, not discussed in Table 3.1) provides knowledge about society–nature interaction mainly through the theory of global change. Older approaches of interdisciplinary thinking in biology include the subjective biology of Uexküll (1926) who constructed a theory of action and interaction of species in sociological terms of social action and lifeworld. This theory and examples for new holistic thinking in biology (Gierer 1998) are not discussed further here;

Table 3.1 Ecological theories of nature–society interaction

Theme	Theories and authors	Main ecological assumptions and theorems	Assessment
(1) Ecological production and reproduction: – biological reproduction (species) – autopoiesis of living systems – primary production of ecosystems – ecosystem functions, maintenance of ecosystems – material/nutrient cycles in ecosystems – ecological models of (circular) economy	Maturana and Varela (1980) Systems ecology (Odum 1983) Ecological economics (see below)	The terms of production and reproduction describe continued processes to maintain life, ecosystems and social systems. Autopoiesis is easily described for living systems as building and renewing from the elements that constitute the system/organism; controversial is whether autopoiesis can be assumed for social systems; in ecosystems autopoiesis can refer to their living components (species), or to processes of material/nutrient cycles	Biological concepts of reproduction and autopoiesis, conceived for living systems, become system metaphors with their transfer to ecosystems or social systems; ecological concepts are not connected to societal forms of reproduction; ideas of a circular economy are simplified models of economic systems in ecological terms

(continued)

Table 3.1 (continued)

Theme	Theories and authors	Main ecological assumptions and theorems	Assessment
(2) Complex adaptive systems – complexity in ecosystems	Lansing (2003), Porter (2006) Norberg and Cumming (2008)	Such systems include large numbers of elements; the interactions between the elements cannot be sufficiently described, explained, or predicted; open systems, far from equilibrium; non-linear development; constant flows of energy to maintain the systems; properties of complexity, self-organisation, emergent order, high adaptive capacity, resilience	Although developing from natural sciences, adaptive systems include many kinds: organisms, ecosystems, social systems, material and symbolic systems; a component of systems theory (to describe similarities and common properties of different kinds of systems) Complexity in ecosystems (Norberg and Cumming): connecting theory of complex adaptive systems and ecosystem/natural resource management
(3) Ecosystem dynamics: – systems ecology – ecosystem-based management	Strategy of ecosystem development (Odum 1969) and system ecology Ecosystem management, adaptive management and governance	Concepts and conceptual frameworks from applied ecological research, formulating criteria and rules for management, regulation, governance of ecosystems	Including heuristics (experience-based rules and techniques) more than theories; varying principles and rules; low theoretical and explanatory value

3 Interaction of Nature and Society in Ecology

(4) Global environmental change	Climate change, loss of species/biodiversity reduction, land use change as main forms of global environmental change—investigated in large research programmes (climate change) and assessments (Millennium Ecosystem Assessment, IAASTD)	No theories of society and nature in the strict sense of the term, although interpreting environmental change as man-made process; the changes observed in environmental research require social-ecological theories to explain the processes in their societal origins, causes and forms	Mainly descriptive, based on ecological research about the consequences of human intervention in natural processes and material cycles; no explicit theory of society and nature beyond the assumption of environmental change as anthropogenic
(5) Planetary boundaries of human resource use	Rockström et al. (2009)	First and preliminary approach to systematise knowledge about the limited availability of natural resources for human use	In difference to carrying capacity of ecosystems (maximum of individual organisms and species an ecosystem can feed), the formulation of global boundaries specifies physical limits to growth

(*continued*)

Table 3.1 (continued)

Theme	Theories and authors	Main ecological assumptions and theorems	Assessment
(6) Co-evolution of social and ecological systems: – co-evolution – biocomplexity	Norgaard (1984), Porter (2006) Pickett et al. 2005	Co-evolution in biology as co-evolution of species, transferred by Norgaard to co-evolution of social systems/economy and ecosystems	The term of co-evolution becomes controversial with regard to kinds of systems that can co-evolve (Weisz 2001). The term build on the biological concept of evolution which is continually disputed for social systems
(7) Coupling of social and ecological systems and social-ecological systems (SES)	Binder et al. (2013)	A theory developing from typologies of concepts and frameworks to describe the connections between social and ecosystems in qualitative and quantitative terms	SES concepts and forms of coupling are hardly theoretically reflected, the theory is not advanced

(8) Ecosystem services for society/humans	Carpenter et al. (2006), Farley (2012), Lele et al. (2013)	Nature has life-supporting functions for humans and society, specified in form of services	Lack of theoretical reflection of the production of services (coproduction by humans and nature?); descriptive form of services derived from knowledge about ecosystem functions; implications for the ecosystem concepts are not reflected further when ecological and biological functions and processes are assumed to become part of social systems
(9) Ecological theories of resilience	Folke et al. (2005), Folke (2006)	Resilience as capacity of ecosystems to buffer disturbance of various kinds, e.g. natural catastrophes	The transfer of the concept to humans, social groups and social systems is controversial (metaphors and analogies)

(continued)

Table 3.1 (continued)

Theme	Theories and authors	Main ecological assumptions and theorems	Assessment
(10) Ecological theories of sustainability	– ecological theory of sustainable systems (Cabezas et al. 2005) – ecologically sustainable development (in difference to the social and economic components of the mainstream idea of sustainable development) – strong sustainability (also in ecological economics) – sustainability in frontier research/SES	Mathematical/information theory. Different theories: varying according to the interpretation of the term sustainability—as ecological integrity, as functional or adaptive coupling of social or ecosystems, as maintaining functions and process in nature and ecosystems; a joint assumption in the variants of ecological sustainability: sustainability is the basis of nature–society interaction throughout all processes of adaptation and transformation and cannot be substituted through other social processes	Limited capacity to integrate social and ecological knowledge

(11) Human, social, cultural, political ecology: – interaction of humans, society and nature – biological and social/societal metabolism – earth system science – world ecology	Several theories converging to an interdisciplinary theory of nature and society in critical social ecology (Frankfurt and Vienna School of social ecology) Ehlers and Krafft (2006) Hornborg and Crumley (2006), Hobbs et al. (2006), Fischer-Kowalski and Haberl (2007), Miao et al. (2009)	Societal metabolism is more than natural resource use in production and consumption: includes complex patterns of connected processes in social and ecosystems	Deliver important knowledge for a new theory of nature–society interaction: – interaction of the social world system and the ecological earth system – global social-ecological change – social-ecological transformation to sustainability – biological and societal metabolism
(12) Ecological economics	Georgescu-Roegen (1971) Daly (1991, 1997) Martinez-Alier 2002	Entropy law and economic processes Steady-state economics Beyond growth: economics of sustainable development Environmentalism of the poor: ecological conflicts and valuation	Ecological economics analyses interaction of economy and nature in specific forms (energy accounting, analyses of material flows) to show the limits to growth; description of an economy of zero growth or degrowth

(*continued*)

Table 3.1 (continued)

Theme	Theories and authors	Main ecological assumptions and theorems	Assessment
(13) Industrial ecology	Frosch and Gallopoulos (1989), Allenby (2006)	Interdisciplinary study of the interaction between the biosphere and the industrial technosphere; society and industry are bounded within nature or the biosphere; industrial metabolism (flows of energy and materials through industrial systems)	Societal components in the interaction of society and nature reduced to technologies; metaphoric use of ecology for a circular model of material flows
(14) Ecological anthropology	Moran (2000), Vayda (2009)	Analysis of complex relations (and changes of relations) between human populations and their natural environment (plants and animals, land, climate, natural resources)	Ecological analysis of specific material components of nature–society interaction

Sources: Own compilation, sources mentioned

they include specific components of nature–society interaction that could become useful for discussing the interactions between humans and animals in modern society. With the social-ecological theory, holistic thinking shares only some conceptual problems, for example, that of possibilities to extend the concept of social action beyond humans. Further areas of interdisciplinary research in historical ecology, environmental history, and environmental psychology are more limited and less relevant for the construction of a theory of nature–society interaction.

The *ecological theories* (themes 1–5 in Table 3.1) add the following components to the social-ecological theory that is mentioned in the Table as part of the broader interdisciplinary discourses of human, social, cultural, and political ecology:

1. *The biological and ecological concepts and theories of reproduction of species, primary production of biomass in ecosystems, autopoiesis of living systems, and maintenance of ecosystems:* These theories complement the sociological theories discussed in Chap. 2 and provide knowledge for the core of the social-ecological theory (theme 1 in Table 3.1). The interdisciplinary knowledge synthesis includes different forms: empirical forms of synthesis, for example, with the analysis of ecosystem services for humans and society, and theoretical synthesis in forms of social-metabolic profiles and regimes from social-ecological research. The biological theory of autopoietic systems is also applied to social systems, but remains controversial in the social-scientific discourse; in biological theory, it has the role of a theory of reproduction in the sense of chemical processes that maintain living cells. An attempt to translate ecological ideas of reproduction into principles for the organisation of human economy and natural resource use has resulted in a simple model of circular economy that is formulated in analogy to circular processes of production, consumption, and reduction in ecosystems: the dysfunctionality of modern economies is seen by their lack of a reduction function as they give waste back to nature. The complexity of economic reproduction in modern capitalism cannot be analysed with this model, but the analysis of reduction functions can be connected with economic analyses of reproduction and accumulation.

2. *Important for a theory of nature–society interaction, but not for its core, are the theories of complex adaptive systems, ecosystem functions and management, or global environmental change from which the interdisciplinary thinking in ecology develops.* These theories describe nature–society interaction insufficiently. The research on these themes makes a significant part of ecological and environmental research, providing knowledge complementary to the systems analysis of SES. The ecosystem processes analysed in this research cannot replace theories of production and reproduction, but become relevant in analyses of global change where the theoretical analyses of system maintenance need to be transferred into analyses of system transformation. These theories of transformation can be divided in two groups: some describe ecosystems, their functions and processes theoretically (global environmental change, complex adaptive systems, co-evolution), whereas others translate ecological knowledge into managerial practices of resource use (ecosystem management and governance). *Complementary to the transformation theories is the theory of planetary boundaries of resources under development.* It is not yet an explanatory theory, rather a theoretical classification of the limits of natural resource use for developing such a theory, based on measurements that deliver information to formulate such a theory.

The *interdisciplinary theories* mentioned in Table 3.1 (themes 6–14) can be divided into four groups of more or less important ones for the social-ecological theory, in different forms connected to the core of this theory:

1. Analyses of *societal metabolism* discussed in human, social, and political ecology develop as an integrative core of the social-ecological theory of nature–society interaction. With the term "societal metabolism" are the conceptual and explanatory components formulated that connect social and ecological theories of production and reproduction and theories of global social and environmental change.
2. *Theories of ecosystem services, of vulnerability, resilience, and sustainability* work with assumptions and hypotheses about the relations and interactions between nature and society at the level of specific ecosystems.

As operational theories, they conceptualise interaction processes between social and ecosystems based on ecological research that is so far badly connected to sociological research. They include preliminary and undeveloped forms of theory and hypothetical assumptions that need to be verified through further research and theoretical codification to dissolve contradicting interpretations of the processes of intersystemic exchange and interaction of SES.
3. *Theories and frameworks of co-evolution and of coupling of social and ecological systems* as specific theories of SES describe forms and conditions of interaction, communication, knowledge exchange, or coupling between different types of social or ecological systems. These theories and frameworks connect to the systemic reproduction and societal metabolism operating through SES.
4. *Ecological anthropology, ecological economics, and industrial ecology* are examples for interdisciplinary approaches combining social and ecological knowledge. They are less used for the formulation of a theory of the societal relations to nature in modern society and more for research on specific economies or cultures and their modes of resource use. This research provides empirical knowledge for the broader theory of nature–society interaction but does not provide knowledge for its conceptual framework.

With this discussion of ecological theories, it seems possible to complete the overview of sociological and ecological components of an interdisciplinary social-ecological theory. The following epistemological reflections describe possibilities of improving the development of social-ecological knowledge integration in ecological research.

Shifting the Limits of Ecological Research on Nature–Society Interaction

Research in ecology shows a series of limits and difficulties that block—as consequences of disciplinary specialisation—the formulation of a broader social-ecological theory. The attempts to formulate with ecological knowledge a theory of nature–society interaction reflect this thematic limitation and specialisation in the concepts used and in the knowledge excluded. These attempts develop from specialised research, using pre-theoretical concepts, models, and frameworks. The following epistemological and methodological reflections refer to possibilities to shift the limits of ecological research on SES through knowledge syntheses.

1. *Interdisciplinary research and theory developed slowly and in limited fields of ecology.* These fields are applied research (such as that on vulnerability, resilience, sustainability) or marginal and heterodox subjects (such as human, social, and political ecology). Interdisciplinary research developed rapidly in the past decades, with the aggravation of environmental problems and the emergence of environmental policy and environmental movements (Jamison 2001). The attempts to integrate knowledge and formulate interdisciplinary concepts and frameworks in ecology began with pre-theoretical reflections of nature–society relationships in models and conceptual frameworks. As the late attempt by Scheiner and Willig (2011) shows, these models did not developed into broader and general theories in ecology. The new theory of Scheiner and Willig does not direct towards an interdisciplinary theory of nature and society; rather it remains within the scope of conventional ecology as natural-scientific discipline that studies the development of ecosystems. Limits of ecological theorising show especially in the few concepts used (e.g. space and time) to describe intersystemic processes in SES. No attempts have been made to work with theoretical concepts from the social sciences in the analysis of humans in ecosystems, although there is some awareness of the limits of ecological or natural-scientific concepts for the analysis of social action or environmental policy as the discussion of vulnerability and resilience below (following section) describes. The first requirement of an interdisciplinary social-ecological theory is to formulate the core concepts and theorems of this theory. This can be done by comparing, systematising, and classifying the ecological and sociological concepts used to describe social and ecosystems.
2. *With the notion of SES the possibilities in ecological research develop to analyse more systematically the forms of the interaction of social and ecological systems.* More complex forms of spatial and temporal relations

are identified, but do not yet reveal the complexity of processes in societal systems and their effects on ecosystems. The spatio-temporal processes studied are identified in ecosystems: phenomena of global environmental change, multi-scale processes of networking and nesting of ecosystem management, transboundary flows of matter, energy and information between systems, material cycles, and the temporal patterns in biological processes of growth—cyclical and seasonal processes in the short run and the longer adaptive cycles. The reflection of knowledge integration ends at the point where it becomes necessary to apply specialised social-scientific terms and knowledge to analyse development and changes in SES in their social complexity (a rare example in ecological research for a methodology to assess the functional (mis)fit of institutions and ecosystems to identify deficits of SES governance: Ekstrom and Young 2009). This deficit includes analyses of the power relations in complex political and economic systems, of human power over nature (called in social ecology "colonisation of nature"), and of socially structured processes of natural resource use in agricultural and industrial production. The advances in ecology happened more through creating new areas of specialised ecological research such as resilience or ecosystem services instead of opening the research for interdisciplinary knowledge synthesis.
3. *The main epistemological problem* in the interdisciplinary knowledge synthesis for SES is that social and ecological processes are of a different nature. Their conceptualisation develops with different ontologies and epistemologies that confirm an irreconcilable separation of the social and natural sciences. A new integration of knowledge requires more than postulates and assumptions as those in the ecological discourse where humans are seen as dependent from nature and the biological necessities of reproduction of life. The theoretical development of the concept of SES requires its systematic connection to social and ecological research and classification of forms of the coupling of social and ecological systems (see Chap. 5). The theoretical analysis begins when the historically varying forms of interaction between society and nature and the social transformation of nature in different forms of society is reconstructed. A historically specified theory is necessary to explain the forms of social and environmental change that

are studied in environmental and ecological research with the general assumption that the changes of climate, biodiversity, land use, landscapes, and ecosystems are man-made. Such a general assumption is not sufficient to understand the manifold and specific forms of interaction between social and ecological systems. This diversity cannot be reconstructed with empirical knowledge from ecological case studies either: these case studies continue to work with pre-theoretical concepts. The interdisciplinary knowledge culture required for a social-ecological theory is challenged to develop knowledge bridges between natural and social sciences with the construction of an epistemological meta-theory of nature–society interaction. Such an epistemological theory develops from older discourses of societal relations with nature and historical materialism.

4. *Not all ecological research is epistemologically reflected* and framed in theoretical concepts as Ostrom's research which developed with the unfolding of social ecology. Often the established terminology of systems theory and conceptual modelling is used, assuming that this is sufficient for interdisciplinary research. The shortcomings can be seen in the example of the conceptual model of the adaptive cycle for analysing ecosystem dynamics (Gunderson and Holling 2002). The adaptive cycle models four phases of ecosystem development, distinguishing between adaptive and transformational change and the assumption that social and ecological systems are continually changing. This model is only a first step towards a theoretical analysis of the dynamics of SES and its generalisation as a conceptual model for SES dynamics; it has the consequence of reducing the complexity of nature–society interaction to the few things that can be found out from the study of ecosystem development. Change is conceptualised in a cyclical temporal structure, describing ecological processes and ecosystem dynamics in which the social system dynamics appear as subordinate processes. Also in the elaborate form of iterative adaptive cycles in a continued sequence of adaptive and transformative change of ecosystems, the model is too simple to capture social system processes. These processes are strongly influenced by two determinants:

- The human capacities of action and agency, anticipation, and planning, and
- the systemic constitution of modern society with complex power relations in political and economic systems.

According to these reflections, the construction of an interdisciplinary social-ecological theory is confronted with two shortcomings:

- *Ecological research underuses social-scientific knowledge* from the complex theories of modern societies.
- *Sociological and social-scientific research underuses ecological knowledge* when it comes to the study of nature–society interaction.

In ecology, Carpenter and Folke (2006) summarised interdisciplinary ecological research in some ideas that show the limited aspiration of ecological theorising as applied research. Ecology is seen as

- improving the understanding of benefits that humans obtain from ecosystems (ecosystem services), contributing to

  - the development of environmentally sound technologies;
  - the development of markets for ecosystem services; and
  - decision-making that accounts for the changing relationship between humans and ecosystems—including research on resilience and reaction to disasters.

A research programme of this kind can easily be adapted to the neoliberal ideas of environmental policy that are not reflected in this description, but seem to have influenced its formulation. The interdisciplinary collaboration among ecologists, social scientists, and decision-makers is envisaged in ecology for the formulation of positive, plausible visions of the relationships between society and ecosystems in the long-term perspectives of sustainability (Carpenter and Folke 2006: 309). This understanding of ecology's role in interdisciplinary research reduces the analysis of the interaction between society and nature to visions that do not require a theory of the kind discussed here. It seems necessary to develop more theoretical forms of SES analysis:

- *An interdisciplinary social-ecological theory requires the use of social-scientific knowledge when it comes to the systems analysis of modern society, economy, and politics.*
- *The nature of the social-ecological theory under construction is not that of a general theory in ecology nor that of a general sociological theory of society. Its form is that of a historically specified theory of nature–society interaction, and its function is that of consolidating interdisciplinary knowledge through methodologically guided syntheses.*

For the further elaboration of the social-ecological theory it is necessary to assess the development of interdisciplinary and theoretical knowledge synthesis in sociology and ecology.

- Simple forms of knowledge synthesis include the use of empirical knowledge from different disciplines as, for example, in cultural and ecological anthropology. Such knowledge transfer does not always imply synthesis in the sense of combining datasets as in statistical meta-analysis. It can also be done by showing complementarity of knowledge from different fields of research in a more complete picture of a problem.
- Another widespread form of synthesis is concept transfer, where single ecological concepts are used for studying social systems, or for connecting social and ecological systems analysis as in the notion of SES.
- Further developed is interdisciplinary synthesis in the form that a theory from one discipline is used in another discipline or more disciplines. Examples for that are the use of thermodynamic physical theory in ecological economics, and older forms of evolutionary theory in sociology. Such attempts of framing a social-scientific discipline with a theory from natural sciences remain controversial in the ecological discourse showing that theory transfer between social and natural sciences has rarely succeeded.
- A more advanced and rare form of interdisciplinary synthesis is the construction of interdisciplinary theories in form of new interdisciplinary theory with knowledge from different disciplines. This is the form of interdisciplinarity followed in the construction of a theory of nature–society interaction.

All these forms of knowledge synthesis are not as epistemologically and methodologically structured as, for example, the method of progressive synthesis described by Ford (2000) for ecological research. They imply preliminary forms of synthesis that can be successively improved through epistemological reflection.

The themes vulnerability, resilience, and sustainability discussed in the following parts of the chapter were developed in recent ecological research. This research fills knowledge gaps in ecology and complements the social-scientific research on dynamics and changes of interacting social and ecological systems. In the theoretical context of a social-ecological theory, these analyses provide relevant knowledge to identify possibilities and hindrances of a societal transformation to sustainability. Improving the search of potential transformation paths towards sustainability requires the theoretical analysis of system mechanisms, of modes of production and reproduction of coupled social and ecological systems, of societal metabolism and of planetary boundaries of natural resource use.

Bridging Concepts in Social-Ecological System Analyses: Vulnerability, Resilience, and Sustainability

The term bridging concept is understood in different ways. Here it is used by describing its epistemic functions in ecological research as that of creating knowledge connections between different disciplines and between science and practice (Baggio et al. 2015). In the reconstruction of the processes that form the changing societal relations with nature it should be found out to what extent these relations

- require interdisciplinary concepts;
- can be regulated by managerial and political decisions;
- require further analyses of non-manageable processes of change to understand the varying relations between nature and society.

The problems in managing and regulating SES are environmental ones in the broad sense, including overuse of natural resources, environmental pollution, modification of ecosystems, and global environmental change. Regarding these problems the environmental historian McNeill (2001) formulated "there was nothing like the twentieth century" in human history—in terms of exponential growth of resource use and environmental disruption. The environmental problems create in their complex interaction the effects that ecological research shows as unexpected and surprising events, as limits of knowledge that make foresight, planning, management and control impossible. This is the widespread view in ecology, repeated by Scoones (1999). Whether these knowledge limits are final or temporary is disputed. With the elaboration of the social-ecological theory, it is assumed that the limits can be shifted

- through further research on vulnerability, resilience, and sustainability that shows potential forms to deal with social and ecological change, adaptation, disturbance, and transformation;
- through interdisciplinary and theoretical synthesis of knowledge about nature–society interaction.

The recent research on vulnerability, resilience, and sustainability in SES provides material for a theory of dynamics to specify the interaction of nature and society. This research adds knowledge to the understanding of global change and to sociological analyses of societal relations with nature. Although most ecological researchers analysing vulnerability, resilience, and sustainability are not aiming at a social-ecological theory of nature–society interaction, it is their research that provides ecological knowledge for this theory. The older theories of human transformation of nature in variants of historical materialism and the newer theories of urban, industrial, and societal metabolism can be connected with this recent ecological research to a theory of transformation to sustainability under conditions of global change.

Analyses of vulnerability and resilience develop through the broadening of risk analyses by attempting to account for potential consequences of human action and events in the unknown future. The terms and their interpretation are seen as confusing and requiring further clarification by Mumby et al. (2014), but the authors do this only for the limited purpose of ecosystem management, bypassing theoretical reflection. In a preliminary differentiation, the terms can be used for the development of a social-ecological theory as follows: vulnerability describes the present situation (or the situation before changes in forms of adaptation or transformation happened), resilience the future in a short-term, and sustainability in a long-term perspective.

- *Vulnerability analysis* shows the sensitivity and exposure of persons, social groups and systems regarding risks and disturbance; it can be seen as supplementary to social-scientific analyses of inequality and poverty and their origins.
- *Resilience analysis* shows the ability of social and ecosystems to cope with disturbance and to adapt to changed situations without collapsing; it can be seen as supplementary to social-scientific analyses of risks and disturbance created by humans in social systems and similar to robustness, a term from engineering and control theory to describe the capacity of a system to maintain a desired state despite fluctuations in internal components or the environment (Mumby et al. 2014: 24).
- *Sustainability analysis,* in a specific meaning as ecological sustainability connecting to the limited availability of natural resources, shows possible pathways for the transformation of coupled social and ecological systems to maintain their functions and development capacity in the very long run; it can be seen as complementary to social-scientific analysis of transformative capacities of social systems in modern economy and society.

Analyses of vulnerability, resilience, and sustainability can in the theoretical perspective of the social-ecological theory be connected to each other: vulnerability analysis creates knowledge for resilience analysis, and resilience analysis provides knowledge to formulate conditions for transition to sustainability. This interpretation of the concepts and their interconnection is discussed in more detail in the following parts of the chapter. The interpretation is compatible with some forms of using these terms in ecological research, but not with all.

Variants of Vulnerability

Social and ecological analyses of vulnerability use this term that exposes its metaphoric character from a concept transfer. The health-related semantic is still visible in the connotations of weakness and exposure to risks and dangers. From fields of applied research as poverty abatement (Cafiero and Vakis 2006) and risk research (Rossignol et al. 2014) the term vulnerability spread in environmental research and so far with little theoretical reflection. Related to environmental problems, ecosystems, and ecosystem processes, the notion of vulnerability implies sensitivity and exposure to natural disasters as a presupposition for understanding and mitigating disasters (Bankoff et al. 2004). Subjects of vulnerability are human or social subjects (individual persons, households, communities, organisations, social groups). In environmental and ecological research not only humans appear as vulnerable to disturbance, for example, through climate change, but also places and areas, ecosystems or biomes, SES, and the whole earth system.

The following three attempts to specify the meanings of vulnerability in ecological research cover relevant variants of the term:

1. *Vulnerability to climate change* is reviewed by Fellmann (2012) and shows that its clarification is reduced to definitions and methodological questions of terminology by ignoring theoretical discussion in the context of SES analysis. The literature provides a variety of definitions for vulnerability in the climate change discussion, mostly depending on the disciplines of their origin. Fellmann concludes: because of the variety of interpretations and concepts of vulnerability, the term should be explicitly defined for each application. Regarding vulnerability to climate change, he found two basic versions of vulnerability in ecological contexts. These include biophysical or ecological forms of vulnerability and socio-economic or social forms of vulnerability, simplified in the formula "biophysical vulnerabilities + socioeconomic vulnerabilities = climate vulnerability + likelihood = climate risk" (Fellmann 2012: 46). This way of interpreting the vulnerability concept is more oriented to application and practices of policy and resource management than to elaborating theoretical concepts.

2. *Further views and meanings of vulnerability are found in analyses and assessments of climate vulnerability.* Vulnerability in climate change literature is underpinned by numerous theoretical ideas from different disciplines resulting in different understandings of vulnerability and different methodological frameworks for its assessment. The many variants helped to frame and shape different understandings of vulnerability and to define the conceptual and analytical elements considered as critical in any climate change vulnerability assessment: *the specific SES for place-based analysis, scales of analysis, key components of vulnerability, causal structures of vulnerability, dealing with uncertainty, multiple perturbations, differential vulnerability across social groups, historical and prospective analysis, and engaging stakeholders* (Soares et al. 2012: 6). This description, which implies that vulnerability cannot be reduced to one version, remains a pluralistic concept.
3. *An example for a widespread ecological application of the vulnerability term is* the calculation of the environmental vulnerability index (EVI) for all countries and geographical areas of the world. This index, which includes subindices (hazards, resistance, damage) and fifty indicators (referring to climate change, biodiversity, water, agriculture and fisheries, human health aspects, desertification, exposure to natural disasters), seems to show all difficulties with the vulnerability concept, its elasticity, and multiple meanings discussed so far. The purposes of applying vulnerability analyses in policy and resource management differ from theoretical reflection of the concept in two points:

– In policy processes, vulnerability analyses aim to create capacities of agency by developing strengths and capacities of actors and systems.
– For that purpose a simple differentiation between vulnerability and resilience is used: vulnerability refers to exposure to risks and the possibility of being damaged, whereas resilience means the opposite—the ability to resist and to recover from damage.

This differentiation between vulnerability and resilience applies to physical entities (people, ecosystems, coastlines) and to social entities (social systems, economic systems, countries). For the construction of the EVI, risks and hazards are measured in the same dimensions, social, economic, and environmental—as in the use of the term of sustainability in international policies, for which a similarly constructed environmental sustainability index (ESI) exists. The results of the EVI calculation are not surprising in the message that nearly all industrialised countries except Australia and Canada are in a fragile state from vulnerable to extremely vulnerable; this mirrors the excessive use of natural resources in the national economies. The fewer countries classified as resilient or at risk are the mainly poor and least developed countries.

Regarding the connection of vulnerability analyses to a theory of nature and society the conclusion is twofold:

1. *The main advantage of applying the term vulnerability can be seen in more differentiated analyses of inequality, risks, and exposures than with the limited concepts of poverty or access to resources.* This applies to the social and ecological meanings of vulnerability and their further specification for different subjects and systems. The empirical knowledge gains from vulnerability analyses in terms of improved diagnoses; differentiated analyses and explanations of interactions between social and ecological systems are theoretically insignificant in vulnerability analyses. The term is useful for a theory of nature–society interaction only in combination with resilience and sustainability analyses.
2. *Vulnerability is a bridging concept for utilisation in applied research, vulnerability assessment, environmental policy, and natural resource management.* The concept is bridging between social-scientific and ecological knowledge, between science and policy or practice, and it helps to specify the action capacities of different actors. Together with the terms of resilience and sustainability, it makes the "ecological trinity" of transdisciplinary concepts in the sustainability discourse. Vulnerability analyses gain explanatory capacity by integration with resilience and sustainability analyses, but vulnerability does not become a general theoretical term that explains the interaction of SES.

Variants of Resilience

Like vulnerability resilience includes social and ecological meanings. The social meanings of resilience are to a large degree disconnected from ecological meanings: resilience in social psychology, for purposes of education, and in the contexts of social work refers to the capacity of humans to maintain balance and regain well-being after psychic shocks or traumata, disease, discrimination, and exposure to violence (Ungar). In ecological research, three variants of the concept are used specifying the basic meanings of recovering from shocks and adaptation: ecological resilience, social-ecological resilience, and social resilience (details: see the Appendix). The critical reflection of these variants has come to the point that the different forms of resilience may be contradicting *and* complementary. For social systems the terms resilience may be inadequate because of reflection capacities of humans and the resulting complexities of social action. At this point of developing towards an interdisciplinary theory, the ecological theorising stops.

Ecological resilience research has been criticised as "disastrous subject" (Reid 2012, with further references) supporting a neoliberal variant of sustainability that favours the deregulation and privatisation of economy and natural resource use. Resilience is interpreted as replacing security in the political discourse. With this broader ecological concept, it seems easier to argue against state-led development and for community-based, local, and self-reliant development *without using political arguments*. The multiple meanings, the inexactness, and the semantic elasticity of the ecological concept of resilience appear as veiling the political consequences of sustainable development. In this reasoning, the ecological debate of resilience "colonialises" the social and political development discourse and needs to be countered by a reflexive strategy of sustainability that defends the political against its neoliberal occupation and instrumentalisation (Reid 2012: 77f). The critical debate is useful for showing possible consequences of resilience and sustainability research in the political discourse and process and for reinterpreting sustainability. Nevertheless, the utility of this critique is limited; it (mis)directs the sustainability and resilience debates towards politics and governance and neglects the understanding of resilience and sustainability as broader social-ecological processes that include managed and non-managed components. A critique of state-dependent modernisation does not necessarily argue with neoliberal ideas of the kind that private enterprise, private property rights, deregulated markets, and the lean state are conditions for the success of further development and modernisation. Also many environmental movements and actors do not support such arguments as Reid shows. This critique has similarities with older forms of critique of political ideology; it does not allow to reconstruct the complex processes in modern society and its interaction with nature for reformulating resilience and sustainability. These processes require broader and interdisciplinary theoretical analyses and reflections of nature–society interaction—beyond politics and governance. The ideological message of neoliberalism, that there is no alternative to the marketisation, monetarisation, and commercialisation of social relations and relations with nature can be countered by showing its selective knowledge use.

The *search for alternatives to the neoliberal mainstreaming of the environmental discourse* is part of the discussion of resilience in social-ecological research that can be described with the following requirements:

1. *Different and contradicting interpretations of the resilience term should be classified* to make visible the pluralistic quality of the concept that causes confusion when the term is applied for different system types, social and ecological systems. To differentiate systematically between the variants of social and ecological resilience and to describe their interrelations would be useful to create further theoretical clarity for the formulation of system-specific concepts of resilience.
2. *Ecological variants as engineering or ecosystem resilience imply limited forms of adaptation.* Reaction and feedback mechanisms create resilience in ecosystems without requiring human subjects and action capacity. Resilience in this meaning cannot be applied for the analysis of societal change and adaptation, unless society is reduced to a self-regulating system of the kind of automatons. Human capacities of knowledge use, action, agency, anticipation, and planning become relevant in the notions of social and social-ecological resilience where they may be described with additional terms as robustness. Social-ecological resilience requires a reconstruction of the interplay of social and ecological mechanisms in resilience.
3. *As long as resilience analysis is limited to ecological and social-ecological resilience, the contradicting requirements of different forms of resilience are not fully discovered*, although between ecosystems resilience and

social-ecological resilience contradictions can be found. Major contradictions appear between social and ecological resilience; ecological resilience can be seen as socially negative or unwanted (Adger 2000). Such critical reflections in resilience research (to deal with contradictions) are limited. Further, theoretical reflection with social-scientific knowledge is required to interpret the contradictions and to connect the analyses of vulnerability, resilience, and sustainability in a coherent theoretical perspective.

Sustainability Research

The theoretical reconstruction of the transformation of modern society to sustainability is discussed in detail in Chap. 4. Here only connections of empirical analyses of vulnerability, resilience, and sustainability are described in methodological perspective to prepare the theoretical synthesis. Sustainability is the most complicated theme and term to deal with in social-ecological research. Yet, most definitions of sustainability are simple and bypass theoretical formulation and contextualisation; they do not highlight the capacities required to maintain social and ecosystems over long time periods, and they are usually formulated in the indefinite form of "future generations", not specified in more concrete time frames and with regard to transformations of social systems and SES. To clarify the concept further requires analyses of complex social and ecological systems and their interaction and criteria to assess sustainability as condition of continued functioning of SES.

For purposes of SES analysis, sustainability can be seen as the overarching concept that directs the use of the concepts of vulnerability and resilience in integrated analyses. This assumption for integrating the three forms of analysis cannot be derived from the practice of using the terms in research and resource management, and it requires theoretical arguments: there is always a plurality of contrasting terms in use. The theoretical argument developed in the theory of social-ecological transformation (see Chap. 4) is that sustainability requires a transformation of the societal and economic system with a new socio-metabolic regime or mode of production which cannot be achieved in the perspective of resilience that requires only adaptation as another kind of and much more limited change (see also Chap. 7). Integrated analyses are mainly practised in global scenarios of possible transitions to sustainability. The scenarios require joint conceptual frameworks and further theoretical reflection of sustainability in terms of transformation of modern society, bringing social-scientific knowledge about modern society in this debate. With the integration of the three kinds of analysis, sustainability can be connected to the theory of nature and society where it has the function to provide a knowledge compass for potential forms and successive steps of transformation under conditions of global change.

From this overarching perspective, resilience and vulnerability can be mapped backwards by specifying the meanings of resilience and vulnerability that are compatible with the social-ecological sustainability perspective. Whereas the function of bridging concepts with multiple meanings can be maintained, the theoretical framing and structuring requires the prioritisation of certain meanings to construct a perspective of societal transformation. Temporal perspectives of action for each term can be specified to avoid, as in some variants of resilience research, the reduction of sustainability to a form of systems change that requires mainly resilience. When the specific meaning of transformation as transformation of society and of societal metabolism is eliminated from SES analysis, the debate of sustainable development tends to become superfluous. The long-term perspective required for sustainable development can be specified further by connecting the social-scientific knowledge about the transformation of societal systems with ecological research on resilience and sustainability. Social systems do not follow the logic and functionality of ecosystems. Also when both system types are seen as coupled and societies as embedded in ecosystems, there remain differences that are specified in the social-scientific theory.

The shortcomings of the sustainability debate that block its further development in interdisciplinary social-ecological perspectives can in an exemplary way be shown in the global scenario debate. This debate reduces the formulation of potential trajectories of sustainable transformation to a methodology of envisioning the future. Scenario analysis indicates the further knowledge requirements in theoretical reflection of sustainability (see Chap. 4). Limits of scenario analysis can be summarised as follows:

1. *Scenarios of global sustainability are often working with a dramatic rhetoric implying warnings of catastrophes as described in the "limits to growth"—discourse.* An example is that of Gerst et al. (2014) arguing that humanity confronts two challenges in the twenty-first century: meeting widely held aspirations for equitable human development while preserving the biophysical integrity of the earth system. Quantifications of trends and future states that address the sustainability challenges do not systematically account for different environmental and social drivers of global change instead relying on quantification methods that exclude fundamental social, cultural, economic, and technological changes that influence the possibilities of transitions to sustainability. For the formulation of several possible trajectories of transformation, three hitherto separate streams of inquiry are connected by Gerst et al.: scenario analysis, planetary boundaries, and targets for human development. None of these are theoretically developed, but at least they advance with regard to interdisciplinary knowledge integration. The progress in the discussion is shown in scenarios that remain within the earth's safe biophysical operating space and achieve a variety of development targets—this seems significant for formulating improved forms and options for transformation to sustainability in the political discourse.

   As the authors argue, dramatic social and technological changes are required to avoid the social-ecological risks of a conventional development trajectory—the zero-variant of most sustainability scenarios in terms of "business as usual". A second narrative, which is predominant in the scenario literature, envisions marginal changes in the social and cultural drivers underlying conventional growth trajectories. This variant requires unprecedented intensity and success of international cooperation, alignment of powerful conflicting interests and conflict resolution, and high political power to direct technological change towards sustainability—assumptions that do not seem realistic regarding established global power relations and vested interests. A third variant is the coupling of transformative social-cultural and technological changes, which formulates more systematically the necessary conditions for transitions towards resilient and sustainable global futures (Gerst et al. 2014: 123).

With the industrial revolution, conditions of discontinuous change have been created for human resource use resulting in exponential growth in modern society. At the beginning of the twenty-first century, the effects of industrialisation on the future development of modern society and on ecosystems can be described with regard to the knowledge about the state of the earth and the limits to growth: global industrialisation is impossible because of limited availability of natural resources. The concrete forms of transformation cannot be sufficiently foreseen. Under these conditions, plausible narratives can be formulated to show the contours of resilience and sustainability of the global society and the earth system:

- Scenarios of policy reform may be possible, but probably disruptive and transformative social and technological changes are required other than in the historical transition from agricultural to industrial society.
- For the future planetary phase of human history Gerst et al. assume the necessity of a great transition: a shift in the development paradigm and a restructuring of the global economy underpinned by a fundamental change in values, a sharp demographic change through reduction of population growth, and a strengthening of institutions of global governance and massive technological change (Gerst et al. 2014: 132).

With this scenario analysis, the authors come close to a social-ecological analysis of transformation of the industrial metabolic regime.

2. *The limits of the scenario method for sustainability* analysis appear in paradigmatic form with the example from Gerst et al.: scenarios use mainly normative information about unwanted, wanted and possible futures, only to a limited degree theoretical knowledge for the construction of pathways of future development. This is comfortable for the policy discourse of sustainability and for decision-makers: it avoids a complicated theoretical analysis and reflection through simpler sets

of coherent assumptions and narratives. However, it reduces the construction of possible transition pathways through selective and fragmented knowledge about modern society. Therefore, improved scenarios should be informed through knowledge about modern society and from social-ecological theory. Yet, scenarios alone do not allow to deal with the transformation problems. The authors discuss relevant questions of sustainability, and those of systems transformation, but they do so without using available theoretical knowledge from theories of society, economy, and ecosystems to discuss the problems, possibilities, and limits for transitions to sustainability. This seems to characterise large parts of the political sustainability discourse. Sustainability or sustainable development is discussed up to the levels of visions and empirical knowledge about the present system state. Excluding important knowledge from theories about nature and society, they show the possibilities and limits of transformation more through guesses than through theoretical information.

The further discussion of sustainability requires

- knowledge and ideas of possibilities to transform industrial society and its societal metabolism;
- methodological integration of vulnerability, resilience, and sustainability analyses (see Appendix);
- developing a theory of transformation to sustainability under conditions of global change.

These three requirements are discussed in the following section as components of the social-ecological theory that develop from ecological research described in this chapter.

Connecting Vulnerability, Resilience, and Sustainability to Social-Ecological Theory

The conditions to connect analyses of vulnerability, resilience, and sustainability for their use in an interdisciplinary social-ecological theory can be summarised as follows:

Vulnerability is a broad and multi-dimensional concept with many specific forms that describe deficits and lack the capacity of social actors, social and ecological systems to deal with social and environmental risks, problems, and turbulences. Integrated analyses can work with various meanings of vulnerability, but the environmental problems and global environmental change need to be specified to describe vulnerability in SES. Such integrated vulnerability analysis provides knowledge for reacting to environmental problems and risks to which analyses of resilience and sustainability add further knowledge about adaptation and transformation of social and ecological systems.

Resilience is specified in SES analysis through the basic meaning of a capacity to cope with disturbances either by maintaining a given system state or by rebalancing a system at another level of functioning that may include less favourable conditions for further development of social and ecological systems. The differences of ecological, social-ecological, and social resilience are maintained in integrated analysis, although not all forms have the same relevance. Ecological and social-ecological resilience imply capacities to cope with environmental disturbance and adaptation to global environmental change; these capacities are required for sustainability.

Sustainability as a capacity of multi-scalar social and ecological systems to maintain the necessary functions and reproductive requirements of both systems in the long run, over generations, was specified in social, economic, and ecological sustainability in the political discourse of sustainable development. These aspects are not complementary. The long-term maintenance of society, economy, and ecosystems requires to maintain different kinds of structures and orders in combined SES: different cultural orders in global society, functioning of global structures and processes in political and economic systems, and maintaining of functions of ecosystems at different scales. The systemic interactions of modern society—its different cultural orders, the modern economic world system, the SES which show the specific forms of interaction of nature and society, and the ecosystems that constitute the earth system—cannot be maintained and regulated in one mode of governance. This complex intersystemic organisation requires integration of environmental governance at different levels to achieve multi-scale governance and to react to continuing global change. For that purpose the indicators of sustainability are of limited significance; they are indicators of the present state of systems. Indicators about the transformability of connected SES are more difficult to construct; they require theoretical analyses of the possibilities of change, adaptation, and transformation of connected systems in society and nature. For that purpose, a systems analysis of modern society and its interaction with nature in the modes of production, reproduction and the societal metabolism needs to be integrated into the social-ecological theory under construction.

The theoretical connection of the ecological research on vulnerability, resilience, and sustainability to the social-ecological theory of nature–society interaction implies to assess the theoretical significance of the three concepts.

1. *Vulnerability* as a dependent variable needs to be connected with resilience and sustainability to develop interdisciplinary perspectives. In the controversy over vulnerability as "contrasting to resilience or a form of it?", a choice needs to be made. *A distinction of the terms as contrasting, where vulnerability marks the lack of capacities and social agency that are required for resilience and sustainability, is coherent with the social-ecological theory and helps to combine the three analyses of vulnerability, resilience, and sustainability.* The further clarification implies the necessity to differentiate theoretically between different forms of vulnerability, resilience, and sustainability.
2. *Resilience and sustainability have different implications for social and ecological systems,* and refer to complex systemic processes in social and ecosystems (see, e.g. the study of climate change resilience of Asian cities by Reed et al. (2014) that highlights the necessities of knowledge integration, joint learning, and cooperation in contrast to conventional policy processes of implementation and mainstreaming that prevail in climate change adaptation). This is not always supported by the political terms in the discourse of sustainable development where the systemic differences are often watered down, allowing interpretation of sustainable development as a variant of adaptive change, and coming close to the meaning of resilience. *Sustainability, interpreted as societal change, requires more than adaptation to and coping with disturbance: a*

theory of transformation of societal systems that is lacking in the resilience research that works with the model of adaptive cycles.
3. *Distinction and connection of the three terms.* Opting against the interpretation of vulnerability as melting with resilience or resilience as melting with sustainability has further consequences. The conceptual differences need to be formulated more clearly to be able to reflect their connections. Vulnerability shows a lack of adaptive capacity that needs to be acquired to achieve resilience. Resilience requires further capacities and functional mechanisms in social systems, ecosystems, and SES to maintain the functions of these systems. In that way it can be distinguished from sustainability that requires still further capacities and possibilities to transform the systems in order to achieve global sustainability.

The difficulty in integrating analyses of vulnerability, resilience, and sustainability in theoretical perspectives is not only that of the multiple meanings of the terms. The limited interdisciplinarity in ecological research and the insufficient use of social-scientific knowledge and theory create more problems. The theoretical systems analysis cannot be reduced to biological knowledge about humans and their communities: it requires use of sociological knowledge and systems analysis of society and economy.

The methodological integration of the three kinds of analysis (see Appendix) supports the elaboration of a social-ecological theory through the production of knowledge that allows to describe the interaction between social and ecological systems more concretely:

- Vulnerability analyses are part of an extended analysis of the social inequalities and risks with regard to environmental burdens created by natural resource use and by global environmental change.
- Resilience analyses regard the dynamics of SES, and their present and future states, by specifying the forms and conditions of adaptation, and less that of transformation.
- To unfold a perspective of societal transformation, theoretical knowledge about modern society needs to be integrated with ecological knowledge; otherwise, sustainability tends to be reduced to a form of resilience.

- A main methodological requirement in all these forms of analysis is the specification of empirical and theoretical knowledge about SES.

With the theories discussed above (Table 3.1), including the integrated analyses of vulnerability, resilience, and sustainability, the ecological components of the theory of nature–society interaction can be summarised (Fig. 3.1), which complements the social-scientific components (Fig. 2.1).

An interdisciplinary knowledge exchange about vulnerability, resilience, and sustainability is hindered through disciplinary and subdisciplinary specialisation of knowledge, methodological difficulties of synthesising incoherent empirical research, different epistemic cultures of natural and social sciences, and through the continuous controversies in theoretical debates. Further reasons for the deficits of interdisciplinary knowledge synthesis are the lack of theoretical concepts, frameworks, theories, epistemologies, and methodologies for knowledge synthesis across the boundaries of natural and social sciences. The difficulties of knowledge synthesis and theory construction need to be dealt with successively in the construction of an interdisciplinary social-ecological theory, and in

Fig. 3.1 Ecological components of a theory of nature–society interaction. *Sources*: Own compilation (theories discussed in this chapter)

the attempts to use empirical and theoretical knowledge from different disciplines more systematically.

Reacting to the difficulties with knowledge synthesis and interdisciplinary theory construction, a methodological procedure is suggested in the Appendix to stepwise advance with the integration of empirically based vulnerability, resilience, and sustainability analyses in the framework of a social-ecological theory that is constructed with the conceptual components discussed in Chaps. 2 and 3. These are preliminary methodological ideas that reflect the early stage of development of this theory. Other methodologies can be developed in the further discourse and elaboration of the theory of nature–society interaction. The approach suggested here is an attempt to follow the theory construction along the knowledge chain of production, dissemination, and application of knowledge in interdisciplinary knowledge creation and synthesis. It proceeds from the combination of theoretical components of systems analysis of SES, connecting these with knowledge from integrated analyses of vulnerability, resilience, and sustainability and ending with the completion of the theory through consolidated research about the change and transformation of coupled social and ecological systems. The integration of empirical and theoretical knowledge in the emerging theory is, with exemplary themes of this theory, described in the following chapters. This methodological procedure cannot deal with all difficulties of integrating and synthesising knowledge for the purpose of theory construction. It has two specific aims in developing the social-ecological theory: that of developing an interdisciplinary theory of nature–society interaction, and that of using the knowledge of this theory to improve the practices of governance and regulation in the transitions to sustainability.

Conclusion: Interdisciplinary Social-Ecological Theory

Under the given conditions of limited interdisciplinary knowledge exchange between social and natural sciences, the social-ecological theory of nature and society develops—with difficulties—from different sources, disciplines, and theories. Chapters 2 and 3 describe the core components of that theory and how these connect with further theoretical components for the elaboration of the theory. This theory includes the following components:

- The *production and reproduction of systems,* of societal systems (material and symbolic reproduction) and of ecological systems (energy-driven circular flows of matter and nutrients)
- The forms of *interaction and coupling between social and ecological systems* structured through the power relations in the national and global economic and political systems
- The *dynamics of interacting social and ecological systems in processes of global change* in terms of vulnerability, resilience, and sustainability, showing adaptation, change, and transformation of the interacting systems
- The *societal metabolism* that connects social and ecological systems in the forms of the basic metabolism and the more complex processes of societal metabolism, configuring in historically differing socio-metabolic regimes for resource use
- The changes in *natural resource flows in and between social and ecological systems* that are necessary for transitions to global sustainability (e.g. potential pathways for societal transformation, hinders of transformation)
- The knowledge transfer in *programmes and processes of the regulation of social-ecological systems,* for example, in climate policy

According to this description, the theory develops at two levels:

- As an overarching theory for the whole development of human societies in a broader historical perspective (this is not discussed further here: see, e.g. Fischer-Kowalski et al. 1997).
- As more specific theory of modern society and the present global social and environmental change and transformation to sustainability (in this specific sense it is discussed further in the following chapters).

What can be gained from a broadened social-ecological theory includes improved knowledge and analyses of

- the possibilities of co-evolution of social and ecological systems;
- the different forms of functional or dysfunctional coupling of SES;
- the natural *and* social resource flows in multi-scale SES;
- the local and planetary boundaries of ecosystem production and functions for human resource use;
- the functions and services of ecosystems *and* social systems for humans and their forms of living, production and consumption;
- SES governance and transformative agency that develops in a critical theory of ecological practice.

According to its epistemological and methodological properties, the interdisciplinary theory of nature and society has open boundaries and requires continuous development through knowledge synthesis. The theory has as its main aim to connect and structure the available knowledge about nature and society. Verification or falsification of knowledge used in that theory requires, further, complex processes of integrated research, theoretical reflection, and knowledge transfer in processes of management, regulation, and governance of natural resource use.

The differences between traditional and critical theory in sociology and the discourse about theory of society have no equivalent in ecology. As similar schism of conventional and heterodox approaches can be seen that between disciplinary ecological research (natural-scientific knowledge) and interdisciplinary approaches (human, political and social ecology). In epistemological terms, the mode of knowledge generation for this theory can be formulated in the expression inherited from the older discourse of critical theory where it was differentiated as "holism ex post" from a "holism ex ante" (Bruckmeier 2014). Holism ex ante can be exemplified by the discourse and movement of systems theory where it is assumed that the concepts and frameworks of systems theory create the explanations by their application in a specific theme or field of research. This theoretical top-down procedure of knowledge integration neglects the *necessity of consolidating empirical research and knowledge production through conceptual codification and theoretical mapping of the knowledge fields to connect—before explanations, systematic theory, and its application in governance processes can become possible.* In the critical theory there developed an alternative in the form of a holism ex post by elaborating theory and explanation in a continuous interchange with empirical research and knowledge.

In the elaboration of such a theory more problems come up that cannot be clarified in the theory itself, but should not be ignored. Problems appear as pitfalls in the transfer of knowledge from science to policy, governance, or resource management. There is an inherent tendency in ecological research, appearing with the governance approaches (see Chap. 5), to make governance practices more and more complicated, and finally replacing decision-makers as practitioners by decision-makers as scientists. The idea of the "reflective practitioner" (Schoen) requires that the social practices of reflection and knowledge use for natural resource management and governance need to be developed, refined, and structured more systematically. If it is not critically discussed how knowledge from scientific research is assessed and applied in social practices of knowledge use, reflective knowledge practices tend to fall back in conventional reasoning of scientific management and knowledge transfer. Such reasoning assumes that the practices of resource use and management require only scientific knowledge and the practitioners and actors need to be "enlightened" through science. In this form, an old Platonic utopia of governance through the wise is renewed—if not governance through philosophers at least through scientists. Socially naïve as it is, it reappears again, such as in the suggestions for managing transitions to sustainability.

Appendix: Connecting Empirical and Theoretical Knowledge—Integration of Analyses of Vulnerability, Resilience, and Sustainability

Reconstructing Social and Ecological Meanings of Vulnerability

In ecological vulnerability studies, two basic variants of the concept can be identified: social and ecological vulnerability. Both of these are related to environmental problems.

1. *Social meanings of vulnerability:* Vulnerability can be caused by social factors (psychic, social, political, economic) or ecological disturbances as climate change; vulnerable subjects are humans, individual persons, or social groups. Also when vulnerability is differentiated in biophysical and social vulnerability (Füssel 2007; Fellmann 2012), humans are the vulnerable subjects, as highlighted in the review by Lundgren and Jonsson (2012):

    – Social vulnerability to environmental hazards is generally dependent upon lack of access to resources (monetary, information, knowledge, or technology).
    – Access to political power and representation, and social capital (including social networks), show important social dimensions of vulnerability.
    – Social variables for vulnerability to climate change are difficult to specify, but age, gender, race, and socio-economic status are generally accepted (Lundgren and Jonsson 2012: 7)

2. *Ecological meanings of vulnerability:* Ecological literature shows vulnerability and resilience as interconnecting and interacting phenomena (Turner et al. 2003), but the connections between vulnerability and resilience are disputed. Also the separation between social and ecological meanings of resilience in ecological literature is not very clear. For non-living and non-acting systems such as ecosystems, it is more difficult for social actors to say in which sense they are vulnerable, except through loss of functions. Vulnerability to deterioration of environmental quality, of ecosystem functions and limits of natural resources for life-maintaining purposes implies the difficult cases of SES where social and ecological meanings of vulnerability are overlapping. That marine and terrestrial ecosystems become vulnerable and may collapse through overuse of their resources is evident. Controversial is whether that loss of functionality requires the term vulnerability that is burdened with organismic thinking, metaphorical meaning, analogous reasoning, and value-loaded connotations that can all be seen as consequences of its derivation from health and life processes.

Reconstructing Social and Ecological Meanings of Resilience

The development of resilience research has been reviewed repeatedly (Adger 2000; Folke 2006; Nelson et al. 2007; Brand and Jax 2007; Bruckmeier 2013), but resilience remains a term with different and contradicting meanings.

1. *Social and ecological resilience:* Resilience concepts are summarised by Folke (2006) in three basic variants:
   - *Engineering resilience* (return time, efficiency, recovery, constancy, vicinity of a stable equilibrium)
   - *Ecological/ecosystem resilience and social resilience* (buffer capacity, withstanding shock, maintaining function, persistence, robustness, multiple equilibria, stability landscapes)
   - *Social-ecological resilience* (interplay of disturbance and reorganisation, sustaining and developing adaptive capacity, transformability, learning, innovation, integrated system feedback, cross-scale dynamic interactions)

   Controversial in this systematisation of the concepts is the use of the notion of resilience for ecosystems and social systems, assuming that there are similarities of functions, interactions, and processes in ecological and social systems that create buffer capacity. In the use of resilience for ecosystems, sometimes anthropocentric or biocentric meanings are ascribed to ecosystems, for example, with the notion of memory of ecosystems or landscapes. Some ecologists discuss the notion of resilience more critically (Carpenter et al. 2001; Anderies et al. 2004; Janssen 2006: 128; Perrings 2006: 217f; Janssen and Anderies 2007), suggesting to differentiate between resilience of ecosystems and robustness of social systems by specifying the notion of robustness through the human capacities to deal with disturbance or coping capacity, based on consciousness, reflexivity, and anticipation.

2. *Epistemological classification:* A classification of resilience concepts in social and ecological research by Brand and Jax (2007) shows the epistemological features of resilience in three groups: (a) *descriptive variants of resilience* (ecological resilience, extended ecological, systemic-heuristic, operational; social resilience as sociological or ecological-economic term); (b) *hybrid variants* (related to ecosystem services and SES); and (c) *normative variants* (metaphoric, sustainability related). As in Folkes classification, the dominant meanings of resilience are ecological. With hybrid and normative variants, resilience becomes a more vague notion. The classification gives the impression that the resilience terminology is inexact, that it glides between different disciplinary applications and meanings, and that it lacks theoretical structuring. This shows also in the overlapping of the terms vulnerability and resilience (in the definition of Eakin (2012) in the aspects coping with hazards, adaptation, and maintaining function) and resilience and sustainability (Nelson et al. 2007: 412, in the aspects adaptation, transformation, agency, and achievement of desired states of social systems). Inexactness, overlappings, and different interpretations seem to be consequences of the abstract and elastic nature of the terms.

Reconstructing Social and Ecological Meanings of Sustainability

Nelson et al. (2007) discuss the nature of adaptation, describing it as a process of deliberate change in anticipation of or in reaction to external stimuli and stress, which is a requirement for resilience and sustainability. They see the dominant research on adaptation to environmental change as social-scientific and actor-centred—conceptualising the agency of social actors to respond to environmental stimuli thus reducing their vulnerabilities. A systematic description of sustainability can start from the classification of sustainability approaches (see Table 4.1).

Sustainability as discussed in the discourse of sustainable development includes conventional and critical approaches where social and ecological meanings of sustainability are combined in many variants. The criterion of differentiation between the two kinds of approaches is whether sustainability is seen to be achievable without changing the modern society and economy, or whether fundamental changes are required.

Conventional approaches include two kinds:

- *Policy-generated ideas, describing aims and policy processes to achieve sustainability*: the Brundtland report (resource sharing, intergenerational solidarity); the mainstreaming of sustainable development through the Johannesburg Summit (social, economic, environmental sustainability); the neoliberal variant of global environmental policy in the Rio+20 process (green economy, market-based policies, payments for ecosystem services).
- *Ideas oriented to the innovation of society, economy, or industry and changes of modern lifestyles and consumption behaviour*: ecological modernisation (MoI) and similar ideas of the greening of the industry; environmental economics (internalisation of negative externalities); certain variants of consumer ethics (e.g. sufficiency, "voluntary simplicity"); certain ideas about social innovation or restructuring of society (e.g. autonomous local communities and regions, nested systems).

In the conventional approaches, the criteria of changes required for sustainable development remain often vague and unclear with regard to the changes of system structures (industrial society, the globalising market economy, social-metabolic regimes, or modes of production). Implicitly sustainability is assumed to be achievable within the present industrial or market economy, through continuous modernisation and policy reforms that aim at reduction of resource use and internalisation of social costs of environmental pollution through taxes and pricing. Changes of lifestyles and consumption are ambivalent. They are seen either as non-political, individual, and cultural changes, or as fundamental changes of mass consumption and change of the growth economy [see Dauvergne (2008) and the analysis of mass consumption based on the Fordist accumulation regime and cheap energy by McNeill (2001)].

Critical approaches can also be divided into two kinds:

- *Ideas originating in environmental movements* aiming at fundamental change of livelihoods which are to be achieved through movement

activities and lifeworld-oriented strategies of change: deep ecology (Ness et al); environmental justice (Schlosberg et al); sustainable livelihoods (Chambers et al); environmental democracy; certain variants of environmental ethics (bio-, ecocentrism); local, small-scale and circular economy; environmentalism of the poor (Martinez-Alier 2002); degrowth (Latouche).
- *Ideas from ecological research* that show the limits of resources through analyses of interacting society and nature at various levels and derive from that the necessity of fundamental change of industrial society and modern economy: carrying capacity; limits to growth (Meadows and Meadows); ecological footprint (Rees, Wackernagel); planetary boundaries (Rockström et al); social-ecological strategies of transforming socio-metabolic regimes.

In critical approaches, the interests of the countries and people in the Global South are more consequently accounted for, arguing in a global perspective, against the exclusive "environmentalism of the rich". However, in many of the critical approaches it is not clear as to how changes of everyday life and consumption behaviour are connected with transformation of societal systems, the modern economy, and its modes of production and reproduction, or the unequal global flows of resources. The question of changing society is either left open or it is assumed that the transformation of society happens through mobilisation and activism of social movements that combine individual changes of life- and consumption styles and collective action.

A Methodology to Integrate Analyses of Vulnerability, Resilience, and Sustainability

In the following description an integrated analysis of vulnerability, resilience, and sustainability for coupled SES is specified that can be connected with the social-ecological theory discussed. Beyond the differentiation and clarification of the concepts (for resilience and sustainability see, e.g. Perrings (2006) who stresses the necessity of understanding the system dynamics to match the concepts) the integration requires methodological reflection and structuring:

1. *Logic of knowledge integration:* The logic of integrated analyses of vulnerability, resilience, and sustainability can be described as a process of

    – *"backward analysis":* going backward from the goal to be achieved (transforming the industrial metabolic regimes to a sustainable societal metabolism) to the present situation (state of societal and ecological systems), followed by
    – *"forward analysis":* going forward by formulating the means to achieve that goal in successive steps of research, knowledge synthesis, knowledge transfer, and formulation of governance and regulation processes for transformation to sustainability (aiming at reducing vulnerability and strengthening resilience and sustainability in social and ecological systems).

2. *Steps of an integrated analysis:*

    – *Classifications and development of conceptual frameworks:* A coherent analysis of SES can be achieved with the help of the classifications of the forms of vulnerability, resilience, and sustainability [see examples for the use of different frameworks in Mumby et al. (2014: 25f)]. The analysis can be done for all forms of coupled social and ecological systems that include use of natural resources, for example, in systems of land use, agricultural production, industrial production, private consumption, local communities, or specific areas. Knowledge integration develops in a "bottom up" process by integrating knowledge from local case studies through comparison of local studies and connecting these to studies at higher levels (using for analyses at national and global levels additional information such as statistical indicators, synthesis reports, global assessments, and simulations).
    – *Beyond classifications: integrated forms of vulnerability, resilience, and sustainability analysis*

    An integrated analysis of *vulnerability* includes two components:

    – *Identification of forms of vulnerability* (e.g. scarcity of and access to specific natural resources, consequences of environmental change,

exposure to natural hazards), subjects of vulnerability (who is vulnerable), the relevant natural resources
- *Dynamic vulnerability analysis:* identification of trajectories and changes of vulnerability over time in a given SES (Magnan et al. 2012), interaction across various scales (Turner et al. 2003), dynamics, and changes of resource use regimes (Young 2010).

An integrated analysis of *resilience* includes the identification of possibilities to build social-ecological resilience of coupled SES:

- *Strategies to reduce vulnerability* through the analysis of *disturbances and their connections*: identifying crucial vulnerabilities through vulnerability assessment; mitigating vulnerability through measures for reducing exposure to hazards and disturbance or compensating for their effects; reducing sensitivity (minimising responsiveness to changes through disturbance); identifying social and institutional capacities to prepare for disturbances and minimise their impacts; projecting changes relevant for future development of coupled systems (Chapin et al. 2009)
- *Strategies to enhance adaptive capacities of coupled systems and coping capacities of actors* [see examples for a differentiated use of frameworks for adaptation and resilience in Nelson et al. (2007)]: maintaining diversity of social and ecological systems; stabilising feedbacks; changes of resource use practices; social and institutional learning in adapting institutions and governance processes to changing environmental conditions; developing multi-level governance through adaptive management/governance

For *sustainability* integrated analysis includes the identification of core elements of a sustainability synthesis beyond the components of adaptation found in resilience analysis:

- *Analyses of the system dynamics of coupled SES:* analyses of societal metabolism and modes of natural resource use in the perspectives of maintaining the reproductive and functional mechanisms of both system components referring to local and global limits of resource use

– *Long-term strategies to enhance the transformative capacity of the SES*: using knowledge from systems analyses of society and economy to identify possible trajectories of transformation; conflict mitigation and strengthening collective action and cooperation of resource users; developing mechanisms of multi-scale and multi-actor governance; developing processes of navigating transformations through different periods of turbulence, uncertainty and crisis (Olsson et al. 2006)

References

Adger, N. (2000). Social and ecological resilience: Are they related? *Progress in Human Geography, 24*(3), 347–364.

Allenby, B. (2006). The ontologies of industrial ecology? *Progress in Industrial Ecology – An International Journal, 3*(1–2), 28–40.

Anderies, J., Janssen, M., & Ostrom, E. (2004). A framework to analyze the robustness of socio-ecological systems from an institutional perspective. *Ecology and Society, 9*(1), Art. 18. http://www.ecologyandsociety.org/vol9/iss1/art18/

Baggio, J., Brown, K., & Hellebrandt, D. (2015). Boundary object or bridging concept? A citation network analysis of resilience. *Ecology and Society, 20*(2), 2. doi:10.5751/ES-07484-200202.

Bankoff, G., Frerks, G., & Hilhorst, D. (Eds.). (2004). *Mapping vulnerability: Disasters, development and people*. London: Earthscan.

Binder, C. R., Hinkel, J., Bots, P. W. G., & Pahl-Wostl, C. (2013). Comparison of frameworks for analyzing social-ecological systems. *Ecology and Society, 18*(4), 26. doi:10.5751/ES-05551-180426.

Brand, F. S., & Jax, K. (2007). Focusing the meaning(s) of resilience: Resilience as a descriptive concept and a boundary object. *Ecology and Society, 12*(1), 23. http://www.ecologyandsociety.org/vol12/iss1/art23/

Bruckmeier, K. (2013). *Natural resource use and global change: New interdisciplinary perspectives in social ecology*. Houndmills, Basingstoke: Palgrave Macmillan.

Bruckmeier, K. (2014). Problems of cross-scale coastal management in Scandinavia. *Regional Environmental Change*. doi:10.1007/s10113-012-0378-2.

Cabezas, H., Pawlowski, C. W., Mayer, A. L., & Hoagland, N. T. (2005). Sustainable systems theory: Ecological and other aspects. *Journal of Cleaner Production, 13*, 455–467.

Cafiero, C., & Vakis, R. (2006). *Risk and vulnerability considerations in poverty analysis: Recent advances and future directions.* SP Discussion Paper 0610. Washington, DC: The World Bank.

Carpenter, S. R., Bennett, E. M., & Peterson, G. D. (2006). Scenarios for ecosystem services: An overview. *Ecology and Society, 11*(1), 29. http://www.ecologyandsociety.org/vol11/iss1/art29/.

Carpenter, S. R., & Folke, C. (2006). Ecology for transformation. *Trends in Ecology and Evolution, 21*(6), 309–315.

Carpenter, S., Walker, B., Anderies, M., & Abel, N. (2001). From metaphor to measurement: Resilience of what to what? *Ecosystems, 4*, 765–781.

Chapin, F. S., III, Kofinas, G. P., & Folke, C. (Eds.). (2009). *Principles of ecosystem stewardship: Resilience-based natural resource management in a changing world.* New York: Springer.

Daly, H. E. (1991). *Steady-state economics* (2nd ed.). Washington, DC: Island Press.

Daly, H. E. (1997). *Beyond growth: The economics of sustainable development.* Boston: Beacon Press.

Dauvergne, P. (2008). *The shadows of consumption: Consequences for the global environment.* Cambridge, MA: MIT Press.

Dyke, C. (1988). *The evolutionary dynamics of complex systems: A study in biosocial complexity.* New York: Oxford University Press.

Eakin, H. (2012). Human vulnerability to global environmental change. *The encyclopedia of earth.* Accessed from http://www.eoearth.org/view/article/153598

Ehlers, E., & Krafft, T. (Eds.). (2006). *Earth system science in the anthropocene.* New York: Springer.

Ekstrom, J. A., & Young, O. R. (2009). Evaluating functional fit between a set of institutions and an ecosystem. *Ecology and Society, 14*(2), 16. http://www.ecologyandsociety.org/vol14/iss2/art16/

Farley, J. (2012). Ecosystem services: The economics debate. *Ecosystem Services, 1*, 40–49.

Fellmann, T. (2012, April 23–24). The assessment of climate change-related vulnerability in the agricultural sector: Reviewing conceptual frameworks. In A. Meybeck, et al. (Eds.), *Building resilience for adaptation to climate change in the agricultural sector.* Proceedings of a Joint FAO/OECD Workshop, FAO, Rome.

Fischer Kowalski, M., & Haberl, H. (Eds.). (2007). *Socioecological transitions and global change. Trajectories of social metabolism and land use.* Cheltenham: Edward Elgar.

3 Interaction of Nature and Society in Ecology

Fischer-Kowalski, M., Haberl, H., Hüttler, W., Payer, H., Schandl, H., Winiwarter, V., et al. (1997). *Gesellschaftlicher Stoffwechsel und Kolonisierung von Natur: Ein Versuch in Sozialer Ökologie.* Amsterdam: G+B Verlag Fakultas.

Folke, C. (2006). Resilience: The emergence of a perspective for social-ecological systems analyses. *Global Environmental Change, 16,* 253–267.

Folke, C., Hahn, T., Olsson, P., & Norberg, J. (2005). Adaptive governance of social-ecological systems. *Annual Review of Environment and Resources, 30,* 441–473. doi:10.1146/annurev.energy.30.050504.144511.

Ford, E. D. (2000). *Scientific method for ecological research.* Cambridge: Cambridge University Press.

Frosch, R. A., & Gallopoulos, N. (1989). Strategies for manufacturing. *Scientific American, 261*(3), 144–152.

Füssel, H.-M. (2007). Vulnerability: A generally applicable conceptual framework for climate change research. *Global Environmental Change, 17,* 155–167.

Georgescu-Roegen, N. (1971). *The entropy law and the economic process.* Cambridge, MA: Harvard University Press.

Gerst, M. D., Raskin, P. D., & Rockström, J. (2014). Contours of a resilient global future. *Sustainability, 6*(1), 123–135. doi:10.3390/su6010123.

Gierer, A. (1998). *Die Gedachte Natur: Ursprünge der modernen Wissenschaft.* Reinbek: Rowohlt.

Gunderson, L. H., & Holling, C. S. (Eds.). (2002). *Panarchy: Understanding transformations in human and natural systems.* Washington, DC: Island Press.

Hobbs, R. J., Higgs, E., & Harris, J. A. (2006). Novel ecosystems: Implications for conservation and restoration. *Trends in Ecology and Evolution, 24*(11), 599–605.

Hornborg, A., & Crumley, C. L. (Eds.). (2006). *The World System and The Earth System: Global socioenvironmental change and sustainability since the neolithic.* Walnut Creek, CA: Left Coast Press.

Jamison, A. (2001). *The making of green knowledge: Environmental politics and cultural transformation.* Cambridge: Cambridge University Press.

Janssen, M. A. (2006). Historical institutional analysis of social-ecological systems. *Journal of Institutional Economics, 2*(2), 127–131.

Janssen, M. A., & Anderies, J. M. (2007). Robustness trade-offs in social-ecological systems. *International Journal of the Commons, 1*(1), 43–65.

Lansing, J. S. (2003). Complex adaptive systems. *Annual Review of Anthropology, 32,* 183–204.

Lele, S., Springate-Baginski, O., Lakerveld, R., Deb, D., & Dash, P. (2013). Ecosystem services: Origins, contributions, conditions, pitfalls, and alternatives. *Conservation and Society, 11*(4), 343–358.

Lundgren, L., & Jonsson, A. (2012). *Assessment of social vulnerability: A literature review related to climate change and natural hazards* (CSPR Briefing No. 9). Linköping: Centre for Climate science and Policy Research.

Magnan, A., Duvat, V., & Garnier, E. (2012). Reconstituer les « trajectoires de vulnérabilité » pour penser différemment l'adaptation au changement climatique? *Natures, Sciences, Sociétés, 20*(1), 82–91.

Martinez-Alier, J. (2002). *The environmentalism of the poor: A study of ecological conflicts and valuation.* Cheltenham: Edward Elgar.

Maturana, H. R., & Varela, F. J. (1980). *Autopoiesis and cognition: The realization of the living.* Dordrecht: Reidel Publishing Company.

Mayr, E., & Provine, P. B. (Eds.). (1980). *The evolutionary synthesis: Perspectives on the unification of biology.* Cambridge, MA: Harvard University Press.

McNeill, J. R. (2001). *Something new under the sun: An environmental history of the twentieth-century world.* New York: W.W. Norton.

Miao, S., Carstenn, S., & Nungesser, M. (2009). *Real world ecology: Large-scale and long-term case studies and methods.* New York: Springer.

Moran, E.F. (2000 (1979)) *Human Adaptability: An Introduction to Ecological Anthropology.* Boulder: Westview Press (2nd edition).

Mumby, P. J., Chollett, I., Bozec, Y.-M., & Wolf, N. H. (2014). Ecological resilience, robustness and vulnerability: How do these concepts benefit ecosystem management? *Current Opinion on Environmental Sustainability, 7*, 22–27.

Nelson, D. R., Adger, N., & Brown, K. (2007). Adaptation to environmental change: Contributions of a resilience framework. *Annual Review of Environment and Resources, 32*, 395–419.

Norberg, J., & Cumming, G. (Eds.). (2008). *Complexity theory for a sustainable future.* New York: Columbia University Press.

Norgaard, R. (1984). Coevolutionary development potential. *Land Economics, 60*(2), 160–173.

Odum, E. (1969). The strategy of ecosystem development. *Science, 164*(3877), 262–270.

Odum, H. (1983). *Systems ecology: An introduction.* New York: Wiley.

Olsson, P., Gunderson, L. H., Carpenter, S. R., Ryan, P., Lebel, L., Folke, C., et al. (2006). Shooting the rapids: Navigating transitions to adaptive governance of social-ecological systems. *Ecology and Society, 11*(1), 18. http://www.ecologyandsociety.org/vol11/iss1/art18/.

Perrings, C. (2006). Resilience and sustainable development. *Environment and Development Economics, 11*, 417–427.

Pickett, S. T. A., Cadenasso, M. L., & Grove, J. M. (2005). Biocomplexity in coupled natural–human systems: A multidimensional framework. *Ecosystems, 8*, 225–232. doi:10.1007/s10021-004-0098-7.

Porter, T. B. (2006). Coevolution as a research framework for organizations and the natural environment. *Organization & Environment, 19*(4), 479–504. doi:10.1177/1086026606294958.

Reed, S. O., Friend, R., Jarvie, J., Henceroth, J., Thinphanga, P., Singh, D., et al. (2014). Resilience projects as experiments: Implementing climate change resilience in Asian Cities. *Climate and Development.* 10.1080/17565529.2014.989190.

Reid, J. (2012). The disastrous and politically debased subject of resilience. *Development Dialogue, 58*, 67–79.

Rockström, J., Steffen, W., Noone, K., Persson, Å., Stuart Chapin, F., III, Lambin, E. F., et al. (2009). A safe operating space for humanity. *Nature, 461*, 472–475.

Rossignol, N., Delvenne, P., & Turcanu, C. (2014). Rethinking vulnerability analysis and governance with emphasis on a participatory approach. *Risk Analysis, 35*(1), 129–141.

Scheiner, S. M., & Willig, M. R. (Eds.). (2011). *The theory of ecology*. Chicago: The University of Chicago Press.

Scoones, I. (1999). New ecology and the social sciences: What prospects for a fruitful engagement? *Annual Review of Anthropology, 28*, 479–507.

Soares, M. B., Gagnon, A. S., & Doherty, R. M. (2012). Conceptual elements of climate change vulnerability assessments: A review. *International Journal of Climate Change Strategies and Management, 4*(1), 6–35.

Turchin, P., & Hall, T. D. (2003). Spatial synchrony among and within world-systems: Insights from theoretical ecology. *Journal of World-Systems Research, 9*(1), 37–64.

Turner, B. L., Kasperson, R. E., Matson, P. A., McCarthy, J. J., Corella, R. W., Christensen, L., et al. (2003). A framework for vulnerability analysis in sustainability science. *PNAS, 100*(14), 8074–8079.

Vayda, A. P. (2009). *Explaining human actions and environmental changes*. Lanham, MD: AltaMira Press.

von Uexküll, J. (1926). *Theoretical biology*. New York: Harcourt, Brace & Co.

Weiher, E., Freund, D., Bunton, T., Stefanski, A., Lee, T., & Bentivenga, S. (2011). Advances, challenges and a developing synthesis of ecological community assembly theory. *Philosophical Transactions of the Royal Society, B, 366*(1576), 2403–2413.

Weisz, H. (2001). Gesellschaft-Natur Koevolution: Bedingungen der Möglichkeit nachhaltiger Entwicklung. Kulturwissenschaftliches Seminar (Dissertation), Humboldt Universität, Berlin.

Young, O. (2010). *Institutional dynamics: Emergent patterns in International Environmental Governance*. Cambridge, MA: MIT Press.

4

Sustainability in Social-Ecological Perspective

The social-ecological theory of nature and society, discussed in Chaps. 2 and 3, has as one aim to clarify the nature of the process called sustainable development. Sustainability is discussed at two levels:

- Of a *discourse* in which the meanings of the notion of the term are continually disputed, interpreted, and re-interpreted in their social, political, economic, and ecological meanings in a scientific and a political discourse.
- Of *a process of change* of the practices of natural resource use in modern society that has started with local, national, and international political decisions in the early 1990s.

To specify and clarify, by way of theoretical interpretation, the forms, and the courses, and the duration of this process of change are attempted in this chapter. In this interpretation, *sustainable development appears as integrated social and ecological process to maintain the functions and processes of societal and ecological systems in a long-term perspective* that is (inexactly) described as intergenerational solidarity of resource use. To achieve this clarification about the nature and perspective of the sustainability process, it is, however, necessary to go through the discourse first, showing how this perspective of an integrated social-ecological process is perceived, reflected, and discussed. This is done in four steps of review:

- Reconstructing *difficulties and failures* in the prior sustainability discourse
- Discussing a renewal and *broadening of the discourse* where the nature of sustainable development as process of change of resource use is specified as a transformation of industrial society
- Evaluating *interdisciplinary analyses of sustainable development* in search of a theory of transformation
- *Integrating theoretical sustainability analyses* in political-economic and social-ecological theories

In attempts to separate the interwoven scientific and political discourses, the nature of sustainable development needs to be discussed more critically. The blending of political and scientific, conceptual and problem-oriented, normative and fact-oriented discussion is a property of the discourse. Attempts of clarification by way of definition and operationalisation resulted in a variety of competing definitions. Two terms have been created to deal with such abstract notions: that of "essentially contested concepts" (Gallie) and that of "floating (or empty) signifiers" (Levi-Strauss). The theoretical reflection of sustainability in social ecology is a way to discuss and assess the different variants of the term and to show how sustainability can be used to analyse global social and environmental change. In this theory-guided discussion, sustainable development is transformed into a scientific concept of social-ecological transformation. The purpose is to study the complex social process of change that cannot be sufficiently understood as a policy and governance process.

Several controversial points about the nature of sustainable development need to be dealt with in the following review of the sustainability discourse:

1. *Sustainable development as political process, as socio-technical process, and as process of global social change:* In the past decade, these three views could be found as salient interpretations of the sustainability process. Sustainability as a *global transition process* is discussed in the

global scenario debate (Raskin et al. 2002). Sustainability as a *political and governance process* shows the qualities described by Meadowcroft (2007) and sustainability as *a socio-technical transition process* the qualities described by Smith et al. (2005). The views of sustainable development as political reform and technical change are limited in disciplinary perspectives, whereas the global transition debate by Raskin et al. seems to approach an interdisciplinary and integrated social-ecological perspective. But also their diagnoses are disputed: that the dynamics of human development are for the first time in human history seen as global, with changes happening in a short historic time (Raskin et al. 2002: 71f) seem to neglect the social and geographical (local, regional, national) consequences of global change. The construction of a common world and future of humankind in the discourses of sustainable development and global environmental change (see Chap. 7) shows the problems of a scientific reductionism that derives its arguments from the physical or ecological description of the world; it cannot sufficiently account for the differences in the social world in terms of wealth and poverty, power and interests, property of natural and other resources.

2. *Sustainable development as sharing and redistribution of resources and as a conflicting process between different interests of resource users:* Much less scientific and political attention is paid to these two social aspects of sustainability that appeared with the discourse-opening policy document of the Brundtland report. These themes come up in the more critical analyses of sustainable development that developed off the mainstream research and debate on sustainability, in social-ecological and political-ecological analyses of large-scale interactions between the biophysical earth system and the social world system. After the initial debate of the ecology of the modern world system by Goldfrank et al. (1999), first syntheses of social and ecological research were that of an emerging earth system science (Ehlers and Krafft 2006) and of interdisciplinary studies of the interacting world and earth systems (Hornborg and Crumley 2006). In the book edited by Hornborg and Crumley a model for the integration of empirical research and theoretical thinking is developed with knowledge from different disciplines.

The new unit of analysis is the interaction of society and nature at global levels, as social world system and ecological earth system, in a perspective of world ecology analysis. This world ecology perspective needs to be elaborated further in social-ecological analyses of global change in the modern world system and its transformation to sustainability. The world ecology perspective of two interacting macro-systems, world and earth system, that operate as an inseparable whole, although with two distinct orders, unfolds in piecemeal theoretical studies of social-ecological systems at different scales, for example, in the studies of Fischer-Kowalski and Haberl (2007, land use change); Miao et al. (2009, methodological questions); Krausmann et al. (2011, socio-metabolic transition); Fischer-Kowalski et al. (2012, transition scenarios); Brand (2015a, b, critical theory of social-ecological transformation). The theoretical discourse is fragmented and dispersed, an integrating interdisciplinary theory not yet visible.

3. *The forms of collective action* in transitions to sustainability are discussed in several theoretical interpretations of sustainability. Meadowcrofts' (2007) analysis of governance of sustainable development under conditions of unequal power of actors and power relations in diverse societal subsystems is insufficient, assuming that sustainable development is an inherently and irreducibly political process. The discussion seems blocked through a clash of two contrasting views of sustainability: that of a political process of steering society (governance), and that of a much more complex process of transformation of society. Transformation includes autonomous social and ecological processes that cannot be managed or only influenced to some degree. Even for formulating political strategies of governance, these complex processes of interaction and change of global societal and ecological systems need to be analysed—to show where policy, steering, and regulation end and need to be reinforced by other processes to become effective. In the social-ecological analysis of sustainability unfolds such a broader, interdisciplinary perspective where sustainable development includes further and autonomous processes of societal change that cannot be reconstructed as governance. An interdisciplinary analysis of interaction and change of social and

ecological systems is not yet achieved by Meadowcroft (2007: 310ff) with the envisaging of structural and systemic underpinnings of societal change and with the description as iterative process of reform over many decades; large parts of social and ecological processes are ignored in Meadowcrofts' analyses.

4. *The implications of constructing sustainable development as a process of global transition or transformation of modern industrial society:* Policy and governance processes are the most visible, public, and communicated, but further processes need to be considered: socio-cultural processes in the lifeworld of people, various forms of change (structural social, economic, technical), processes of global social and environmental change (as globalisation and land use change) that should be influenced through sustainable development. Examples of such autonomous processes are population growth and demographic transition, urbanisation and land-use change, economic and physical resource flows, transport and communication processes and their technical infrastructures, environmental pollution, energy development, or the global material cycles in ecosystems as the carbon, nitrogen and water cycles, changes in atmospheric and ocean circulation in biological diversity, and others. The complex and interwoven processes cannot be managed, coordinated, or steered in one big process of global governance. But in governance strategies, it needs to be found out which processes can and need to be steered in the transition to sustainability. The process of transforming industrial society towards sustainability has just started with the transformation of the industrial energy regime based on fossil fuels (see Chap. 8). The future success of sustainable development depends to a large degree from the success of these efforts to transform the global economy into a low carbon economy with significantly reduced CO_2 emissions in the atmosphere. Local and national strategies for transition to sustainability are parts of the global transformation. Continuous deliberation and negotiation, iterated and revised decisions and evaluation are necessary to maintain a process at different levels of policy, in social and ecological systems, for which the forms and levels of integration and regulation are continually matched and rescaled.

The following review of the prior discourse of sustainable development is not about the evaluation of the implementation of national or international policy programmes, for example, the global programme "Agenda 21" or the Convention on Biodiversity. These forms of policy analysis are part of empirical evaluations and assessments of environmental policies; they cannot show the problems and difficulties in the sustainability process. The discourse is reviewed to show the advances, changes, failures, and modifications in the sustainability process as an erratic process of approaching the complexity of social transformation.

The Prior Discourse of Sustainable Development

Reviews of the sustainability discourse since its beginning in 1987 show: only gradually the discourse approached the complexity of the societal transformation process with a more coherent picture of the dynamics of social and ecological systems in modern society. The interpretations of the idea of sustainable development in the discourse have been reviewed many times (e.g.: Moffatt 1995; Barry 1999; Lee et al. 2000; Sneddon et al. 2006; Bruckmeier 2009). The main views are described below (Table 4.1), compared and classified to identify the components of strategies to regulate social-ecological change and to initiate societal transformation. Two aspects of the sustainability discourse seem important for its development: a political debate to develop, discuss, and adopt ideas, and a controversy about economic growth in relation to sustainability.

1. *The political debate of sustainable development* started before scientific research on global change was carried out. The report of the North–South Commission "Our Common Future" from 1987 marks the beginning of international policies and strategies for sustainable development. In the report, sustainable development was formulated in a vague global perspective as intra- and intergenerational solidarity in resource use. Sustainable development remained a vague idea of a common future, a platform concept for organising the policy process.

4 Sustainability in Social-Ecological Perspective   131

Table 4.1 Sustainable development—changing interpretations

Core concepts/themes	Approaches and authors	Main ecological assumptions and arguments	Assessment
(1) Intra- and intergenerational solidarity of natural resource use	Report of the North–South Commission 1987 as policy document, subsequently developing global ("Agenda 21"), national and local policies (Rio-conference 1992)	Sustainable development requires global management, sharing and redistribution of natural resources	No consensus was found in the policy processes how to realise intergenerational solidarity and implement sustainable development; policies adopted reflect the ideas and interests of powerful "global players"
(2) Mainstreaming sustainability: social, economic, and environmental sustainability	Johannesburg Summit 2002	Sustainability as pluri-dimensional process where ecological principles of resource management need to be matched with contrasting social and economic principles	How to find compromises between the three contrasting components requires continuous discussion
(3) New ecological paradigm/ecological embeddedness of society and economy	Human ecology (Catton and Dunlap 1978), ecological economics Georgescu-Roegen (1971), Daly (1991)	Human dependence upon the natural environment and functions of ecosystems, society cannot be decoupled from nature	Ecological principles for framing the sustainability concept and adapt it to the ideas of environmental movements

(continued)

Table 4.1 (continued)

Core concepts/themes	Approaches and authors	Main ecological assumptions and arguments	Assessment
(4) Carrying capacity Limits to growth/ external natural limits on human economic activity Planetary boundaries of resource use	Ecolgy (Odum 1993) Neo-Malthusianism and Systems analysis of "Limits to Growth" (Meadows et al. 1972; Ehrlich and Ehrlich 1970) Rockström et al. (2009)	To stay "within carrying capacity and therefore (being) sustainable" (Daly 1991) To prevent overuse of natural resources requires ecological criteria for limiting resource use: first ideas to formulate global limits ecologically: planetary boundaries (connected to global cycles of water, nitrate etc., and ecosystem, functions)	Reformulation of Malthusian ideas Inexact use of the term carrying capacity—differs for humans and other species: for humans to be calculated for the culturally varying levels of consumption, not as biological fix (Rees) The formulation of planetary boundaries of resource is inexact (measurement problems) and remains controversial
(5) Ecological modernisation, greening of industry, clean production, technological innovation ("triple helix"), green economy	Ecological modernisation (Mol) Greening of industry (Speth) Technological innovation in cooperation of government, industry, science ("triple helix": Leydesdorff et al.)	Variants of sustainable industrial production according to ecological rationality principles (research, technology improvement, saving material and energy in production)	Ideas that support the interests and positions of the industrialised countries and private corporations

(6) Environmental economics: maintaining natural, economic, and human capital Ecological economics Degrowth	Internalisation of negative externalities in economic production costs Payments for ecosystem services Georgescu-Roegen (1986), Daly (1997) Latouche, Martinez-Alier	The scarcity of goods and the pollution of the environment can be measured and managed by attributing monetary value to goods and "bads"/damages Zero growth of economy and population to maintain natural capital	Adapting the idea of sustainable development to principles of capitalist market economy Critical economic variants of the idea of limiting growth and resource use, connected with the neo-Malthusian discourse
(7) Environmental politics based on economic arguments: weak and strong sustainability	Weak sustainability Strong sustainability	Substitutability of human and natural capital Human and natural capitals are complementary	The different views of sustainability correspond to the differences between mainstream environmental economics and critical ecological economics
(8) Environmental ethics: ethics of care, deep ecology, redistribution of resources, sufficiency, environmental justice	Connects to different variants of environmental ethics, mainly bio- and ecocentric variants and conservation ethics	Formulating ideas of fair/just distribution of resources and environmental burdens with different ethical arguments	Requires consequently global solidarity, referring to all humans and all countries: necessary to argue with different cultural values/worldviews

(continued)

Table 4.1 (continued)

Core concepts/themes	Approaches and authors	Main ecological assumptions and arguments	Assessment
(9) Livelihood changes/ sustainable livelihoods "Environmentalism of the poor" Environmental democracy	Connecting resource use and sustainability to human/private consumption (households) and basic human needs: Chambers and Conway (1991), Scoones (1997) Martinez-Alier (2002) Lee et al. (2000)	A livelihood "comprises the capabilities, assets (stores, resources, claims, and access) and activities required for a means of living: a livelihood is sustainable which can cope with and recover from stress and shocks, maintain or enhance its capabilities and assets, and provide sustainable livelihood opportunities for the next generation; and which contributes net benefits to other livelihoods at the local and global levels and in the short and long term". (Chambers and Conway 1991)	Ideas supporting poor (rural) populations in the Global South Ideas to give rights and priority in resource use to local and indigenous populations who need to live from the resources provided by the local ecosystems

4 Sustainability in Social-Ecological Perspective

(10) Ecology and economy: – small-scale production and local, circular economy – eco-economy – ecological footprint	Schumacher: "small is beautiful" Brown (2001) Rees, Wackernagel	Variants of the argument to reconnect production, distribution and consumption of resources more closely, rebuilding small-scale systems and global resource flows, supported from ecological and anthropological research	"Ecological utopias" of local society/economy, influenced by the ideas of Schumacher Brown's description of the eco-economy is based on thorough ecological analysis, but the transition is described within the present economic system, with insufficient policy instruments Renewal of the ideas of local economy through arguments for reducing global resource flows and "resource/food miles"

Sources: Own compilation, sources cited

In the discourse about principles of natural resource use ideas of social and environmental justice were adopted and the support and cooperation of many governmental and non-governmental actors was sought. The success of the political discourse was visible in the mobilisation of many actors with different interests in a global debate and in the formulation of action programmes, especially "Agenda 21". The interaction between political and scientific debates of sustainability generated further possibilities to develop the vague idea, but no authoritative institution as that of the Intergovernmental Panel on Climate Change (IPCC) in the global climate discourse provided scientific knowledge syntheses for the political debates. Also the sharing and redistribution of resources at global levels, the social components of sustainable development, need to be negotiated in more complicated governance processes than most actors imagine. The critique of policies of sustainable development that unfolded in the controversy about economic growth and possibilities of a non-growing economy is the key to understand the complexity of the governance and transformation processes.

2. *The idea of a non-growing economy* is old. Ecological economists and other discourse participants refer to the concept of stationary or steady state economy of the classical political economist John Stuart Mill (1848). He saw that economic growth once will come to an end, followed by a stationary economy. With the ecological and sustainability discourse the question is taken up again, whether a non-growing economy is one of misery on an environmentally devastated earth, or whether it is a higher form of economy, better satisfying human needs than the ecologically primitive industrial growth economy. The argument supporting the critique is that industrialisation, with all its scientifically based technologies, destroys the natural resource base of human society and undermines society's future. In this critique, sustainable development is an ecological imperative to live within the resource limits of the earth system through appropriate forms of sharing and redistribution of natural resources. Mill was inclined to see the stationary economy as one of social and moral progress with improved quality of life, although he saw the possibility of overuse of resources and environmental disruption.

Main Ideas of Sustainable Development

In the policy-driven discourse of sustainable development, it remained an open question how scientific knowledge is used in the political processes—although it became clear, that sustainable development cannot be limited to environmental policy and natural resource management processes. Broader social processes of change in the societal systems of politics and economy are part of sustainable development. Furthermore, the interaction of society and nature needs to be analysed, not only normatively constructed as in the value-based and ethical thinking that influences the sustainability discourse. A weak echo of the broad view of sustainable development is found in the mainstream variant with the differentiation of social, economic, and environmental sustainability. The broadening brought new theoretical debates and controversies on the nature of sustainable development. A major controversy is that between conventional ideas assuming transition to sustainability is possible within the existing global economic order through a series of policy reforms, and more critical approaches aiming at fundamental transformations of modern society and economy (Table 4.1).

Many of the interpretations of sustainability in Table 4.1 are derived from ecological worldviews, visions, and normative ideas that take up ecological concepts as carrying capacity or ecologically embedded resource use. In most ideas of sustainable development, the future is vaguely formulated in normative forms as wanted states or ideal states of resource use. How to achieve a sustainable future is controversial. The necessary changes of society and economy on the way to global sustainability are only gradually discussed with the help of theories connecting social- and natural-scientific research. In the early debates of sustainable development theoretical reflections were overshadowed by controversial interpretations of the sustainability idea as one in the interest of the Global North and the rich countries, or the Global South and the poor countries in the global economy, as shown in the review of the discourse by Lee et al. (2000) and in the analysis of the "environmentalism of the poor" (Martinez-Alier 2002).

The ideas in the sustainability discourse can be divided in conventional and critical variants (as described in the Appendix of Chap. 3). Conventional variants see sustainability as achievable within the industrial society and its global order, critical variants argue for changes of lifestyles and societal, political, and economic systems. In both variants, sustainability is not sufficiently discussed with regard to possibilities of transforming modern society and economy.

Social-ecological theory connects to the sustainability discourse at the point where the discourse, in conventional and critical approaches to sustainability, ends. The theory argues that sustainability is achieved with a new mode of production that can be theoretically described in terms of a societal metabolism. Social ecology provides one of the possible theoretical underpinnings of sustainability as great transformation of industrial society, or as "Promethean revolution" as it is called in ecological economics. The dominant view of sustainability is that of a process with three different forms of social, economic, and ecological sustainability between which compromises need to be found in the policy process. A practical difficulty is that the three dimensions and their interrelations are understood differently and tend to generate incoherent changes of policies and governance strategies.

Assessment of the Sustainability Discourse

The global sustainability discourse developed, since the Brundtland report in 1987, in complicated interaction between science and policy processes, where the notion of sustainable development became an overburdened idea. Although continually discussed, the idea remains controversial; the multiple interpretations move between normative, practical, and scientific knowledge forms. The political process can be seen as one of articulating and matching different interests. Further knowledge from scientific analyses of societal, economic, and ecological systems cannot be adequately transferred into interests of actors, also not in the form of advocacy policy by speaking for the interests of the non-actors (future generations, the ones excluded from policy processes, nature, other species, ecosystems). To understand the process complexity, it seems necessary to identify and assess the difficulties in the sustainability discourse that is stuck in contradicting and competing interpretations of the idea of sustainable development:

- The *consequence of the pluralist nature of the discourse* is that the bridging concept of sustainable development cannot be reduced in meaningful ways to one common concept: no consensus is achieved by the many participants with their varying interests. The global spreading of the debate is seen to make a consensus unrealistic. This implies that sustainable development is not a scientific concept, but a bridging term to allow a variety of interpretations and approaches under a guiding idea.
- The *consequence of the perception of sustainability as implying normative ideas* is that it cannot be clarified through scientific, methodological, and theoretical discussion, but need to be negotiated in policy processes where heterogeneous interests and goals of social actors need to be articulated and defended.
- The *nature of sustainability and the derived term sustainable development is that of "essentially contested concepts"* (Gallie 1956; Collier et al. 2006) that develop with a variety of meanings, in continuous controversy and discussion, without ever achieving the status of concepts and criteria verified through scientific or practical knowledge.
- *Compromises between different interests* that are in conflict with each other can only be achieved to some degree in the sustainability process, through norms and strategies on which many actors with different interests and worldviews can agree.
- The *consequence of a lack of critical reflection of the idea* of sustainable development in the policy process is that knowledge from research and theoretical analyses of modern society and its interaction with nature is ignored.

All of these reasons for the difficulties in the sustainability discourse may be justified to some degree, but only the last one shows a possible way out of the dilemma with contested concepts and interpretations. This interpretation is used here: the lacking theoretical reflection and underpinning of the ideas of sustainability requires a renewal of the discourse with knowledge about the dynamics of social and ecological systems that is integrated with the help of social-ecological theory. It is suggested to connect the notions of sustainability and sustainable development with theoretical analyses of societal and ecological systems in social ecology and similar critical discourses.

Three points seem important in the further discussion and clarification of the social properties of sustainability: the discursive nature of the idea, the operationalisation, and measurement of sustainable development, and the neoliberal mainstreaming of the discourse:

1. *The discursive nature of sustainability appears as diffuseness of the notion and the debate*: The most widespread idea of sustainable development is from the Brundtland report, referring to intra- and intergenerational solidarity as the normative compass of global economic development. The clarification of the term and the ways towards sustainability went back and forth between science and policy when the political actors were seeking scientific knowledge, support, and justification for their ideas.

    Described in terms of a power-dependent debate, the political discourse of sustainable development tended towards a consensus of the powerful actors, achieved by selective knowledge use and insufficient search for alternatives, in the "disjointed incrementalism" of the political debate. After the Johannesburg summit of 2002, the mainstream variant became that of differentiating between social, economic, and environmental sustainability as three dimensions that need to be balanced in sustainability strategies and political programmes. This managerial reasoning emerged from political debates, in attempts to catch some of the complexity of the processes in interacting social and ecological systems. The complexity was seen in the contradictions between the three forms that can be dealt with by compromises. After the mainstreaming and the neoliberal occupation of the idea, the political sustainability process shifted from debates about goals towards implementation of (international) sustainability policies. The search for "another development" that was less economically driven, more taking into account environmental criteria for the use of natural resources, was finally channelled in an economic reasoning supporting economic globalisation through the neoliberal consensus. The conference "Rio+20" of 2012 supported a green economy building on market forces and commercialisation of nature or ecosystem services. The debate of a great transformation vanished from the political agendas. It continues in critical discourses, after the millennium ecosystem assessment especially in the global scenario debate.

Described in terms of a scientific debate, the sustainability discourse did not find to scientific clarification and approval of the contested idea. Scientific knowledge from several disciplines influenced the discourse selectively. Most of the heterogeneous ideas of sustainable development classified above in conventional and critical approaches found some scientific support. The policy processes at national and international levels turned out to be "immune" against the adoption of certain forms of scientific knowledge. More complicated theoretical ideas and critical social-scientific analyses of societal systems found less interest in environmental movements and among the political actors where simple, vague, and normative ideas were preferred. The lack of theoretical reflection supported illusions to achieve sustainability in the short run, through technologies and ecological modernisation, within the present societal, political, and economic systems, in "win–win" situations, within national policies of "splendid isolation", by building sustainable systems at local, regional, or national levels, without changing the globalised economic system.

Assessing the process of discussing sustainable development since 1987 it can be said: The idea of sustainable development created at the beginning enthusiasm, as many ideas appealing to common values and interests, especially that of "our (mankind's) common future". This enthusiasm faded away with the difficulties of the sustainability process that became visible in the continuing process and the inefficiency of most policy programmes. These programmes were not based on systems analyses of modern society and economy, also not on thorough analyses of problems and hindrances, more dictated by the necessities to act under time pressure and generating some change. Different political interests and expectations of the discourse participants should be matched in participatory processes built into political programmes. The early debate unfolded unexpected success in mobilising people, governmental and non-governmental organisations, and social movements for global action. With the conceptual vagueness and the confusing debates the notion of sustainable development was quickly adopted by governmental and non-governmental actors and spread in a global discourse, in many national and international policy programmes: it became the symbolic idea guiding local, national and international environmental policies, engaging in many countries scientists and practitioners in the fields of politics, economy and culture. This can be seen as widespread awareness that sustainability is required to maintain the systems and processes in nature and society that changed during modern society in risky ways, but consensus as found for the idea is not achieved for the process and for methods, measures, and interim goals.

The weakness of the sustainability process can be attributed to several factors: that of a political process distorted by interests of powerful actors and institutions; of a process in which science and research did not find consensus about ways of system transformation; of a process where the ambitious and morally justified goals faded away on the way in power fights and compromises. The last point can be connected to the necessity of a global debate for which the global arena of the government-dependent institutions of the UN system is not an optimal platform. The UN does not provide sufficient possibilities to initiate critical debates about transformations of societal and economic systems; critical voices are not necessarily suppressed in the policy process but ignored in policy programmes. The externalisation and the shifting of burdens of environmental disruption to weak social groups continue in forms of "managed decline" in national and international economic policies, also in the industrialised Western countries. The maintenance of economic growth and new industrialisation imply that the burdens of resource use and environmental pollution are to a large degree that of future generations and people in the Global South.

2. *Attempts to operationalise and measure sustainability and the progress of sustainable development:* The policy-driven sustainability process generated many attempts to measure sustainability and the implementation of sustainable development policies through indicators and criteria for the valuation of political strategies and programmes at

national and international levels. The indicator debate is stuck in the complexity of constructing quantitative and qualitative indicators for each dimension of social, economic, and environmental sustainability, based on available public statistics. Attempts to measure the whole process aggregate the indicators in indexes as the EVI, the ESI, and the environmental performance index (EPI). The results are abstract figures without clarification of possibilities to measure societal transformation. The indices seem to measure the exact position of a country or national economy at certain points in time, as its higher or lower progress towards sustainability. They leave the impression that important information is missing. Statistical indicators can measure the quality of life and wellbeing in the countries, their economic development, the manifold environmental problems and damages, but summarising these measurements in indices does not yet give guiding ideas of how to achieve sustainability. The advantage of indicators and indices is that they are constructed with empirical data. However, the difficulty with indicators for sustainability is, that they often use data for measuring other, less complex phenomena than sustainabiliity.

3. *The neoliberal sustainability discourse:* The manifold ideas that appeared in the political debates did not show, rather mask, the real course of the economic and policy processes under the influence of neoliberalism. The neoliberal consensus that marks the present deadlock of the sustainability discourse is not a consensus of the majority but of the powerful. This power-based steering of the process works for several reasons: there are too many and heterogeneous interests that cannot be integrated; there is not sufficient critique of and resistance against the fragmentation, segmentation, and segregation of society through market-based policies; the powerful vested interests that contradict sustainability are articulated as forms of sustainability and ecological reforms of the economy, for example, as ecological modernisation.

The core ideas of neoliberal sustainability policy (deregulation of markets and international capital flows, monetarisation and commercialisation of natural resources) were not in the foreground in the early discourse. The idea of sustainable development was launched as a new idea and a moral appeal to change the economy and economic resource use practices. In the debate about a sustainable economy emerged three paradigmatic ideas:

- The idea of the *bio-economy* (Enríquez-Cabot 1998) is oriented to (bio-) technology as a source for industrial innovation and growth.
- The idea of the *eco-economy* (Brown 2001) is based on a thorough analysis of the environmental problems caused by the global economy, but suggests insufficient and inefficient policy instruments for a transformation to sustainability.
- The idea of the *green economy* is influenced by ecological thinking, but developed in practice as supporting monetary and market-based instruments (see the critiques of green economy by Lander 2011; Hoffmann 2011; Spash 2012).

Although none of the ideas is programmatically following a neoliberal reasoning, they do not develop alternatives to growth and market-based development. In practice, the attempts to measure the natural capital that should be maintained in the sustainability process ended in the neoliberal ideas of green economy, reformulating sustainable development as economic growth and environmental responsibility as reinforcing growth through the valorisation of nature and ecosystem services (Daily 1997). The economic valorisation supports further inequalities in an economic system that is already characterised through inequalities. The ecological goals of protecting and maintaining natural resources are distorted through the financial mechanisms and the incoherent measures for achieving the goals (see Chap. 5). To manage a neoliberal consensus under the label of "green economy" implied that a part of environmental movements and NGOs accepted the ideas that through monetary valorisation of natural resources, and in practice through payments, nature is protected better than through ecological criteria for resource management and the natural resource base can be maintained for sustainable use in this way. The "economisation of ecology" is not new, discussed in environmental economics, theories of natural capitalism (Hawken, Lovins) and environmental finance (Sandor, Daily). What is new is that the practice of sustainable development, neglecting the ethical goal of intra- and intergenerational solidarity is connected with historically unprecedented forms of privatising and commercialising natural resources.

Following this assessment, a collective learning and renewal of the sustainability discourse require more interdisciplinary knowledge integration, critical discussion, and theory-guided re-interpretation of sustainable development. Many of the shortcomings in initiatives and policies of sustainable development are consequences of a narrow view of the process in terms of a policy process and as one of development, not of renewal and transformation.

Ways of Broadening and Renewing the Sustainability Discourse

In the social-ecological discussion, "sustainable development" becomes "societal transformation to sustainability" or "social-ecological transformation", understood as a global process of transforming the societal metabolism of industrial society, or, in more traditional terminology, developing a new mode of production. In the discourse of social-ecological transformation in social ecology and critical theory (Demirovic 2012; Brand 2015b) two themes are highlighted:

- *Theoretical analyses* of the system of modern society and its economic logic of development
- *Practical ideas* for changing modes of life and consumption under conditions of global social and environmental change

Sustainability as societal transformation is a broader, more complex process beyond market processes and political governance. Further, autonomous social and ecological processes that cannot be politically and economically managed or regulated are part of the overall transformation. The established term of transition management (Fischer-Kowalski and Rotmans 2009) can be used further on to specify the management and governance components of the overall transformation process that includes managed and other, autonomous processes (see above, introductory discussion). Risks and disturbances identified through vulnerability and resilience research (Chap. 3) appear as signals in the managed and non-managed processes. The autonomous processes can be influenced in governance processes only indirectly or to some degree. Indirect transformative governance as a broadening of environmental governance and a way of social-ecological regulation requires additional, interdisciplinary, and theoretical knowledge about the systemic properties of society, economy, and ecological systems.

The broadening of the governance perspective for sustainability implies

- to clarify the complicated and changing forms of interest that need to be matched in participatory governance and through conflict management;
- to become aware of the limits of environmental governance: how can these limits be shifted, and how can non-manageable social and ecological processes be connected to governance processes?
- to become aware of other processes of collective action in society that do not or cannot adopt political forms and not become part of public policies: how do they influence the possibilities of transformation to sustainability?
- to become aware of the symbolic and material relationships between spatiotemporal patterns of global resource flows and modes of production, metabolic regimes and modes of consumption: what kind of theoretical analyses and systems analyses of societal, economic, political systems do they require?

To approach the complexity of sustainable development in this broad perspective requires a series of changes of knowledge practices in science and in policy:

1. *Renewing the idea of sustainable development:* Sustainable development can be seen as an example of an "essentially contested concept", for which it is doubtful whether it can be formulated in scientific terms. Sustainability is often reduced to a political notion in the sense of normative goals as intergenerational solidarity in resource use, to achieve through environmental policies and natural resource management. Also, the mainstream variant of connected social, economic, and environmental sustainability is a vague idea that social, economic,

and environmental processes of change are interconnected. How the processes are interconnected cannot be found out through reconstructing sustainable development in a perspective of social-ecological transformation where the unclear political notion is replaced through one that can be interpreted more systematically, with theories and scientific knowledge about SES, not mainly with data and quantitative indicators. The notion of sustainability was (over)used in a long debate for different purposes, which tends to lose its communicative value in science and politics. Still no adequate new idea to replace it appeared in the ecological discourse. Environmental justice cannot take this place: it covers only part of the sustainability process with the reduction to normative ideas for distribution and sharing of resources and environmental burdens. It can be seen as an idea for the solution of problems inherent to sustainability, for seeking ways to fair distribution of resources and environmental burdens. In the renewal of the idea and discourse of sustainable development as social-ecological transformation, it needs to be assessed which processes and paths are promising and why. Sustainable development becomes an indicator for a joint learning process in the continuing discourse, improving the idea, the knowledge used, and the strategies. This has similarities with the ideas and strategies discussed in the ecological approaches of adaptive management or governance. Although these are designed for more limited purposes and simpler tasks, they may work in more complex strategies of sustainable transformation that built on such interim steps and processes as adaptive management, analyses of vulnerability and resilience (see Chap. 3), and interdisciplinary knowledge syntheses.
2. *Restructuring the sustainability discourse with the help of interdisciplinary knowledge and new practices of knowledge use:* A critical, theory-guided analysis of possibilities and ways of transformation of modern industrial society is developing with the interdisciplinary social-ecological theory. The presently incomplete theory, the complexity of the transformation process, and the unknown distant future make this transformation difficult to foresee and to design in form of potential pathways. The process cannot be prognosticated and planned, but qualitatively described in its many components. Also scenario analysis

is only a method providing limited insights, however, the one from where to begin to think about further methods of thinking through, projecting, envisioning, modelling, and simulating of changes in complex social and ecological systems. Global climate change is the model process for constructing processes of social-ecological transformation. Significant changes of the global economy in terms of decarbonisation and transformation of energy regimes are necessary to mitigate climate change and to achieve sustainability (see Chaps. 7 and 8). Although the future society and mode of production are unknown, they are influenced and constructed by the decisions made presently, and in policies and governance processes that can have consequences of reducing or maintaining possibilities of future development. These decisions require social-scientific and ecological knowledge, knowledge about the dynamics of social and ecological systems and their reproduction and interaction, besides the knowledge about environmental problems and their technical solution.

The dynamics in society and nature imply manifold and specific processes at different spatial and temporal scales. The social consequences of global environmental change—described in simple terms of temperature rise, species reduction, or land-use change—differ between countries and regions. Whether societal development is now only driven by these global environmental changes can be doubted. The hypothesis of humans as geological force in the era of the anthropocene needs to be specified and interpreted more carefully, with historical data about environmental and social change and social-scientific knowledge. Much of that knowledge seems to be ignored in the simplified and generalised concept of anthropogenic change. The constituting processes of global change, as main social process that of economic globalisation, show contradicting forms and consequences in local social systems (Sassen), through the fragmentation of local society in global cities or villages that profit from globalisation and others that become the "new hinterland". Beyond economic globalisation further interactions of SES in sub-global systems and national political or societal processes show contradicting effects that influence transitions to sustainability in unforeseeable ways. Global and ecological processes of change seem to become more important in shaping the future of modern society. These processes generate new inequalities and differences between countries, economies, and social groups. The transitions to sustainability cannot be understood without further knowledge and new knowledge practices.

3. *Clarifying the implications of the process of transformation:* In the social-ecological debate sustainability is interpreted with the theoretical term of transformation of the socio-metabolic regimes of industrial society. This term implies an analysis of the interaction of social and ecological systems in industrial resource use regimes and their combinations of material and energy resources. The theoretical reflection starts from the assumption that *sustainable development is not an evolutionary process of incremental change, as its connotation seems to be, but a rupture of path-dependent development by way of a new great transformation in the search for a future sustainable society.* This society and its mode of production differ certainly from the ideas and visions of scientists and political actors today, but they are influenced in their genesis from these ideas. Such a transformation happened for the last time in modern history with the rise of the capitalist market economy and the transition to industrial capitalism, for which the term "great transformation" was used by Polanyi (1944). Industrialisation became in the capitalist economic world system the core process of the modernisation of national economies, spreading at first only in few Western countries. Ecologically seen, it is a process based on growth and intensification of natural resource use under conditions of rapid population growth. Industrialisation made possible further social and economic transformation processes of accelerated urbanisation, mass consumption, and industrialisation of agriculture and food production that brought the environmental problems of today.

In the processes of modernisation, science-based technologies play an increasing role in transforming nature and ecosystems. Science and technology are used for specific purposes and interests of a minority of the global population, interests organised economically in terms of private property-based capital accumulation. The temporal and spatial dynamics of modernisation, in a sequence of several accumulation regimes, are complex: technology and energy-intensive processes of resource use, dependent on changing fossil energy sources, first mainly coal, later oil. The present changes through globalisation imply a restructuring of the economic processes of production and exchange with growing dependence from financial capital and its risks and crises. The commercialisation process has during the twentieth century reached large parts of non-valorised natural resources that were in earlier phases of capitalist accumulation regimes available in high quantities and with low extraction costs. The mechanism of unequal ecological exchange established during modern colonialism did not vanish in later modernisation, but changing its forms and dimensions; it is part of the structural separation of national economies in such of the core and the peripheral countries. The protection and conservation of nature and the maintenance of the natural resource base are, finally, becoming part of market-based regulation mechanisms. An ecological analysis, if it is not connected with an economic systems analysis, can say little about these changes and how they affect the environment and ecosystems. The new transformation towards sustainability becomes a power struggle between the capitalist "market ecology" or the world ecology of capitalism that develops through acceleration and intensification of resource use, and the "ecological ecology" of sustainability that asks, in one or another form, for the reduction of natural resource use, saving, redistribution, and de-commercialisation of resources.

4. *Systems analyses of societal and economic systems to understand the interaction of nature and society:* As a synthetic theory constructed from several theories, the social-ecological theory of nature and society works with concepts, theorems, and explanations from other theories. Large parts of knowledge from systems analyses of modern society and economy are already available and do not need to be carried out again in social ecology, only syntheses and complemented. World system theory, political economy, sociological, economic, and cultural-anthropological theories provide a stock of knowledge that can be reviewed and used in social ecology. The social-ecological theory is elaborated in three main components or sub-theories renewing earlier discourses: social relations with nature, societal metabolism, and colonisation of nature

4 Sustainability in Social-Ecological Perspective

(through human transformation of nature, ecosystems, and natural resources). These components have been described in detail elsewhere (Fischer-Kowalski et al. 1997; Becker and Jahn 2006; Bruckmeier 2013) and can be used for the systems analysis of the interaction of modern society and nature. The core process of societal metabolism is specified in the socio-metabolic regimes of industrial society and the accumulation regimes connected with that; both regime types of the capitalist mode of production and economic reproduction require theoretical analyses of natural resource use.

The self-destructive system dynamics of the modern economic world system unfolded through its programming for capital accumulation and economic growth that show the externalisation of social costs of production through private enterprises as negative concomitants of development. Externalisation, the shifting of burdens of natural resource use like environmental pollution to other social groups, countries and to ecosystems, is growing with the quantitative growth and intensification of natural resource use during modernisation. The trend towards externalisation is through the environmental policies—that started, nationally and internationally, in the 1970s—not yet significantly changed. In the history of modern capitalism, externalisation was practised through the colonisation of the Global South that became the cheap resource base of the capitalist world system. The historical externalisation processes included destruction of nature in the colonies, ruin of civilisations, massacres of indigenous populations, and slavery. The colonial economy supported the breakthrough of industrialisation in the European core countries as a societal transformation through unequal exchange and colonial exploitation of humans and nature. The later development of industrial society included modifications of the externalisation process with relocation of polluting industries from the industrialised countries to the newly industrialising countries and the Global South. The self-destructive mechanisms of the modern world system have been analysed first in the critical political economy, in the analysis of previous accumulation and of the exploitation of humans and nature in modern capitalism. Later Polanyi and others described this as the self-destruction of the growth society, and more recently Beck as environmental consequences of the risk society (see Chap. 2).

The dynamics of societal system transformations are summarised by Fischer-Kowalski et al. in a historical perspective, interpreting the new transformation towards sustainability as a rupture of path-dependent development of modern society and its industrial socio-metabolic regimes (based on the use of coal, oil, gas). This can happen in partially managed and regulated transformation or in unwanted chaotic, violent, and catastrophic processes where the transformation is done instead through human action, through natural hazards and catastrophes following from climate change, lack of natural resources, and pollution of the environment. Fischer-Kowalski and other social ecologists (Fischer-Kowalski 2007; Fischer-Kowalski and Haberl 2007) formulate a theoretical model of the transformation process with ecological and evolutionary ideas. Socio-ecological regimes are constructed in abstract terms as systems with adaptive and transformative dynamics:

– Processes within a regime (e.g. the present industrial metabolic regime) include gradual change, adaptation, and path dependence during the regime development that can be interpreted as "maturation".
– The system comes into crisis through external or internal factors (the industrial regime through the present overuse and scarcity of natural resources in general and fossil energy resources more specifically).
– The crisis interrupts the development path and results in a transformation towards a new regime.

The historical transformation to the industrial metabolic regime happened with the industrial revolution; the new transformation to come is not the digital revolution of information technology that is expected to allow the economic system to continue without dramatic ruptures, but a new great transformation triggered by the resource and energy problems.

The theoretical core concept for a systems analysis of SES is that of socio-metabolic regimes with different system components—the first three of the following components are core components of the social-ecological theory (Fischer-Kowalski et al.), including the societal modes of production and reproduction:

- *A socio-metabolic profile* is a system of energy and material flows connected to economic production and consumption (flows that can be measured per capita of human population).
- *A certain pattern of use and change of nature/environment* is connected with the societal metabolism: land use, resource exploitation, pollution, effects for biological evolution—the specific forms of colonisation of nature in a given society.
- *A resource management system is* organised with the help of infrastructures (transport and communication systems) and specific technologies (in agricultural, industrial production).
- *Specific institutions for economic and political governance* (market order, law systems, national and international political systems) frame the resource-management processes.
- *A pattern of demographic reproduction* structures human forms of life, life time, gender relations, and forms of human labour and employment (the components four and five transfer the systemic components of a society or socio-economic system into specific forms of public action, government and governance, and into the organisation of social and biological reproduction in the social lifeworld).
- It is assumed that between the different components of a social-ecological regime, *between the socio-economic system and its natural environment, positive or negative feedbacks are possible*; all regime components include interaction of social and ecological systems.

With the social-ecological regime concept societal development is not interpreted as a linear process, in contrast to older evolutionary concepts, or as repeating a limited number of cycles, as the model of the adaptive cycle in ecology. Development can be interrupted and the system can fall back in earlier stages of the same path. System transformation is not a determined process; the future appears as open, better: as unknown. This view of a non-deterministic, contingent development is compatible with Wallerstein's analysis of the modern world system and its genesis as a result of the system crisis of medieval European feudalism. It is also compatible with the ecological description of ecosystem dynamics, for example, in resilience research. The abstract conceptualisation of social-ecological transformation makes sense when it is connected to a theoretical systems analysis of the modern industrial society. The model of the transformation process described by Fischer-Kowalski is based on historical comparison of societies and knowledge about prior societal transformation processes in human history. In its abstract form and evolutionary logic, the transformation process is not sufficiently contextualised. The social-ecological change in the perspective of the temporal perspective of "longue durée" in history can be described as one of connecting two time scales in coupled social and ecological systems, where the specific processes in each system type influence the processes in the other type (e.g. the capitalist accumulation regimes influence the processes of ecosystem development and its adaptive cycles and the other way round). Planning processes in public policy and resource management have a time perspective of less than a decade, scenario analyses of several decades. The transformation of industrial society towards a more sustainable society is cautiously estimated as a process of several generations (but probably underestimated, it may take hundreds of years). Beyond that all views of the distanced future are indefinite.

The social-ecological view of transformation to sustainability and the theoretical core concept of societal metabolism require further analyses of societal transformation:

5. *National and international policy programmes for sustainable development need to be assessed:* Governmental sustainability policies have been implemented by nearly all states and international institutions, but global assessments show their lack of effectiveness and success. After two decades of policies under the guiding idea of sustainable development the policy in some, especially European, countries is wrongly interpreted as success. Rice (2007) called this the "rich country illusion effect": improvements of environmental quality in early industrialising countries imply the shifting of pollution to other, newly industrialising and Southern countries. The Millennium Ecosystem Assessment (2005) showed mainly continuing deterioration of ecosystems, and necessary policies to halt the negative trends are not on the way. Also climate policy has, after some early success, since the first decade of the twenty-first century failed. The conference

"Rio+20" in 2012 resulted in a non-binding declaration which is seen as failure. All that sums up to lack of success of policies of sustainable development, with two possible conclusions:

– To give up policies of sustainable development, for which an argument seems: sustainable development is impossible because of the complexity of the global processes to manage for which never sufficient knowledge will be available (Benson and Craig 2014). This can be seen as an attempt to avoid the other conclusion.
– To improve the processes of environmental policy and governance with a diagnosis of policy failures and choice of new or other policy instruments and approaches.

The first conclusion is rather an intellectual capitulation facing the contingencies and complexities of processes to analyse. In ecology, the argument that humans cannot understand and manage the complexity of ecosystems is easily mobilised; it is part of a tradition of ecology as science that is aware of its knowledge limits. With the second conclusion, it is possible to argue more consequently for a renewal and improvement of the sustainability process and seeking knowledge and methods for that. This is also possible with the conviction "it is already too late", often argued in critical assessments of climate policy that is a key to further success of sustainable development. Sustainability can be achieved only when the transformation process is continuing, even under deteriorating environmental conditions. Transformation seems necessary as the difficult process of changing the societal use of natural resources. Arguments for a transformation include the limited availability of natural resources, the overuse of natural resources and the overshoot of carrying capacity of ecosystems, the functional disturbance of ecosystems and natural cycles, and the social consequences of global social and environmental change. Global sustainability, seen with regard to these processes, requires a change of the mode of production or the societal metabolism of industrial society. This change—what it means for the people in the affluent and for that in the poor countries—is not sufficiently understood with the vague idea of intra- and interdisciplinary solidarity in resource use as the core of sustainable development.

6. *Transformation to sustainability is dependent on—not determined by—the system structures and functions of modern societies and the globalising economy.* Structures that block sustainable development are such that they keep modern societies on their path of economic growth under conditions of inequality: markets, capital, and private property; specific social and class structures: inequalities between social groups, regions, and countries in terms of political and economic power; unequal global flow and exchange of natural resources; unequal exchange and distribution of resources (Rice 2007, 2009). These structures of modern society have been created through the capitalist world system (Wallerstein 2000); its continuous development, modernisation, and innovation depend on the maintenance of power asymmetry and unequal access to resources. Most of the problems emerging in the interaction of society and nature as environmental pollution, environmental change, deteriorating ecosystem functions, and overuse of natural resources are consequences of such inequalities: the problems are not human made in an anthropological sense of being species-specific behaviour of humans, but consequences of the modern industrial society and its forms of societal metabolism.

7. *Power asymmetries and inequalities in the modern world system make transitions to global sustainability difficult.* Transitions require changes of the political institutions and the mechanisms that direct the policy process and decision-making to achieve sustainability (see Chap. 6). In the global scenario debate, there is some awareness of the problems resulting from power structures and social inequality, but no adequate theory-based analysis. According to Stutz (2009: 49) transition to sustainability requires reduction of environmental impacts and reduction of economic growth, which makes the process difficult. But the only means to deal with that is an "income transition" which does not reflect the complexity and the system contradictions in the modern economic world system. A further question is the change of political and economic power structures in the global system. This change can be approached through the analysis of social movements that initiate changes of lifestyles, outside the public policy process and in specific forms of collective action. In many of these social processes, knowledge

and experience relevant for the transformation to sustainability are generated. It seems necessary to analyse, for purposes of sustainability governance, the social spheres and forms of collective action and knowledge generation:

- The forms of collective action that are described in the social sciences as social movements, formal organisations, enterprises, networks, groups, and coalitions of these different actor types
- The forms of knowledge generation described in social research as new inter- and transdisciplinary forms of knowledge production through cooperation of different social actors and knowledge bearers

These processes are to a large degree outside policy, governance, management, and planning processes. Global governance and regulation of societal relations with nature require analyses of these broader processes that affect the transitions to sustainability. Not all of these analyses are possible with social-ecological theory of nature–society interaction, but important parts are analyses of interactions of societal and ecological systems, systemic structures and constraints in the global economy, possibilities of environmental movements and civil society action, changing forms of lifestyles, and public and private consumption.

The changes of knowledge practices in the seven points above are all to be realised with the help of interdisciplinary knowledge generation and use of available knowledge from a variety of theories described below (Table 4.2).

Interdisciplinary Analyses of Sustainable Development

The badly understood complexity of interacting SES, the wicked problems, and the fact that the transition to sustainability requires global governance are not the only reasons of limited success of sustainability policies. Insufficient analysis of the interaction between social and ecological systems that results in simplified views of the transitions to sustainability counts as more widespread practice for disorientation and

Table 4.2 Transformation to sustainability—research and scientific discourse

Core concept	Theories and authors	Main ecological assumptions and theorems	Assessment
Transition to sustainability	Dutch transition approach: Fischer-Kowalski and Rotmans (2009)	Transition management; intervention experiences; connected social, cultural, economic changes; timeframe of several decades	Complementary to the social-ecological analysis of transformations of socio-metabolic regimes that has a longer timeframe
"Great transformation" Transformation of socio-metabolic regimes Social-ecological transformation	Social-ecological research on global transformation to sustainability Haberl et al. (2011) Brand (2015b)	Sustainability requires the rebuilding of the economic system towards a non-industrial system	Core idea of critical sustainability analyses
Global assessments of social-ecological systems	Millennium Ecosystem Assessment, IAASTD	State of the global environment, specified for different ecosystems, positive and negative trends	Knowledge synthesis and expert-based assessment of global trends
Global sustainability scenarios	Tellus Institute Raskin et al. (2010)	Formulating different transition strategies: scenario variants—policy reform, ecological modernisation, sustainability transformation	Normative formulation of scenario alternatives on the basis of visions, reduced use of scientific knowledge about the society and the environment
Global environmental governance	Biermann (2004, 2011) Critical approaches: political and social ecology (climate change governance)	Integrating environmental concerns in all political decisions, connecting people and ecosystems	The dominant approaches are policy-focused, limited integration of ecological knowledge

4 Sustainability in Social-Ecological Perspective

Zero growth, degrowth	Ecological economics, political ecology Daly, Martinez-Alier, Latouche	Ideas from ecological economics (steady state, zero growth) in strategies of changing consumption cultures and lifestyles (reducing overconsumption in industrial countries)	Integrating ideas from heterodox and ecological economics in action of social and environmental movements (Local Exchange and Trade Systems: LETS, Degrowth)
Local/urban sustainability Analyses	Rees (2003) Schulz (2005) McDonald and Patterson (2007)	Urban metabolism (Wolman) Ecological footprint of cities "Urban sustainability multiplier" (Rees)	Reducing global resource flows through local (food) production as basic idea: cannot solve all urban sustainability problems
Theoretically guided analyses of societal transformation to sustainability	Emerging in social ecology	Combining analyses of social metabolism, governance of transition, scenario analysis	Integrating ideas in the transition discourse, multi-scale analysis

Sources: Own compilation; sources quoted

difficulties in sustainability practices that are often covered through normative ideas and visions that seem to justify the action sufficiently. Why much of the available knowledge about modern societal systems is not used in the sustainability discourse can, to some degree, be explained with the competing diagnoses and knowledge offers from different disciplines and theories that keep the transformation debate in continuous controversies about the relevant knowledge and the adequate forms of governance. The failures of global environmental policies in the past and the continuing difficulties indicate the necessity of renewals of the governance process, especially the slowdown of global climate policy in the governmental negotiations in the past decade. The neoliberal consensus about the green economy is delaying or preventing important policy reforms that could open possibilities of transformation. Further difficulties result from

- the lack of success and slowdown of many local and national initiatives and policies aiming at sustainable development;
- the rapidly increasing multi-scale conflicts that have been called economic and ecological distribution conflicts (Martinez-Alier);
- the slow processes and difficulties in rebuilding the national energy systems based on the metabolic regime of industrial society with finite fossil energy resources;
- the decarbonisation of national economies to combat climate change and achieve sustainability (carbon sequestration, capture and storage of atmospheric carbon dioxide, additional to natural processes, in sinks like saline aquifers, oil fields, or ocean water, wetlands);
- the continuing economic growth of the global economy annihilating effects of resource saving and dematerialisation of production.

The renewal of global environmental governance and efforts to knowledge synthesis and global assessment require complicated and conflicting processes of developing transformative agency:

1. *Integrated analyses of vulnerability, resilience, and sustainability* (discussed in Chap. 3) are first steps of creating an interdisciplinary knowledge base for transitions to sustainability. They are based on

data from local or regional social-ecological systems, analysing global change by making visible its consequences and effects at sub-global levels. Local or regional analyses can be networked to enable multi-scale environmental policy and governance. Analyses of local, regional, and national transition processes using data from global resource flows support the reconstruction of connections between local and supra-local systems and processes. Knowledge from social-ecological systems analyses can finally help to understand the lacking success of sustainability policies guided by normative, visionary, and illusionary ideas of sustainable development.

2. *Global assessments of interacting SES:* The analysis of interacting global social and ecological systems in the *Millennium Ecosystem Assessment from 2005* has not yet used much social-scientific knowledge about the modern society and economic world system. Nevertheless, it showed that sustainability requires further changes than the ones appearing in policies. The Millennium Assessment identified many maladaptive processes of societal interaction with nature directing towards rapidly deteriorating environmental conditions, resource scarcity, and disturbance of ecosystem functions. The conclusion from the assessment is that most of the changes in economics, politics, and in forms of production and consumption required for global transition to sustainability are not on the way. Further examples for critical global assessments include the critical report of IAASTD (2009) assessing agricultural development and the possibilities to change it towards sustainability, or the global scenario debate. In the expertise of the German Scientific Committee for Global Change (WBGU 2011), three priority areas of action for a great transformation to sustainability identified include climate change, land-use change, and urbanisation. This prioritisation is not based on an elaborate theory, but the three megatrends with closely interacting processes of global change show important problems to deal with in transformation to sustainability. They imply further analyses of processes that are part of the societal metabolism of modern society: population growth, food production, and production of waste and toxic substances as by-products of industrial production.

3. *Critical assessment of path-dependent societal and economic development:* Following the description above, global transformation requires environmental governance and regulation of societal relations with nature that direct towards a rupture of path dependency in the development of modern society, a transformation of global power structures and the economic growth mechanism. Knowledge about modern society and economy and about the global environmental change analysed in ecological research is available from interdisciplinary system analyses of SES in social and political ecology, especially regarding climate change and transformation of energy regimes (see Chaps. 7 and 8). Reducing socially and environmentally ruinous exponential growth becomes the key for developing transformational agency and is meanwhile more intensively discussed in the public, among scientists and political actors (see "New Scientist" 2008, special issue "The Folly of Growth"), not only in the international "degrowth" movement that developed in the past decade.

Theoretically based societal systems analyses (as suggested in Chap. 2) have not been carried out in the global assessments, but these included at least relevant knowledge about resource use practices, so that the difficulties of successful global transition to sustainability became clearer and improved transformation strategies can be formulated. Economic growth is continuing with the neoliberal consensus as mechanism to deal with the crises it generated. Continuing growth, intensifying use of natural resources, pollution and ecosystem degradation, and loss of biodiversity and cultural diversity show that the modern world system and its supporting political institutions are not transforming, only adapting.

The progress of transitions to sustainability depends on improved knowledge and governance practices to deal with the inherent conflicts, the transformation of the global socio-economic system, and the blocking of this transformation through vested interests of powerful actors and institutions in global policy. The mechanisms of self-blockade of the sustainability discourse are visible in small efforts to initiate and implement sustainability policies in many countries and in international policies, in wrong and inadequately designed policy programmes (e.g. in European countries the failure of strategies of integrated coastal zone management: see Chap. 7), in non-addressing of ecological distribution conflicts, in lacking institutional change. Under continuing economic crises, the short-term requirements to stimulate economic growth annihilate efforts and success of sustainability initiatives. Inherent weaknesses in the sustainability process are studied in critical analyses of system transformation (described in Table 4.2) that show further requirements of research.

The approaches described in Table 4.2 can be used in social-ecological analyses of coupled social and ecological systems, although they are not fully coherent. Three forms of critical analysis are important for *the construction of coherent strategies for a global transformation to sustainability:*

1. *Analysis of factors blocking sustainable development:* Sustainability is less hindered through ignorance or unwillingness, more through complications in the processes of transformative governance. These complications are of different kind:
   - The heterogeneous and changing interests of actors and the conflicts in the resource-management processes at various levels require complicated integration processes.
   - The power structures in established institutions, in political and economic systems (including powerful vested interests), are not easily changed.
   - Insufficient knowledge is used in policy and governance about the structures, functions, and processes of interacting social and ecological systems.

   It seems impossible to achieve sustainable resource use without changes of the economic world system that is programmed for global inequality, further economic growth, and intensification of natural resource use. The transformation to a resource-saving economy and the fair distribution of natural resources between countries make changes of economic and political institutions necessary. Some of these changes are shown in global scenario analyses. Changes of institutional structures and reduction of economic growth need to be enforced against the resistance of powerful actors and the short-term interests of maximising welfare for presently living generations. The transformation to sustainability requires other institutions and forums of debate and further knowledge generation, integration, and transfer. A widening of the discourse includes new forms of public discussion, collective action, and knowledge integration.

2. *Analysis of contradictions and conflicts in the sustainability process:* A doubtful construction of the sustainability process is the idea of a "win–win-process", assuming economic advantages for all actors and ignoring unequal access to resources or unequal distribution of environmental burdens. Analyses of dilemmas and conflicts in natural resource use and management are not well developed in environmental research and resource management. Ostrom's collective choice theory, arguing in a public policy perspective, has works without sophisticated forms of conflict analysis and conflict resolution, although she formulates this as a necessary point in her criteria for local approaches to sustainable resource management (see the Appendix). More useful with regard to conflict mitigation as part of strategies of sustainable resource management are the theoretical analyses of ecological distribution conflicts by Martinez-Alier (1995, 2004) and further forms of cultural or economic distribution conflicts (Escobar 2006). The sustainability process can be seen as full of conflicts that need to be solved in the resource-management process. These conflicts emerge because of different interests of actors, incompatible institutional goals, and internal contradictions in societal and economic systems. Some of the "system contradictions" come into view in social-ecological analyses of unequal exchange and global resource flows in which ecological distribution conflicts appear, but redistribution of resources as conflict reducing or preventing process does not yet exist in significant degrees. Redistribution policies work better in the European countries than between rich or core countries and countries in the periphery of the world system, for example, the Global Environmental Facility for funding projects worldwide to improve the global environment. Global sustainability requires stronger redistribution mechanisms and prevention of the shifting of environmental problems and burdens from rich to less developed countries, from productive to extractive economies. This phenomenon called by Rice (2007) as "rich country illusion effects" is a source of additional conflicts.

3. *Analysis of multi-scale processes in the transformation to sustainability* include autonomous processes, not only managed change and governance. The processes that affect transformation can be systematically described in systems analyses of interacting SES as manageable, indirectly manageable, and non-manageable components. Non-manageable components include such that are too complex or follow a specific system logic of development: modes of production and reproduction in social and ecosystems, many of the risks and unplannable events and disturbances analysed in resilience research, especially systemic risks resulting from global social and environmental change. Complex interacting systems show the limits of governance and public policy processes. The debates about global environmental change in the anthropocene brought ideas of managing complex systems as the global climate or the earth system, for example, in ideas of geo-engineering, but only few ideas to deal with ecosystem complexity in "governance beyond management".

Advancing from studies of local sustainability to more complex systems and multi-scale processes requires—beyond empirical research and conceptual modelling of systems complexity—new forms of knowledge generation. Epistemological and methodological procedures to integrate local and global analyses of SES and to structure knowledge syntheses are not yet advanced (see the Appendix).

Integration of Sustainability Analyses in Political-Economic and Social-Ecological Theory

Further progress in understanding the problems, difficulties, and possibilities of global transformation to sustainability depends on the availability of extended systems analyses: connections of political-economic and social-ecological systems analyses of the modern world and earth system, analyses of the inherent contradictions and blocking mechanisms in both systems, and theory-guided analyses of possibilities of degrowth and de-commercialisation of natural resource use. Preliminary forms of such extended systems analysis developed

- in political economy that was renewed in human ecology (Schnaiberg 1980; Schnaiberg and Gould 1994);
- in social ecology that adopts several theory components from critical political economy, critical theory of society (Brand 2015b), world system analysis (Wallerstein 2000), and analysis of global unequal exchange (Rice 2007);
- in ecological research on interaction and coupling of social and ecological systems (Janssen 2006; Janssen and Anderies 2007.)

The analysis of transformation of energy regimes (see Chap. 8) can be seen as a core component of the transformation of modern society and economy to sustainability. The industrial society depends on high levels of energy throughput that are programmed from its societal metabolism and growth mechanisms. Nature–society interaction in modern society implies an intensity of human use of material and energy resources that is only possible through strong modifications of ecosystems and global material cycles (water, nitrate, phosphate, etc.). These modifications disturb the ecosystem functions and tend to exceed the planetary boundaries of resource use. This trend towards intensification and growth of resource use cannot continue long time without triggering ecological and economic collapses. The future mode of production needs to be one reducing global resource use to lower levels, going away from growth-dependent development. The knowledge available for the formulation of transformation strategies that aim to leave the path of growth is synthesised in social-ecological analyses of societal metabolism of historical and modern societies (Fischer-Kowalski and Haberl 2007; Krausman and Fischer-Kowalski 2010; Haberl et al. 2010; Fischer-Kowalski et al. 2012). The social-ecological perspective of the transformation to sustainability is described by Haberl et al. (2010) in four main points:

- A historical perspective for the analysis of long processes continuing over hundreds or thousands of years (modes of production or sociometabolic regimes that remain stable for long time through adaptation, whereas transitions between regimes are seen as great transformations of nature–society interaction in shorter times).

- A view of sustainability as a temporally unlimited exchange between nature and society (this seems impossible in the industrial system that is overusing the natural resource base; examples for sustainable systems can rather be found in land use and agriculture).
- The future forms of a sustainable society or mode of production are unknown (cannot be sufficiently projected with the help of ideas of an ecologically ideal economy, e.g. as local systems of circular economy).
- Transformation is a rupture of present path-dependent development and growth (visions of wanted futures in global scenarios are to some degree useful to trace different transformation possibilities).

The transition theory is elaborated further by Fischer-Kowalski et al. (2012), where a framework for the analysis of socio-ecological transitions between different energy regimes and concomitant ecological changes is outlined. According to this analysis, European countries have completed the historical transition into the fossil fuel-based industrial regime and reached an energetic and material stabilisation phase at high levels. The new transition to sustainability, away from fossil fuels, has just begun in Europe. Many countries of the Global South (where a large part of the human population is living) are not yet in the industrial system; for the industrial countries they are only important as extractive economies that deliver natural resources. The late industrialisation in some large countries (Brazil, Russia, India, China, South Africa: BRICS) may cause further delays and difficulties in the global transformation. The dynamics of global transformation include dis-simultaneity of social-ecological transformation. Referring to the historical transition to industrial society with its profound economic and technical change of the forms of human labour, Fischer-Kowalski et al. discuss potential conditions and forms of transforming labour in the future sustainable society. To estimate the quantitative dimensions of the future transformation, six global megatrends are extracted from a literature review:

- Three societal megatrends (population dynamics, shifting economic and political centres of development, new information technology and knowledge sharing)

- Three megatrends in the natural components of SES (transforming energy systems, problems of resource security, increasing climate change impacts)

Most of these trends imply contradicting information about possibilities of transformation. The conditions under which system transformations can happen vary between mediated (soft) forms under favourable conditions and non-managed (hard) forms of catastrophic global social and environmental change.

To develop transformation strategies for European countries, the authors discuss three ideas that illustrate different degrees of governed transformation: "no policy change", "ecological modernisation", and "sustainability transformation". The scenario "business as usual" (no policy change) fails under favourable and unfavourable conditions. The ecological modernisation scenario is for Europe successful under favourable conditions where market-based development and gradual change can continue. The sustainability transformation scenario is working under both conditions, but can better deal with a complex and quickly changing society and nature.

In the discussion of the new transformation the following question arises: *what can be learned from earlier societal transformations in human history for transition to global sustainability?* The transition to sustainability is not a return to historically older forms of society or resource use. Analyses of earlier "great transformations" do not help much to understand the process of a global transition to sustainability. The course of this transition is to a large degree unknown. This supports the idea of *partial and indirect regulations and governance, seeking stepwise solutions and ways to the future in a variety of knowledge practices: knowledge syntheses, scenario analyses, simulations, policy experiments, systems analyses, and theoretical reflection.* Analyses of temporal structures in human societies and social action are useful to connect the past and the future. For Tilly, human history is a sequence of futures built by humans with the help of their visions of the future (Tilly 1997: 583). How the near future appears in social and economic action is visible from risk, vulnerability, and resilience analyses, and from the attempts to protect against negative future events by way of assurance, contracts, planning, organisational development, social learning, and adaptation. This is the temporal perspective of the future in modern society since its beginning. Reducing the time horizon of sustainability to such short-term perspectives of return-of-investment and planning individual courses of life would imply to give up the perspective of transformation to sustainability in favour of "muddling through" and living with awareness about risks, catastrophes, and unexpected events. This seems the alternative when sustainability is replaced through resilience as the guiding idea for formulating pictures of the future. The view that the future is unknown—influenced through collective and individual decisions made today but unforeseeable—seems to suppress further and systematic thinking and knowledge practices to imagine possibilities of transforming societal systems.

Knowledge from critical systems analyses of the global economy and of ecosystems is useful for thinking through the conditions of transitions to sustainability and identifying limiting and enabling factors, potentially successful or failing strategies. This strategy of knowledge seeking and of balancing knowledge and ignorance is open for changes, corrections, and improvements. A theory of nature–society interaction is one form, not the only one, to collect and synthesise the knowledge available and make better-informed guesses about transitions to sustainability. Other possibilities, not discussed further here, include more widespread attempts: mathematical modelling global development or climate processes, scenario analysis as pre-theoretical construction and envisioning of possible futures, and global assessments of the interaction of nature and society (as in the Millennium Ecosystem Assessment).

Some discussion can be found about the scope of a social-ecological theory of transformation in earlier social-ecological research. It cannot be a theory of co-evolution of nature and society which provides a biologically preformed view of development and change. Biological and social system components that can co-evolve are specific and limited, which need to be identified through theoretical reflection (Weisz 2001: 115f). This discussion needs to be elaborated, asking at which levels of biological organisation (genes, organisms, species, ecosystems) and of social organisation (individuals, social groups, social communities, societies) evolution and co-evolution of coupled social and ecosystems are possible and in which forms. The co-evolution debate is controversial, similar to the whole sustainability debate, because of the heterogeneous definitions and interpretations of the concepts. Transferring the biological theory of co-evolution of species to a social-ecological theory of co-evolution of society and nature (Norgaard 1984) seems one of many forms of analogous reasoning that can be criticised with the social-ecological theory. The other question to elaborate by way of theoretical reflection is: how to connect co-evolution with strategies of sustainable transitions? At this point the clarification of limiting factors in biological and socio-cultural evolution is necessary. Weisz (2001) suggests time as measurable factor. The adoption of the term evolution in the social sciences for analysing socio-cultural change shows a variety of temporal structures in social systems that do not match with these in ecosystems. The differences of biological and societal time scales and regimes are significant. What the differences of temporal structuring of processes in nature and society (different types of short-term and long-term, cyclical and linear structures, disruptive and stable time regimes) imply for the coupling of social and ecological systems is not sufficiently analysed. The theoretical systems analysis of SES in political-economic or social-ecological perspectives adds to the conceptualisation of space and time in physics a more complex picture of time, social and ecological differentiation, and fluidity of time and space in societal development.

Conclusion: Social-Ecological Transformation to Sustainability

With the concepts and theories discussed in Chaps. 3 and 4, several steps of the development of a social-ecological theory of transformation to sustainability have been described:

- *Preparatory analyses of vulnerability, resilience, and sustainability* as part of ecological analyses of the systemic dynamics of coupled social and ecological systems
- *Formulating a broader interdisciplinary perspective of sustainable development beyond normative and policy-centred perspectives* to describe the processes of social and ecological change that are part

of sustainable development (but often neglected in the ecological discourse)
- *Reformulating strategies of sustainable development in policy and resource management,* using theoretical and social-ecological knowledge to identify potential problems and hinders in the transformation process
- *Formulating* potential *trajectories of transformation towards sustainability with* knowledge from a social-scientific and ecological research.

The theoretical re-interpretation of sustainability thinking has three major consequences regarding knowledge use in research and policy:

1. *Re-formulating the perspectives in the policy-centred sustainability discourse:* The sustainability discourse shows inclinations of political actors, environmental movements, but also of researchers, to reduce sustainable development to a political and politically steered process for which mainly knowledge about the power structures and processes in political systems is required. The more complicated processes of transforming societal, economic, and social-ecological systems are neglected. More complex, process-oriented views of the future, beyond political visions and normative ideas about the good society, help to formulate realistic transformation strategies. Integrated knowledge from social and ecological research can be used to design strategies and to prepare for the changes to be expected on the way towards sustainability. The "learning of passages into the future" happens in similar forms as discussed by Ostrom et al. in the critique of panaceas as universal forms of solving environmental problems. Panaceas indicate lack of knowledge and can be replaced through better strategies with improved knowledge about the processes of change in social and ecological systems. Whether and how such learning happens among the actors, how much time it requires, is difficult to foresee. The alternative of not continually improving strategies of sustainable development or governed transformation to sustainability is that of waiting for catastrophic forms of environmental change where disasters and resource scarcity "act" and enforce sustainability in unwanted forms

through collapse of the modern societal and economic systems. In the debates of limits to growth and now of climate change are such catastrophic scenarios possible variants of the transition to sustainability.
2. *Re-assessing the sustainability of coupled social and ecological systems:* This requires interdisciplinary knowledge integration to improve the understanding of possibilities of transformation. Sustainability as an overarching term for social-ecological research needs to be re-analysed with regard to forms, coupling, and functioning of different types of social and ecosystems, for *contradictions, controversies, and conflicts.* These "three c's" appear at local, regional, national, and global spatial levels of analysing SES, caused by social, political, and economic structures, and asymmetric power relations and incompatible interests. The first conclusion from the existence of systemic tensions is global sustainability requires prevention of shifting of environmental problems and burdens from rich to less developed countries, or from productive to extractive economies. Sustainability seems impossible within a divided economic world; social-ecological analyses of transformation ask for more theoretically reflected knowledge than most social and ecological theories offer. To synthesise knowledge is not an aggregation of always larger datasets, but an abbreviating methodology, including the use of available knowledge from different disciplines.
3. *Re-conceptualising the generation and co-production of knowledge in sustainability analyses:* Analyses of knowledge practices in policy and governance processes are more found in social-scientific research, as the theories discussed in Chap. 2 show (reflexive modernisation, ecological modernisation), whereas in ecological research the functionalist thinking in terms of adaptation, cyclic processes, and resilience prevails. In the social sciences, knowledge appears as generated in social processes of interaction and cooperation, in dialogical and discursive forms, in processes of education and learning, work and research, in varying culture- and language-bound processes. Knowledge production is context-dependent or situated and relational as, for example, shown in the sociology of knowledge (Jamison 2001). In environmental, biological, and ecological research such arguments are not self-evident. Power

asymmetries and inequality appear in processes of communication and knowledge use as disturbing and distorting communicability, acceptability, and matching of interests. Some new forms of social and collective production of knowledge (participatory research, knowledge syntheses) came with recent development of inter- and transdisciplinary science.

Initial processes of the complicated and long (several generations) process of transformation to sustainability are described in the following chapters: the operationalisation of nature–society interaction in terms of SES and ecosystem services for the formulation of local and multi-scale transition strategies; the managerial forms of adaptive management and global environmental governance; global adaptation to climate change discussed in exemplary way for coastal areas; and transformation of the industrial energy regimes that are based on fossil resources. Other phenomena of global change and environmental governance can be discussed in similar ways and connected to the sustainability discourse and process: land-use change, urbanisation, global food production, ocean management, modification of ecosystems through humans, biodiversity loss, limits of growth, and scarcity of industrial key resources.

Appendix: Connecting Local and Global Strategies of SES—The Significance of Empirical Knowledge

Knowledge about the complexity of local and global social and ecological systems: The difficulties to advance from local to global complexity and from empirical studies of sustainable resource use to global strategies of sustainability are visible in the attempts to connect local and global social processes. They require a science of complexity to integrate knowledge from empirical research systematically with theoretical knowledge, maintaining discursive forms of knowledge generation and application to support learning, corrections, and improvements in all knowledge processes.

(1) Ostrom: The Complexity of Local Resource Management—Principles of Sustainability

Management principles: clearly defined boundaries; monitoring; costs and benefits should be proportional and fairly distributed between users; creating rights to organise; power of resource users for rulemaking ("collective choice arrangements"); graduated sanctions; mechanisms for conflict resolution; nested enterprises—for larger resource-use systems (Becker and Ostrom 1995: 119)

Learning from local systems analysis for global environmental governance:

- Dealing with fast social and biophysical processes at all scale levels of SES (e.g. carbon emissions, population increase and migrations, overharvesting and pollution, loss of species)
- Strong diagnostic methods for analysing the diversity of processes and the multiplicity of potential social and biophysical solutions to cope with the variety of processes
- Avoiding simple solutions to complex problems (often resulted in worse outcomes than the problems addressed)
- Building an interdisciplinary science of complex, multi-level systems to match diagnoses, solutions, and social-ecological contexts
- Analysing non-anticipated effects of policy interventions and developing multi-tier governance systems (Ostrom 2007)

(2) Ecological Analyses of Urban Sustainability: Integrating Contrasting Results of Research

A theorem of impossibility of urban sustainability
Eugene Odum: great cities are planned and grow without regard for their nature as "parasites on the countryside"; the countryside supplies food, water, air, and degrades wastes. Odum (1993: 263): cities are "parasites on the biosphere" which is not meant to belittle them but to be realistic.

A theorem of ecological advantages and disadvantages of cities
Rees (2003: 116): cities have ecologically seen advantages as well as negative consequences—advantages as saving of space for settlement,

reducing transport, disadvantages as large ecological footprints resulting from the production of resources they need.
Possibilities to approach urban sustainability—analyses of urban metabolism
Wolman (1965): calculating the demand of energy, food, water, and emissions (sewage, refuse, and air pollutants) for a hypothetical city of one million. Urban metabolism provides a basis for assessing the sustainability of cities (Schulz 2005). Extended Urban Metabolism (Alberti, Newman, Kenworthy, Newton) includes the physical, biological, and human basis of the city, including economic and social aspects of sustainability as the provision of social amenity, health, and well-being by the city (McDonald and Patterson 2007: 181).
Material and energy flow accounting for urban areas
As components of global resource flows, material and energy flow analyses of cities (Kennedy et al. 2015) complement the calculation of ecological footprints of cities (Rees 2003).

(3) Local Systems as Part of Global Systems

The empirical social-ecological research on sustainability by Ostrom et al. has advanced to the point where a science of global complexity requires more knowledge than that from empirical social and ecological research to prevent simple solutions to complex processes that have frequently led to worse outcomes. The ideas to develop a science of complexity from the empirical research, embracing local and global sustainability, are limited as the reflections of Ostrom (2007, see above) show. Such complexity cannot be sufficiently analysed through empirical research and requires combinations of empirical research, modelling, and theoretical analyses of interacting SES.

With the dominant trend towards urbanisation and to mega-cities with many millions of inhabitants, cities become unsustainable. They are part of the process of global modernisation that develops with ecologically unequal exchange (Rice 2007) and in multi-scale networking of resource-use processes. The natural and other resources for large cities cannot be provided from the surrounding environment, but through global trade and exchange of goods and services. Cities become dependent on their development on a globalised economy.

(4) Knowledge Synthesis as Integration of Different Forms of Knowledge Production

The analyses of complex and interacting social and ecological systems require combinations of several forms of knowledge production (empirical research, modelling, theory construction, knowledge synthesis), and a variety of methods to make visible global complexity in local case studies and the other way round. A kind of "empirics of totality" (Bonss 1983) is described in the integration of analyses of vulnerability, resilience, and sustainability (Chap. 3). Different forms and degrees of knowledge integration are, for example, described in the method of "progressive synthesis" by Ford (2000) for ecological research. Interdisciplinary and global knowledge syntheses develop with the global assessments of the kind of the Millennium Ecosystem Assessment, or the IAASTD reports about agriculture (cooperative syntheses of many researchers from different disciplines).

In the knowledge syntheses for governance, collective action and strategies for sustainable development other forms and models of synthesis are important—"action syntheses". These include knowledge practices in rudimentary forms applied in the ecological approaches of adaptive management and governance (Chap. 6) and forms of transdisciplinary and participatory research. More of such practices develop in the social sciences: action research and practices of social movements as knowledge producers and knowledge users. Action syntheses include a large number and variety of forms of cooperation and social knowledge practices applicable in environmental governance. Scientifically they are supported through analyses like that of Baecker (2007) about network synthesis of social action, and of the discussion of "societal synthesis" in the tradition of critical theory [theory-based analyses of the contradicting, inequality, and conflict-based forms of system integration in modern capitalist society and economy (Sohn-Rethel 1985; Armanski 2015; Demirovic et al. 2015; Martin et al. 2015)]. Action syntheses can be methodologically described as triangulation of different forms of knowledge generation, dissemination, and application, including theoretical analyses of the spatiality of knowledge production and of knowledge practices, the "spaces of global knowledge" in historical constellations (Finnegan and Wright 2015).

References

Armanski, G. (2015). *Monsieur le Capital und Madame la Terre*. Münster: Westfälisches Dampfboot.
Baecker, D. (2007). The network synthesis of social action I: Towards a sociological theory of next society. *Cybernetics and Human Knowing, 14*, 9–42.
Barry, J. (1999). *Environment and social theory*. London: Routledge.
Becker, E., & Jahn, T. (Eds.). (2006). *Soziale Ökologie. Grundzüge einer Wissenschaft von den gesellschaftlichen Naturverhältnissen*. Frankfurt am Main: Campus.
Becker, C. D., & Ostrom, E. (1995). Human ecology and resource sustainability: The importance of institutional diversity. *Annual Review of Ecology and Systematics, 26*, 113–133.
Benson, M. H., & Craig, R. K. (2014). The end of sustainability. *Society and Natural Resources, 27*(7), 777–782.
Biermann, F. (2004). The global governance project. *IDGEC News, 8*, 10–11.
Biermann, F. (2011). *Reforming global environmental governance: The case for a United Nations Environment Organisation (UNEO)*. Amsterdam: Earth System Governance Project and VU University Amsterdam.
Bonss, W. (1983). Kritische Theorie als empirische Wissenschaft. Zur Methodologie 'postkonventioneller' Sozialforschung. *Soziale Welt, 1*, 57–89.
Brand, U. (2015a, June 10–12). *How to get out of the multiple crisis? Contours of a critical theory of social-ecological transformation*. Paper presented at the conference The Theory of Regulation in Times of Crises, Paris. Accessed November 10, 2015, from https://www.eiseverywhere.com/retrieveupload.php?
Brand, U. (2015b). Sozial-ökologische Transformation als Horizont praktischer Kritik: Befreiung in Zeiten sich vertiefender imperialer Lebensweise. In D. Martin, S. Martin, & J. Wissel (Eds.), *Perspektiven und Konstellationen kritischer Theorie*. Münster: Westfälisches Dampfboot.

Brown, L. (2001). *Eco-economy: Building an economy for the earth*. New York: W.W. Norton.
Bruckmeier, K. (2009). Sustainability between necessity, contingency and impossibility. *Sustainability, 1*, 1388–1411. www.mdpi.com/journal/sustainability.
Bruckmeier, K. (2013). *Natural resource use and global change: New interdisciplinary perspectives in social ecology*. Houndmills, Basingstoke: Palgrave Macmillan.
Catton, W. Dunlap, R. (1978) 'Environmental Sociology: A New Paradigm', *The American Sociologist*, 13: 41-49.
Chambers, R., & Conway, G. (1991). *Sustainable rural livelihoods: Practical concepts for the 21st century*. IDS Discussion Paper 296. Institute of Development Studies, University of Sussex.
Collier, D., Hidalgo, F. D., & Maciucean, A. O. (2006). Essentially contested concepts: Debates and applications. *Journal of Political Ideologies, 11*(3), 211–246.
Daily, G. (1997). *Nature's services: Societal dependence on nature*. Washington, DC: Island Press.
Daly, H. E. (1991). *Steady-state economics* (2nd ed.). Washington, DC: Island Press.
Daly, H. E. (1997). *Beyond growth: The economics of sustainable development*. Boston: Beacon Press.
Demirovic, A. (2012). Reform, revolution, transformation. *Journal für Entwicklungspolitik, 3*, 16–42.
Demirovic, A., Klauke, S., & Schneider, E. (Eds.). (2015). *Was ist der 'Stand des Marxismus'?: Soziale und epistemologische Bedingungen der kritischen Theorie heute*. Münster: Westfälisches Dampfboot.
Ehlers, E., & Krafft, T. (Eds.). (2006). *Earth system science in the anthropocene*. New York: Springer.
Ehrlich, P., & Ehrlich, A. (1970). *Population, resources, environments: Issues in human ecology*. San Francisco: W. H. Freeman.
Enríquez-Cabot, J. (1998). Genomics and the world's economy. *Science, 281*, 925–926.
Escobar, A. (2006). Difference and conflict in the struggle over natural resources: A political ecology framework. *Development, 49*(3), 6–13.
Finnegan, D. A., & Wright, J. J. (Eds.). (2015). *Spaces of global knowledge: Exhibition, encounter and exchange in an age of empire*. Farnham: Ashgate.

Fischer Kowalski, M., & Haberl, H. (Eds.). (2007). *Socioecological transitions and global change. Trajectories of social metabolism and land use.* Cheltenham: Edward Elgar.

Fischer-Kowalski, M. (2007, September). *Socioecological transitions in human history and present, and their impact upon biodiversity.* Presentation to the Second ALTER-Net Summerschool, Peyresq, Alpes de Haute-Provence. Accessed August 25, 2015, from https://www.pik-potsdam.de/news/public-events/archiv/alter-net/former-ss/2007/11-09.2007/fischer-kowalski/presentation_fischer-kowalski.pdf

Fischer-Kowalski, M., Haas, W., Widenhofer, D., Weisz, U., Pallua, I., Possanner, N., et al. (2012). *Socio-ecological transitions: Definitions, dynamics and related global* scenarios. Project NEUJOBS, European Union. Accessed August 25, 2015, from http://www.neujobs.eu

Fischer-Kowalski, M., Haberl, H., Hüttler, W., Payer, H., Schandl, H., Winiwarter, V., et al. (1997). *Gesellschaftlicher Stoffwechsel und Kolonisierung von Natur: Ein Versuch in Sozialer Ökologie.* Amsterdam: G+B Verlag Fakultas.

Fischer-Kowalski, M., & Rotmans, J. (2009). Conceptualizing, observing, and influencing social-ecological transitions. *Ecology and Society, 14*(2), 3. http://www.ecologyandsociety.org/vol4/iss2/art3/

Ford, E. D. (2000). *Scientific method for ecological research.* Cambridge: Cambridge University Press.

Gallie, W. B. (1956). Essentially contested concepts. *Proceedings of the Aristotelian Society, 56,* 167–198.

Georgescu-Roegen, N. (1971). *The entropy law and the economic process.* Cambridge, MA: Harvard University Press.

Georgescu-Roegen, N. (1986). The entropy law and the economic process in retrospect. *The Eastern Economic Journal, XII*(1), 3–25.

Goldfrank, W. L., Goodman, D., & Szasz, A. (Eds.). (1999). *Ecology and the world system.* Westport, CT: Greenwood Press.

Haberl, H., Beringer, T., Bhattacharya, S. C., Erb, K.-H., & Hoogwijk, M. (2010). The global technical potential of bio-energy in 2050 considering sustainability constraints. *Current Opinion in Environmental Sustainability, 2,* 394–403.

Haberl, H., Fischer-Kowalski, M., Krausmann, F., & Martinez-Alier, J. (2011). A socio-metabolic transition towards sustainability? Challenges for another great transformation. *Sustainable Development, 19,* 1–14.

Hoffmann, U. (2011). *Some reflections on climate change, green growth illusions and development space.* United Nations conference on Trade and Development. Discussion Paper 205, 28 pp.

Hornborg, A., & Crumley, C. L. (Eds.). (2006). *The World System and The Earth System: Global socioenvironmental change and sustainability since the neolithic*. Walnut Creek, CA: Left Coast Press.

IAASTD. (2009). Agriculture at a crossroads. In B. McIntyre et al. (Eds.), *International assessment of agricultural knowledge, science and technology for development (IAASTD): Global report*. Washington, DC: Island Press.

Jamison, A. (2001). *The making of green knowledge: Environmental politics and cultural transformation*. Cambridge: Cambridge University Press.

Janssen, M. A. (2006). Historical institutional analysis of social-ecological systems. *Journal of Institutional Economics, 2*(2), 127–131.

Janssen, M. A., & Anderies, J. M. (2007). Robustness trade-offs in social-ecological systems. *International Journal of the Commons, 1*(1), 43–65.

Kennedy, C. A., Stewart, I., Facchini, A., Cersosimo, I., Meleb, R., Chenc, B., et al. (2015). Energy and material flows of megacities. *PNAS, 112*(19), 5985–5990.

Krausmann , F., ed. (2011) 'The socio-metabolic transition. Long term historical trends and patterns in global material and energy use'. Social Ecology Working Paper 131, Vienna, Institute of Social Ecology: Social Ecology Working Paper 131.

Krausman, F., & Fischer-Kowalski, M. (2010). *Gesellschaftliche Naturverhältnisse: Energiequellen und die globale Transformation des gesellschaftlichen Stoffwechsels*. Social Ecology Working Paper 117. Vienna: Institute of Social Ecology.Lander, E. (2011). *The green economy: The wolf in sheep's clothing*. Transnational Institute. Accessed August 1, 2015, from www.tni.torg

Lee, K., Holland, A., & McNeill, D. (Eds.). (2000). *Global sustainable development in the 21st century*. Edinburgh: Edinburgh University Press.

Martin, D., Martin, S., & Wissel, J. (Eds.). (2015). *Perspektiven und Konstellationen kritischer Theorie*. Münster: Westfälisches Dampfboot.

Martinez-Alier, J. (1995). Political ecology, distributional conflicts and economic incommensurability. *New Left Review, 211*, 70–88.

Martinez-Alier, J. (2002). *The environmentalism of the poor: A study of ecological conflicts and valuation*. Cheltenham: Edward Elgar.

Martinez-Alier, J. (2004). Ecological distribution conflicts and indicators of sustainability. *International Journal of Political Economy, 34*(1), 13–30.

McDonald, G. W., & Patterson, M. G. (2007). Bridging the divide in urban sustainability: From human exemptionalism to the new ecological paradigm. *Urban Ecosystems, 10*, 169–192. doi:10.1007/s11252-006-0017-0.

Meadowcroft, J. (2007). Governance for sustainable development in a complex world. *Journal of Environmental Policy and Planning, 9*, 299–314.

Meadows, D., Meadows, D., Randers, J., & Behrens, W. W., III. (1972). *The limits to growth: A report for the club of Rome's project on the predicament of mankind*. New York: Universe Books.

Miao, S., Carstenn, S., & Nungesser, M. (2009). *Real world ecology: Large-scale and long-term case studies and methods*. New York: Springer.

Mill, J. S. (1848). *Principles of political economy with some of their applications to social philosophy*. London: Longmans, Green and Co.

Millennium Ecosystem Assessment. (2005). *Current state and trends*. Global Assessment Reports, Vol. 1. www.millenniumassessment.org

Moffatt, I. (1995). *Sustainable development: Principles, analysis and policies*. Carnforth: The Parthenon Publishing Group.

Norgaard, R. (1984). Coevolutionary development potential. *Land Economics, 60*(2), 160–173.

Odum, E. (1993). *Ecology and our endangered life-support systems*. Sunderland, MA: Sinauer Associates.

Ostrom, E. (2007, February 15–19). *Sustainable social-ecological systems: An impossibility?* 2007 Annual Meeting of the American Association for the Advancement of Science, Science and Technology for Sustainable Well-Being, San Francisco.

Polanyi, K. (1944). *The great transformation*. New York: Farrar & Rinehart.

Raskin, P., Banuri, T., Gallopin, G., Gutman, P., Hammond, A., Kates, R., et al. (2002). *Great transition. The promise and lure of the times ahead*. Report of the Global Scenario Group, Stockholm and Boston.

Raskin, P. D., Electris, C., & Rosen, R. A. (2010). The century ahead: Searching for sustainability. *Sustainability, 2*, 2626–2651. doi:10.3390/su2082626.

Rees, W. E. (2003). Understanding urban ecosystems: An ecological economics perspective. In A. Berkowitz et al. (Eds.), *Understanding urban ecosystems*. New York: Springer.

Rice, J. (2007). Ecological unequal exchange: Consumption, equity, and unsustainable structural relationships within the global economy. *International Journal of Comparative Sociology, 48*(1), 43–72.

Rice, J. (2009). The transnational organization of production and uneven environmental degradation and change in the world economy. *International Journal of Comparative Sociology, 50*(3–4), 215–236.

Rockström, J., Steffen, W., Noone, K., Persson, Å., Stuart Chapin, F., III, Lambin, E. F., et al. (2009). A safe operating space for humanity. *Nature, 461*, 472–475.

Schnaiberg, A. (1980). *The environment: From surplus to scarcity*. New York: Oxford University Press.

Schnaiberg, A., & Gould, K. A. (1994). *Environment and society: The enduring conflict.* New York: St. Martin's Press.

Schulz, N. B. (2005). *Contribution of material and energy flow accounting to urban ecosystem analysis: A case study Singapore.* UNU-IAS Working Paper No. 136. Institute of Advanced Studies, United Nations University, Tokyo.

Scoones, I. (1997). *Sustainable rural livelihoods: A framework for analysis.* IDS Working Paper 72. Institute of Development Studies, University of Sussex.

Smith, A., Sterling, A., Berkhout, F. (2005). The governance of sustainable socio-technical transitions. *Research Policy, 34*: 1491–1510.

Sneddon, C., Howarth, R., & Norgaard, R. (2006). Sustainable development in a post Brundtland world. *Ecological Economics, 57*, 253–268.

Sohn-Rethel, A. (1985). *Soziologische Theorie der Erkenntnis.* Frankfurt am Main: Suhrkamp.

Spash, C. L. (2012). Green economy, red herring. *Environmental Values, 21*(2), 95–99.

Stutz, J. (2009). The three-front war: Pursuing sustainability in a world shaped by explosive growth. *Sustainability: Science, Practice, & Policy, 6*(2), 49–59.

Tilly, C. (1997). *Roads from past to future.* Lanham, MD: Rowman and Littlefield.

Wallerstein, I. (2000). Globalization or the age of transition? A long-term view of the trajectory of the world-system. *International Sociology, 15*(2), 249–265.

WBGU. (2011). *Welt im Wandel – Gesellschaftsvertrag für eine Grosse Transformation. Hauptgutachten.* Berlin: WBGU.

Weisz, H. (2001). Gesellschaft-Natur Koevolution: Bedingungen der Möglichkeit nachhaltiger Entwicklung. Kulturwissenschaftliches Seminar (Dissertation), Humboldt Universität, Berlin.

Wolman, A. (1965). The metabolism of cities. *Scientific American, 213*(3), 179–190.

5

Social-Ecological Systems and Ecosystem Services

Social-Ecological Systems and Ecosystem Services: Genesis and Use of the Terms

The terms social-ecological system (SES) and ecosystem services (ESSs) are used in ecological research and in environmental policy. *In this chapter, the theoretical assumptions and implications about nature and society in both terms are critically reviewed for the purpose to transform the terms into theoretically elaborate concepts.* To use the terms in theory-based interdisciplinary research and knowledge integration requires methodological, epistemological, and theoretical reflection and their modification in differentiated forms of application. The assumptions and hypotheses about relations and connections between social and ecological systems that guide the use of the terms show that they lack coherence and theoretical reflection. In the elaboration of the social-ecological theory discussed in the preceding chapters, the terms need to be specified and differentiated:

- An SES is a heuristic device to construct, by way of preliminary analysis and with insufficient empirical and theoretical knowledge, social, and ecological systems as interconnected systems. The notion

SES is based on the assumption that social and ecological systems are always interconnected. This assumption does not hold when the dynamics of social and ecological systems, their interaction and coupling, and causes of maladaptation in the development of coupled systems are analysed.
- With the term ESSs, connections between human society and nature are constructed, assuming that humans depend on a life-maintaining, producing, and resource-providing nature. This dependence is specified in various benefits humans receive from ecological and life-sustaining functions of ecosystems as providing food and other, also cultural, services. Such services are not produced from nature alone. ESSs depend on varying degrees from human capacities, knowledge, and technologies in complex forms to co-production of services.

The following discussion is guided by the assumption that *the theoretical elaboration of SES as overarching term helps to reformulate the notion of ESSs for its use in an interdisciplinary perspective*. Difficulties with the use of the terms in ecological research and in a social-ecological perspective are discussed in the following seven points:

1. *The construction of SES and ESSs requires in-depth analysis of the systems dynamics of social and ecological systems and of the possibilities of their co-evolution which is not done in the ecological research where the terms are used.* The concepts of co-evolution, of social and ecosystems are not easily combined and integrated as their review and interpretation with the help of interdisciplinary social-ecological theory shows. Naïve and speculative evolutionary theories as that of Spencer in the nineteenth century, assuming that the laws of evolution are the same in society and nature, are no longer in use. Yet, attempts to understand society in physical or biological terms continue, in systems theory and neo-evolutionism [see the ecological-evolutionary theory of Lenski (2005)]. The concept transfer from biological evolution theory to sociology in the systems theory of Luhmann can be seen as a limited and controlled experiment with the concepts of evolution, selection, and variation, more metaphorical use of the biological terms and

creative technique to reformulate sociological knowledge in non-sociological terms than theoretical analysis and explanation; the historically changing relations between society and nature are not analysed, but their consequences remain unclear. The manifold differences between biological evolution and societal development can rather provide arguments against the use of the biological term evolution for social systems: the differences in time perspectives and scales, the significance of culture and symbolic interaction for societal development, the differences of social action from biological forms of behaviour, and the different forms of development and change of social and ecological systems. In new ecological research on resilience, SES, and ESSs, the differences between social and ecological systems are not analysed in depth; the expectation seems that the lacking clarification and reflection of the concepts will happen somehow with the progress of empirical research. The new research areas create their own terminology for specialised research to become independent from prior research and its theoretical concepts.

2. *In ecology, the connection of social and ecological systems is assumed, not theoretically analysed.* To become a theoretical concept, SES requires more than stating by definition a coupling of social and ecological systems: to show the different and changing forms of coupling of both system types in historically specific modes of production, socio-metabolic regimes, and systems of society. In ecological research with the notion of SES, the primacy of nature is assumed in an ahistorical and ontological sense. The ontological assumptions about the constitution of the reality of social and ecological systems imply that there is only one reality that counts: the material world, inexactly described as nature, scientifically analysed in physics, biology, or chemistry. To interpret the material reality as non-social is one form of such reductionist thinking, others are the reduction of knowledge to information flows in physical and living systems that do not require culture, or the view SES as spatial or bio-geo-physical systems with specific social actors and institutions connected to it. Society or social systems are in such forms of naturalistic reductionism conceived as part of nature and investigated with a simplified and inexact terminology, bypassing social-scientific terminology. Thus, large parts of social-scientific

knowledge about modern societies and their relations with nature can be ignored in ecological research. Society is seen as secondary entity, abstract system, and product of the human mind. In the final analysis, society is reduced to a consequence of human brain development and a symbolic reality. The continuous materialisation of society in human action, natural resource use, economic production and construction of technical artefacts, or the social structuring of space and time and the manifold forms of social system types described in sociology are ignored in the reductionist view of SES. The reduction of societal systems and social reality to forms that can be described in natural-scientific terms cannot be done consequently and consistently; the relations and interactions between social and ecological systems are too complicated to be explained in a simple way through a priority monism. A more interdisciplinary knowledge culture that can integrate different methodological, epistemological, and theoretical perspectives has in the practice of ecological research hardly begun. The paradigm change that Scoones described (see Chap. 3) seems more to block than to support interdisciplinary knowledge synthesis; it helps to cut back ecological research to forms of analysis, interpretation, and explanation that are compatible with the ecological repertoire of concepts and methods.

3. *Both terms of SES and ESSs require the identification of the heterogeneous forms of coupling of social and ecosystems to improve their analytical capacities.* For that purpose, further implications of the distinction between social and ecosystems need to be discussed. Does the assumption that the systems are coupled in forms that can be empirically described imply that their analytical distinction as social and ecological systems is wrong? Or does each system type represent a specific sphere of reality, specific in its constitution and development, implying that the conceptualisation of social and ecological systems requires different concepts, theories, and languages? With the second assumption, the coupling of systems cannot be reduced to a simple statement of embeddedness of society in nature; the forms of (dis)embeddedness need to be specified for different historical forms of societies and their natures, in terms of ecologically favourable or maladaptive change. When different types of interaction between the systems are theoretically classified and the

static dichotomy of two system types is dissolved, the integration of sociological and ecological knowledge can be done in other ways than the selective knowledge use in ecological research. The differences between nature and society cannot be directly read off from empirical research and data, but require a theoretical reconstruction of the types of coupling of social and ecological systems.

4. *To conceptualise the coupling between social and ecosystems requires a typology of the historically varying forms of connectedness and adaptive or maladaptive coupling of nature and society in SES.* Classifications are an important method in ecology for the purpose of systematising available knowledge, and in this function they can be used to connect empirical research to theory. Without classifications of the forms of interaction between society and nature, the hybrid forms of nature and society in modern society, constructed as socio-natures or technological natures, remain unclear. Assuming that social and ecological systems are interconnected, the historical process of colonisation of nature by humans needs to be described and theoretically codified to show the changing forms of interconnection, the processes of constructing, shifting and lifting the barriers between the spheres of society and nature in the history of human society, or the possibility of emancipation of society from nature through human culture and science. Hybrid forms of socio-natures are discussed and sometimes formulated as historically changing relations to nature in modern society. Brondizio et al. (2009: 253), for example, conclude from ecological research growing connectedness of resource-use systems and growing functional interdependencies of social and ecological systems. This may be seen as based on empirical observations, but shows the necessity of methodological and theoretical reflection to understand the distinction between society and nature. The epistemological dilemma can be described as follows:

- Social-scientific research and theoretical reflection tends to separation and assuming emancipation of human society from nature.
- Ecological research tends to confirmation of increasing connection and interdependence of society and nature.

How both forms of specialised research can be connected is one of the tasks of the social-ecological theory in progress.
5. A theory-based analysis of interactions between ecological and social systems can generate *more precise concepts of SES and ESS and more convincing arguments for the production and transfer of services* than found in ecological research. The assumption of "ontological primacy of nature over human society" in the construction of SES and ESSs supports a biological reductionism in conceiving the coupled systems, underestimating the role of humans in them. A similar reductionism and biased terminology was discussed by Weisz (see the Introduction) as blindness of social scientists for natural-scientific research and the other way round. The concept of ESSs developed from ecological research; for its development, it is necessary to ask how the relations between social and ecosystems are thought in this concept and which assumptions and normative criteria it uses for the reconstruction of social systems. The assumption that the services are produced by nature becomes unclear when these services vary according to the human perception and to the types of ecosystems, most of which are significantly influenced, modified, changed, or manipulated by human technologies and labour in the colonisation of nature. A conception of ecosystems that includes humans in their double biological and social nature, as species who affected, modified, and transformed nature throughout their history, can come to a more adequate view of production of ESSs. In the production of services, the involvement of nature and society can vary on a continuum of services produced by nature and those produced by human society, with differing degrees of co-production through social and ecosystems, human labour, and technology in natural resource use. ESSs, so the assumption guiding the following discussion, can in the last analysis not be explained by attributing to nature or ecosystems the exclusive quality of a producer.
6. *The theoretical implications of the term ESSs are practically relevant for ecosystem management*: How (far) does the idea of pricing nature and payments for ESSs modify the idea of ESSs from an ecological to an economic construction with consequences that cannot be judged in ecological research? An economic conceptualisation oft ESSs implies

assumptions about the systemic properties of modern economy, commodification of nature, the ways ESSs can be maintained and managed by humans, and how nature can be protected-this is not sufficiently clarified in the ecological debate of ESSs. Modifications of the concept are critically discussed by some ecological economists, although the theoretical implications remain unclear when the inclusion of ESSs into markets and payment schemes is discussed. Gomez-Baggethun et al. (2010: 1209f) argue that the trend towards monetarisation and commodification of ESSs is the result of an economic paradigm change in the history of economics, that is, between classical economics (where benefits from nature were seen in terms of values of use) and neoclassical economics (where they are seen in terms of values of exchange). The current practice of focusing on monetary valuation of ESSs seems ambivalent, as well supporting ideas for conservation as the commodification of nature.

Interpreting ESSs with the theoretical concepts of value of use and of exchange does not offer much better understanding of the economic and ecological problems with the services but leads back to the dilemma with incompatible paradigms of classical and neoclassical economics. The paradigm change in the history of economic theory is insufficient to explain the distortions and inconsequent reasoning in applications of the concept of ESSs, given also the meanwhile manifold variants of the economic term of value. The term "value of use" can be understood to imply a reconstruction of benefits for humans from nature in physical, biological, and social terms of well-being; so far it appears as an alternative to monetary valuation and commodification of natural resources. Yet, the commodification of nature is not a consequence of the conceptualisation of services, but of the application of the ESSs term for a specific purpose in the neoliberal policy of a green economy. To assess different variants of ESS requires a more critical analysis and theoretical reflection of the interaction between modern society and nature and of resource-use processes. The modifications of the concept that happened with payments for ESSs evoke questions as to who makes decisions about payments, who has to pay, who can pay, paying for what purposes, and what are the intended and non-intended consequences of payments? With these questions come

up the power relations in modern society, in policy, and governance processes, relations that are to a large degree ignored in the construction of ESSs.
7. *Through their theoretical reflection, SES and ESSs can be connected with the sustainability debate, by construction of indicators measuring limits of natural resource use in a sustainability perspective* (Kulig et al. 2010). Global limits need to be translated in or supplemented through ecological criteria for natural resource management, nature protection, and ecosystem restauration at local and regional levels. A large number of indicators are necessary to assess resource flows between social and ecological systems and to specify planetary boundaries of human resource use (Rockström et al. 2009). These indicators include material and energy flow accounting and further ecological indicators of human resource use: carrying capacity, ecological footprints, human appropriation of net primary production of ecosystems, energy return of input. Additionally, social and human indicators are required to measure human development and well-being or unequal exchange of natural resources. The complex and historically changing forms of material and energy flows are insufficiently analysed in the sociology of flows (Castells, Mol, see Chap. 2) where its implications for the interaction of society and nature are more veiled than analysed, assuming that the complexity of flows makes the identification of causes and measuring of consequences difficult. To account for the limits of growth a number of indicators need to be assessed with regard to their relevance and combination in the social-ecological theory. Doing that requires two kinds of indicators:

– Indicators showing the absolute limits of natural resources in terms of their physical quantity and availability
– Indicators showing the social and economic distortions of resource use, including the unequal exchange between countries and national economies in terms of material and energy flows

The combination of both indicator types gives a more adequate picture of consequences of economic growth and resource use in modern society. The carrying capacity of ecosystems (a variable, not a constant factor) can be exceeded in absolute terms. But for the assessment of sustainability of natural resource use it becomes as important to show the socially unequal forms of resource use that result in "overshoot".

These seven points show how further theoretical reflection of SES and ESSs in a theory of nature–society interaction can proceed. For this theory, SES classifications become an important methodological component, showing

- different types of SES according to the forms of natural resource use and economic production;
- different forms of coupling of the social and ecological components of SES;
- historically varying relations between "wild" ecosystems and ecosystems modified or dominated by humans; and
- different qualities of ESSs according to their varying forms of co-production by humans and nature.

With the classifications the concepts need to be assessed epistemologically: how far are they heuristics and how far can they adopt explanatory quality, how far are they based on empirical research and how far on normative assumptions, under which preconditions can they be used to explain forms and problems of nature–society interaction? The following analysis of the SES debate has implications for ESSs: *the differentiation of connections between social and ecosystems influences the construction of ESSs that are so far discussed in naturalistic perspectives as benefits which humans and society receive from nature.*

Theoretical Knowledge Synthesis Through Classifications of SES

The following theoretical reflection comprises attempts of conceptualisation of SES emerging from the discussion above where the use of the notion in ecological research has been reviewed: classification of different types of such systems, of different forms or their coupling, and of historically varying relations between social and ecosystems.

The core concept for a theoretical reconstruction of forms of coupling of SES, elaborated in social ecological theory, is that of societal metabolism specified in socio-metabolic regimes. These regimes include various components that are not taken into account in the construction of the SES term. The important components are the modes of production and reproduction of societal, economic, and ecological systems. That social and ecosystem are interconnected and co-evolving, as said by resilience researchers (Folke 2006; Gerst et al. 2014), seems to simplify the notions of evolution and co-evolution and to ignore significant differences between biological and sociocultural evolution that are neither synchronic nor parallel processes, directed by heterogeneous principles, implying possibilities of maladaptive change in society. The normative idea to bring society (again) under control of nature or embedding it in nature seems insufficient for interpreting the interactions between society and nature with many different forms of interconnections between social and ecological systems that developed, differentiated, and changed since the beginning of human history. With the concept of socio-metabolic regimes such historically varying and complex interconnections between social and ecosystems can be theoretically codified, specified, and systematically described. The complexity can be conceptualised through combinations of theoretical terms of modern society (see Chap. 2), of modes of production and socio-metabolic regimes. Socio-metabolic regimes show that SESs are more complex than the spatio-temporal connections appearing in geographical and ecological descriptions: with cultural specific symbolic and knowledge components. The comparison of local SES, as, for example, in the ecological research of Ostrom (2007), shows casual constellations of social actors, social groups, institutions, and their interactions with ecosystems in given forms of resource use and production, but not the background variables of the societal and ecological systems and the conditions of their reproduction. Such background variables include social and class structures, societal division of labour, and socially unequal appropriation and distribution of natural resources as far as they are caused by structures and functions of societal systems, of economic systems determining natural resource use, and political systems organising the governance of coupled SES. The forms of interaction and coupling of SES include more factors and processes than ecological research and empirical case studies of SES can show.

Analysing the historically specific socio-metabolic regimes in the development of modern industrial society is a way to find theoretical explanations for the, ecologically seen, unsustainable interactions between society and nature. The connections and interactions between modern society and nature cannot be described in one form. Relevant for social-ecological analysis is not that social and ecological systems are somehow interconnected, but the human activities, institutional forms, and societal systems that structure the connection and coupling of SES, and the consequences of each forms of coupling for modern society and nature. Different possibilities of conceptually and theoretically specifying the term of SES are shown in Table 5.1. All these variants seem relevant for the development of a social-ecological theory of nature and society.

The different theoretical constructions and interpretations of SES cannot be combined in one coherent theory, but they require systematisation (Herout and Schmid 2015) through critical reflection of the practices of constructing SES and their consequences for research and knowledge use. SES analyses can be connected to many forms of research and practice of resource management, to empirical research or theoretical reflection about nature–society interaction, and to practices of environmental governance and resource management:

- The themes 1 and 2 in Table 5.1 show theoretical concepts and reflections based on empirical ecological research on SES
- The themes 3–6 show theoretical forms of analysis in which the notion of SES is connected with further theoretical and philosophical reflections about the relations between nature and society
- The themes 7 and 8 show examples of social and interdisciplinary research that do not cover systematic analyses of nature and society, only limited and specialised analyses of production and natural resource use relevant for a theory of SES

The theoretical elaboration of the term SES can to some degree be done with earlier theories and research in natural and social sciences. Older theories referred to in Table 5.1, political-economic and critical theories of society, give examples for that. Insofar the concept of SES appears as an innovation it is ignoring or selectively discussing the knowledge available for the interaction of social and ecological systems. With the construction of the new term SES, much of the theoretical knowledge created in

Table 5.1 Social-ecological systems (SES) in different theoretical constructions

Core themes and concepts	Approaches, theories, authors	Main ecological and social assumptions	Assessment
(1) Social-ecological systems/SES	Ecological research and connecting areas as – Limits to growth – Planetary boundaries of human resource use (Rockström et al.) – Ecosystem services – Resilience research (Folke et al.) – Sustainability science (Kates, Clark et al.) – Earth system science (Ehlers and Krafft)	Social and ecological systems are always interconnected	Heuristic, hypothetical assumptions; natural-scientific perspective of social systems, focus on spatio-temporal relations and natural limits of resource use
(2) Coupled systems of humans and nature	Social-ecological research (Ostrom et al.)	Similar to SES, analysis of intersystemic coupling	Conceptual frameworks for interdisciplinary social-ecological research, focus on resource use and resource management
(3) Interaction of society and nature in theoretical analyses: – Societal metabolism – Socio-metabolic regimes – Modes of production – Social-ecological change	Social-ecological theory (Becker, Fischer-Kowalski et al.), analyses of socioenvironmental change (Hornborg and Crumley 2006) Political-economic theory, agricultural and industrial societies (many authors)	Interrelations of society and nature are forming in complex metabolic regimes that connect social and natural processes in the human use of natural resources	Societal metabolism as core of social-ecological theory: theoretical explanation of historically specific forms of nature–society interaction

(4) Global environmental change/GEC: – Syndromes of environmental change – Global change in ecology	Systems theory/sustainability science (Schellnhuber et al.) Ecological theories of global change (Bolin, Crutzen et al.; International Geosphere-Biosphere Programme, International Human Dimensions Programme)	– Three types of man-made syndromes of environmental change: development syndromes, pollution syndromes, sink syndromes – Three forms of man-made GEC: climate change, biodiversity reduction, land use change	Descriptions of environmental change connected to human resource use and its consequences, using ecological terms (GEC) or a metaphoric notion of "syndrome"
(5) Societal relations with nature	Marxist and critical theories of society (many authors) – Philosophical and sociological theories of societal relations with nature – Historical materialism – Historical-geographical materialism	Societal relations with nature are historically specific and changing: transformation of nature through human labour and with that also transformation of social relations in society, specified in the epistemological theory of historical materialism	Concepts and theorems influencing the analysis of society–nature relations in social ecology

(continued)

Table 5.1 (continued)

Core themes and concepts	Approaches, theories, authors	Main ecological and social assumptions	Assessment
(6) Philosophy of nature – Ontology – Ethics	In the ecological discourse important: – Nature as producing/productive force ("natura naturans") and produced nature ("natura naturata") – Ethics in anthro-, bio- and eco-centric variants – Ethics of caring for nature – Ethics of human domination/power over nature	Principles of environmental ethics: – stewardship and responsibility for the environment (Leopold: land ethic; Carson: ethics of the sea; Holmes Rolston, ethics for animals, plants, species, ecosystems) – Jonas, ethics of responsibility – Bloch, hope and social emancipation in alliance with nature	Connecting to SES analysis with broader terms of nature and society from which to develop conceptual frameworks and theories for relations between humans and nature in different cultures and societies; relevant for normative and theoretical framing of ecological research
(7) Analyses of primary production in economics, sociology, anthropology, agricultural sciences: agriculture, fishery, forestry	Research on natural resource use in historically changing modes of production Special theories, for example that of the peasant mode of production and economy (Cajanov)	– Social assumptions about the social structures/division of labour that shape human use of natural resources – Economic assumptions about effectiveness of production and maximum of yield or sustainable yield	Specialised research in the social sciences, selective use if ecological knowledge

| (8) Environmental sociology and human ecology | – Social paradigms of nature (Catton and Dunlap)
– Political economy (Schnaiberg, Gould: treadmill of production)
– Structural human ecology (Dietz, York, Rosa)
– Unequal ecological exchange (Bunker, Rice, Jorgensen) | Interdependence of society and nature (new ecological paradigm)
Global complexity of capitalist production and reproduction in socio-economic and ecological terms | Reformulating interaction of society and nature in sociological perspectives
Conceptualising the global complexity of nature–society interaction and SES in critical theoretical analyses |

Sources: Own compilation; sources quoted

the social sciences about nature–society interaction is ignored. With this concept appears in ecological research a late effort to connect social and ecological sciences, as before in the interdisciplinary subjects of human, social, cultural, and political ecology. In difference to the concepts used in these heterodox forms of ecological research, the notion of SES has rather limited capacity to grasp the complexity of interacting and interwoven society and nature.

With the new ecological term of SES, the interaction between society and nature is analysed with doubtful assumptions about the interaction between both system components. The assumption that social and ecosystems are always connected is inexact, bypassing the theoretical knowledge about modern society, drawing upon evidence from limited ecological research of ecosystem processes and human use of natural resources. To study the interaction between society and nature with results from empirical research in ecology (understood as applied science) as in the ecological research on vulnerability, resilience, sustainability, and ESS, has as consequence to exclude knowledge about the system dynamics of SES. This seems a similar limitation as in sociology the empirical research on attitudes, perceptions, communications, and everyday action of humans. In both cases, the understanding of social and societal systems is reduced to that what can be empirically observed and measured, without clarifying the object of measurement through theoretical definition. In sociology there are, however, complementary theories that show the influence of structures and processes of societal systems on empirically studied processes. These theories are also relevant for the analysis of societal interaction with nature—in modes of production, in socio-metabolic regimes, in societal and economic reproduction, in biological regimes of reproduction and societal energy regimes. *To develop SES analysis in theoretical perspectives, three methodological requirements can be formulated that are not always taken into account in the research traditions described in Table* 5.1.

1. *The normative assumptions and implications of SES analysis should be laid open.* SES analyses in ecological research follow more or less explicitly formulated ecological world views or paradigms, also such discussed in popularised forms found in environmental movements as "to follow nature's lead". Such views are based on normative assumptions, ethical convictions, and visions of society as embedded in nature.

The strong influence of these assumptions seems to be that they make the ecological discourse "immune" against certain forms of empirical and theoretical knowledge and selective in the interdisciplinary knowledge used. Biology provides—not with all its research—a knowledge basis for environmental action that allows for ignoring much of social-scientific knowledge when it is assumed that biology has epistemologically privileged knowledge about humans and nature, of higher value than that the knowledge social sciences can provide. Biology is, beyond its disciplinary specialisation, also part of the interdisciplinary knowledge base for analysing the interaction between humans, society and nature. Biological knowledge needs to be connected to knowledge about the human modification of nature through labour, technologies, and scientific knowledge, the societal transformation of nature that are required in SES analyses, when these should say something about possibilities and pathways towards sustainability. Social ecology deals in theoretically elaborate forms with the interdisciplinary combination of social and natural-scientific knowledge. Other interdisciplinary forms of ecological research, resilience research, and sustainability science do less systematically reconstruct the nature–society interaction.

2. *To develop the concept of SES for the use in a theory of nature and society, classifications are an important step. From classifications, a theoretical taxonomy of SES can develop that covers the wide range of historically varying forms of interaction between society and nature.* Two forms of classification help to refine the SES concept:

   – Classifications of types of SES (Table 5.2) based on the theoretical conceptualisation described above (Table 5.1)
   – Classifications of different forms of coupling of social and ecological systems (Table 5.3)

   The classification of SES types discussed in this chapter is formulated for modern society, whereas historically earlier and different relations between social and ecosystems are not discussed further; they are in some detail described in the social-ecological literature (e.g. Fischer-Kowalski et al. 1997; Fischer-Kowalski and Haberl 2007). The classifications show the historical variations of intersystemic interactions between nature and society from the comparison of modern and earlier forms of society.

Table 5.2 Typology of social-ecological systems in modern society

---
1. *Systems of primary economic production where ecological relations dominate:* agriculture, fishing, forestry, extracting of natural resources (mining)—mainly rurally located economy
2. *Systems of secondary economic production where ecological and social relations are blended:* artisanal production, manufacturing and industrial production, energy conversion systems—mainly urban economy and urban social systems
3. *Systems of symbolic reproduction of society where social and cultural relations dominate:* education, science, culture, social services ("third sector economy")—mainly urban location; ecological relations are not ignored but of limited relevance
4. *Ecosystem types differentiated according to their degree of transformation through humans and during human history:* marine ecosystems, arctic ecosystems—weak transformation; agro-ecosystems, forests, urban ecosystems—strong transformation through colonisation of nature; desert ecosystems, industrial ecosystems—ecosystems resulting from human destruction of nature/ecosystems (deforestation etc.). With global environmental change it becomes difficult to estimate degrees of transformation (also remote ecosystems where no humans are found, in the arctic zone or in the deep sea become strongly influenced through human activities), but the transformation can still be seen in the phenomenology of landscapes.
5. *Social-metabolic regimes:* complex regimes conceptualised in social-ecological theory to analyse the interaction between society and nature in natural resource use (sometimes also called biological regimes of a society): metabolic regimes of agricultural and industrial societies, specified according to different forms and subregimes, for example, the industrial energy regime based on fossil fuels developed first with the coal regime, than with the oil regime.
6. *Multi-layered (synthetic) social-ecological systems:* systems that include combinations of the forms described above (1–5).

All types of systems classified here can be described in further detail through their spatial and temporal relations and their socio-cultural and historical specificities.

Sources: Own compilation

Societies are differentiated systems that cannot be sufficiently analysed with a general definition of society. As older theories from critical political economy show, theoretically more relevant than definitions are combinations of theoretical concepts and analyses of the mode of production (in the specific variant of social-ecological theory: the socio-metabolic regime), of societal relations with nature, and of culturally

Table 5.3 Coupling of social and ecological systems

Types of coupled social-ecological systems:
1. *Social systems with dense and direct coupling to ecosystems:* coupling through extraction and processing of natural resources—all types of primary and secondary economic production (historically and culturally specific forms of agriculture, fishery, forestry, industry)
2. *Social systems with loose or indirect coupling to ecosystems:* non-productive systems—all systems for symbolic reproduction of society (education, science, culture, norm formulation, social services)
3. *Systems with maladaptive coupling:* all coupled systems overusing and destroying their natural resource base—mainly systems of economic production (historically and modern forms of agriculture and manufacture/industry)
4. *Human-dominated ecosystems with dense material and spatio-temporal coupling to societal subsystems in modern society* (urban and industrial ecosystems, agro-ecosystems)
5. *Nature-dominated ecosystems with loose material and spatio-temporal coupling to social systems:* ecosystems where no humans are living (arctic/antarcticecosystems, deep sea ecosystems, certain mountains, and desert areas; these systems may be influenced through human resource use)

Forms of intersystemic coupling of social and ecological systems:
1. *Forms (spatial and temporal) of coupling: as uni- or multi-scalar systems and systems with specific time regimes:* for example, marine reserves that are nested in larger systems of marine governance, or similar networking and nesting of land-based areas of conservation
2. *Processes of coupling, interaction and exchange*: human labour and production; extraction, appropriation, production and processing of natural resources; human-made forms of coupling that change material and energy flows in ecosystems, natural cycles of water, nitrate, and so on. in economic production; "biological coupling" of humans and nature transformed into social-ecological coupling where the specific social and ecological forms need to be specified (production, processing and consumption of food in hunting, fishery, and agriculture)
3. *Media of intersystemic coupling*: flows of information/symbols, matter, energy
4. *Formal descriptions of intersystemic coupling:* loose/tight, intensive/extensive, strong/weak, soft/hard, symbolic/material, functional/dysfunctional, uni-scalar/multi-scalar

Sources: Own compilation; Bruckmeier (2014)

framed interactions between humans and nature. The differences between political-economic and social-ecological theories correspond to different times of their origin and purposes of formulation. The critical political-economic theory of modern capitalism developed since the mid of the nineteenth century in attempts to reconstruct the societal relations of power and exploitation of humans and to reflect how these can be transformed. The new social-ecological theory of nature–society interaction developed since the end of the twentieth century for purposes of analysing historically changing societal relations with nature to describe and explain the consequences of natural resource use for society, nature, and their sustainable coexistence. Explanations from both theories can be combined without neglecting their differences. Political-economic theory described the mode of production in modern societies as that of modern capitalism as an economic world system, whereas social-ecological theory, following the ecological tradition of classifying modes of production, described it as that of industrial society specified through social-metabolic regimes that are characterised through historically dominant forms of energy sources: for the industrial society the coal regime, then the oil regime that lasts until today. The differences can be seen as different perspectives of social-scientific and ecological research, of human society as organising the survival and well-being of humans, or as modifying and disturbing the functioning of ecological systems.

3. *The analysis and reconstruction of the systemic dynamics and complexity of interacting nature and society require a methodological reflection of the interplay of empirical and theoretical analysis in knowledge synthesis.* The conceptual apparatus of social ecology is constructed for the analysis of such complex intersystemic relations that developed in three interconnected theoretical components: the theories of societal metabolism, of societal relations with nature, and of colonisation of nature, all of them to describe and explain facets of the historically changing interaction between nature and society. This social-ecological research programme cannot, as ecological research does to a large degree, ignore what is known about changing societal forms of power and domination in human history: historically specific power relations between humans in societal, political, and economic systems, and power of humans over nature. The manifold power relations are not always manifest as exercise of power and control; they are incorporated and veiled in the organisation of societal division of labour, in

property rights, in forms of appropriation of natural resources, in market relations, in technologies and large technological systems, in the influence and power of science and scientific knowledge. Complex, socially structured metabolic regimes and power relations shape the interaction between society and nature and cannot be reduced to that what can be observed in empirical case studies on the extraction and use of natural resources. Social forms of natural resource use tend to be simplified in ecological analyses when the differentiations of concepts, knowledge types, and levels of interaction between empirical and theoretical knowledge production are not envisaged. Especially differences between (limited) theories that can be directly proven through empirical research and other, broader ones that develop through a series of interim steps and knowledge syntheses are important in interdisciplinary theorising. In ecological research, interacting social and ecological systems are reconstructed in naturalistic perspectives, as similarly in the political discourse of environmental movements. The naturalistic thinking seems to ignore the analyses of complex modes of production, productive forces, and relations of production that have been carried out in research in political economy and in the newer interdisciplinary fields of human, social, and political ecology. The interdisciplinary research in these fields supports, furthermore, open, analytically differentiated forms of reconstructing nature–society relations with a plurality of connected perspectives.

In the following discussion, first conclusions are drawn from the critique of SES constructions with regard to theoretical synthesis of social and ecological knowledge and to forms of coupling of social and ecological systems.

Using modern society as the paradigmatic case and the theoretical and explanatory components discussed in Chaps. 1 and 2, a classification of SES becomes possible in two forms, as a typology of SES (Table 5.2) and of their coupling (Table 5.3). The types of SES, identified from social-scientific and ecological research, are connected to modes of natural resource use, modes of production, and socio-metabolic regimes found in modern society. Through the theoretical embedding of concepts in social-scientific and ecological theories, the term of SES differentiates in historically, culturally, and economically varying forms of interaction between society and nature. In Table 5.2, the term SES is reconstructed with theoretical denotations in theory of society. This does not necessarily imply to give up the simple notion of SES as geo-biophysical systems connected with actors and institutions, but to reduce it to a subordinate notion that needs to be connected with further theoretical terms to identify trajectories and trends of development of coupled social and ecological systems.

The typology above is limited to the interaction of social and ecological systems in modern industrial society. With this typology, the term of SES which is as diffuse and abstract as the constituting concepts of social and ecological systems is modified: not formulated as a single theoretical concept, but differentiated in a number of theoretically specified SES types to be considered in a broader theory of nature–society interaction. The progress of integrating the term of SES in the formulation of such a theory follows the forms and principles of constructing an interdisciplinary theory that can be described in methodological and epistemological terms as follows:

1. *The conceptual reconstruction of SES does not create a single analytical framework or a closed theory, but develops the heuristic device of SES in ecological research into a series of historically and geographically specified theoretical concepts.* It can be assumed that interdisciplinary social-ecological research needs to work for longer time with heterogeneous knowledge components, classifications, concepts, and theories of disciplinary origin, and with limited analytical and explanatory capacity before new explanations develop. More elaborate and coherent interdisciplinary theories develop gradually through progressing knowledge synthesis in subsequent steps of empirical syntheses, of formulating concepts and frameworks, of theoretical syntheses and epistemological reflection. Classifications of the form above characterise the early stage of a theory of interaction between nature and society, describing relevant differences between SES that develop from research and knowledge about natural resource use in modern society which guides and limits the possibilities of theoretical explanation. Although the further development of the theory may result in other classifications, it seems relevant in the present phase of development

of modern society (described in Chap. 2) to construct a differentiated concept of SES and not to exclude a priori certain forms of SES from research and from the synthetic theory to develop. Showing the relative validity and explanatory value of each type of SES for the theory of nature and society is a precondition for the further development of the theory.

2. *To elaborate the classification of SES types to theoretical description and explanation of historically specific types of societies and their relations with nature*, it seems promising to use from the theoretical knowledge in sociology and ecology three concepts where nature–society relations are reconstructed in different analytical perspectives and with different forms of abstraction:

- *Social systems*, including society, where societal relations to nature are described and analysed in symbolic forms of views of nature, humans, and society and in the analysis of material production and reproduction of society in its economic forms.
- *Ecosystems*, where the interaction of nature and society is described and analysed in a biological frame of reference for the analysis of living systems or species, their development, and biological reproduction.
- *Coupled systems or SES* where the interaction of nature and society can be described and explained more systematically, in the theoretical terms of social ecology, as societal relations with nature, sociometabolic regimes of societies, and colonisation of nature.

To differentiate between these three *analytical* perspectives in the development of the theory of society and nature can help to create a variety of rich theoretical analyses and possibilities of explanation that enlarge the knowledge about nature–society interaction beyond the empirical research on SES. Such typological differentiations can, furthermore, help to prevent the levelling of theoretically and practically meaningful conceptual differences in quickly and inexactly formulated hybrid terms to describe the interaction of society and nature, for example, "socio-natures" or "techno-natures". Such de-differentiation of theoretical concepts and systems analyses does not support the reconstruction of the historical variations of nature–society interaction it may be looking for in attempts to describe the blending of society and nature. To work with three analytical perspectives as described above seems useful to develop an interdisciplinary theory of nature–society that can connect and maintain several theoretical perspectives and knowledge from different disciplines. In such a theory, different parts of societal relations with nature can be conceptualised from the study of their historical variation and hybrid concepts are not constructed before theory-guided analysis, but as a result of knowledge syntheses. *All three perspectives are needed in the theory of nature and society as an interdisciplinary and intertheoretical construction, for purposes of identifying relevant knowledge for social-ecological research, for analysis, explanation, and for knowledge synthesis.*

3. *The construction of a theory from elements of different theories implies to deal with heterogeneous cognitive components, contradicting assumptions, epistemologies, and explanations in different parts of the theory that cannot always be dissolved in a hierarchy of progressive theoretical explanations and generalisations.*

Heuristic rules of the thumb and methodological devices for developing the analysis of SES in ecological research argue with the idea that differences and contrasts between SES forms appearing at lower levels of abstraction can be integrated at higher levels of theoretical abstraction and generalisation (Olsson et al. 2006). Theoretical codification becomes more difficult when concepts and knowledge from other disciplines than ecology are used, for which little methodological support is provided from epistemology or theories of science. In the present forms of applying the term SES, theory construction and theoretical reflection is bypassed through shifting the reflection from theory or epistemology to paradigms or ontology. Knowledge production is directed through normative assumptions and guiding ideas, in form of views of nature and society, values, norms, assumptions, premises, and postulates. For SES analyses that require knowledge from the social and natural sciences other rules would be needed that develop in interdisciplinary research and knowledge syntheses. Interdisciplinary theories work with several perspectives, theories, and explanations in knowledge syntheses from different fields of research.

Complex theories as that of nature–society interaction can use various forms and levels of explanation in open and pluralistic theoretical configurations where different explanations emerge at various stages of knowledge generation, in empirical research, theory formulation, and interdisciplinary knowledge syntheses. Interdisciplinary theories of this kind require, furthermore, critical reflection of the possibilities of connecting strong universally valid explanations from deductive reasoning and weak, historically and culturally relative explanations achieved through induction or historical and intercultural comparison.

With an interdisciplinary social-ecological theory develops—beyond the two prominent forms from twentieth century, systems theory and critical theory in sociology—a new form that is more balanced in the use of knowledge from different disciplines than these prior examples that remained either within a natural-scientific or a social-scientific logic and form of explanation. Systems theory illustrates variants of natural-science-based thinking and knowledge synthesis ex ante, where theoretical explanations are generated with the general concepts of the theory applied for the description and analysis of special types of systems. In sociology, this resulted in theories as that of Parsons and Luhmann where empirical research is rather neglected for theory construction. Critical theory provides examples of knowledge synthesis ex post, based on social-scientific knowledge, where explanation emerges successively with theory construction, in several steps of combining and reflecting empirical and theoretical knowledge from earlier and new research.

4. *The main conclusion from the discussion of methodological and epistemological problems of theory construction is for the social-ecological theory of nature and society the model of a general, universally valid theory is inadequate.* More adequate seems the idea of a historically situated and specified theory reconstructing nature–society relations specifically for each form of society and mode of production in human history. This theory of nature and society is not identical with a theory of society, but requires such a theory to reconstruct societal relations with nature for modern society. The levels and forms of abstraction of this theory depend on the historical specificity of society and its three social-

ecological determinants of societal relations with nature, societal metabolism, and colonisation of nature. These determinants generate descriptions of the societal system structures and the state of nature modified through humans at a given historical time.

For modern society

- *the state of society* is that of a global system according to its dominant mode of production and the modern world system;
- *the state of interaction of social and ecological systems* is that of man-made global environmental change, described with the hypothesis of "mankind as geological force";
- *the state of nature* is one of disturbed global material cycles and overshoot of carrying capacity as consequences of the quantity and intensity of human resource use.

These general descriptions need to be detailed and deepened in the integrated theory of nature and society. Analyses of the transformation to sustainability cannot be done with such general and inexact diagnoses of non-sustainable states. This deepening requires also further epistemological discussion and interdisciplinary knowledge synthesis. Epistemologies and methodologies that developed throughout the past century, mainly from disciplinary perspectives or abstract philosophical approaches (Schneider 1999), neglect the cognitive problems of border crossing and interdisciplinary knowledge synthesis.

In ecological research, limited theorising is found within the new research on vulnerability, resilience, and sustainability. Only in rudimentary forms are the difficulties and problems of interdisciplinary conceptual integration and knowledge synthesis reflected for analyses of the interfaces of nature and society, for example, by Young et al. (2006), ending without theoretical clarification of the connections and interactions between social and ecological systems. Although the relations between man, nature, and society are the main topic of interdisciplinary social, cultural, and human ecology, a systematic theory of nature–society interaction has not yet developed from these interdisciplinary approaches. The elements of theory developing in social ecology include theories of societal relations with nature, of societal metabolism, of colonisation of nature, and—more recently—the theory of socio-ecological transformation.

The work with heuristic devices and pre-analytical assumptions of the kind that social and ecological systems are always interrelated and co-evolving seems to become a dead end for the term SES. The arguments and justifications for new areas of specialised empirical research need to be revised when the term SES is theoretically reflected and used in interdisciplinary theory. The emerging social-ecological theory needs to develop the argument that the construction of interdisciplinary theory happens in several steps of knowledge integration. The construction of the theory of society and nature develops through analyses of interrelations, forms of coupling, and of co-evolution of social and ecological systems in different perspectives: in theoretical triangulation, keeping possibilities open for further integration and synthesis, creation of new concepts, and further explanations. A plurality of system perspectives helps to explain society–nature interaction through progressing knowledge generation and synthesis. Such a synthesis is predetermined by knowledge about the historical specificity of modern society and needs to deal with competing theories of that society (described in Chap. 2).

There are several epistemological and methodological problems to solve in this synthesis. Social and natural-scientific knowledge from several disciplines, such as sociology, biology, ecology, physics, is generated with heterogeneous ontologies, epistemologies, and methodologies. These knowledge forms do not add in a continuum of knowledge and cannot be made coherent. Interdisciplinary theories require methods to deal with or to bridge epistemological ruptures and gaps between different knowledge systems. In advance, there are no safe criteria to find out which knowledge is relevant and can be synthesised, which not. Therefore, interdisciplinary knowledge synthesis progresses through a series of steps that are not yet methodologically safe and include trial-and-error, require further reflections on the way of theory construction. Only the beginning of a theoretical synthesis is possible with the three steps discussed here:

- Identifying different theories potentially relevant for the theoretical codification of SES (Table 5.1)

- Specifying different types of social and ecological systems for which the interaction can be described in a broader typology of SES (Table 5.2)
- Specifying the systems interaction further by classifying the concrete forms of coupling of specific social and ecological systems (Table 5.3)

The last step marks the interim target for the operationalisation of an interdisciplinary social-ecological theory of nature–society interaction in the present stage of elaboration.

A typology of forms of coupling of social and ecological systems (Table 5.3) is not fully visible from the typology of SES (Table 5.2), and not sufficiently elaborated in ecological research. With such typologies, the reductionist view of SES as "geo-bio-physical systems with actors and institutions connected to them" can be broadened and connected with theoretical analyses of societal systems and their interaction with nature.

To specify the concept of coupling for SES and to differentiate it from further concepts to describe intersystemic relations, several criteria from the classification in Table 5.3 can be combined. Two forms of coupling that seem dominant in SES are those of material and spatio-temporal coupling:

- *Material coupling* can be described in social-scientific terms through socio-economic links, based on human labour, production, extraction, appropriation, and processing of natural resources; or in ecological terms through biophysical links, based on flows of information, material, and energy between the systems and through the biological and social metabolism. The forms of material coupling connect knowledge on the production and reproduction of social and ecological systems and the interaction of society and nature through biological and social metabolism.
- *Spatio-temporal coupling* refers to the interrelations between social and ecosystems that can also be described, in social-scientific and ecological terms: time and space as constituents of social reality; as constituting the forms of human perception of the world, nature, and society; as social structuring of time and space are forms of socially transformed time and space to frame social action. Spatio-temporal relations in ecosystems, based on physical concepts of

time and space are of another kind, assumed to be objective qualities of nature, not constructions of the human mind. Both forms of spatio-temporal coupling can be connected to material coupling, to system structures and processes, and to social action.

Material and spatio-temporal coupling include informational and symbolic coupling, but this is less specific for social-ecological coupling, exists in all types of systems, ecological, technical, cybernetic, symbolic, and social systems. Societies as complex social systems include further forms of informational coupling: communicative coupling through power relations, organisational coupling, and knowledge use which are also relevant for SES. The term coupling is not exclusively applied to analyse interrelations between social and ecological systems, but this term should not be overburdened with theoretical meaning: Forms of coupling of systems can be differentiated from other forms of interrelations to which they can be connected: functional interdependence, interaction, feedback, exchange, networking, and further, more specific forms of intersystemic relations. Lack, disturbance, or loss of coupling can be conceived, for example, with the concepts of "noise" or "resonances" used in some forms of systems theory and ecological research, or in forms of violent and non-violent conflicts.

With these steps of reflection of SES, the concept seems sufficiently prepared for theoretical use and for connection to an interdisciplinary theory of nature and society. The heuristic construction of SES can be transformed in theoretically reflected concepts that specify nature–society interaction, which are no longer dependent on doubtful assumptions, for example, that social and ecosystems are always interconnected and co-evolving. The interconnections can be investigated in their different and historically varying forms and with the problems they generate. In this way can be described the cognitive programme for the further elaboration of a theory of the interaction of nature and society in modern society. Furthermore, with that differentiation of SES types it should become possible to discuss more critically and theoretically the concept of ESSs and its significance for a theory of society–nature interaction.

ESSs

ESSs are the benefits people obtain from ecosystems. This definition from the Millennium Ecosystem Assessment (2005) shows an unclear blending of terms with multiple meanings. ESSs require a more detailed analysis than with the functional terminology of ecosystem functions and services.

- An anthropocentric perspective appears in the statement that humans benefit from nature—not necessarily, but with the practical purpose of maintaining human well-being and quality of life with this idea. That human beings benefit from ESSs does not exclude that other living beings, animals and plants, also benefit from them, but this is practically irrelevant. It is only a side effect of ecosystem governance that aims primarily to support human well-being and implies to protect ecosystems from disturbance or destruction through humans. This perspective differs from the conventional anthropocentric argument of human supremacy that makes nature an object of human use, modification, or improvement, but both views do not exclude each other.
- An ecocentric perspective appears in the assumption that the benefits are obtained from nature or ecosystems, not co-produced by humans and nature, and should help to maintain the functions and processes of ecosystems. The unclear point in both anthropocentric and ecocentric interpretations is, how humans are integrated in ecosystems and how in society.

Without analysing further, more systematically and critically, the interaction of social and ecological systems and the contribution of humans or society to service production, the term remains unclear in its scientific and practical applications. *From the four kinds of ESSs—supporting, provisioning, regulating, and cultural services (Table 5.4)—it is not clear, whether this classification is meant to be exhaustive and universally valid, how it is derived from the theoretical construction of ecosystems or empirical research, how far it needs specification for different types of ecosystems, and how far ecosystems include social components.* These deficits can be seen as indicating insufficient elaboration of ESSs as a theoretical concept. Theoretical

Table 5.4 Classification of ecosystem services: Millennium Ecosystem Assessment

The Millennium Ecosystem Assessment differentiated between the following types of ecosystem services:
1. Supporting services (nutrient cycling, soil formation, primary production, and others)
2. Provisioning services (food, fresh water, wood and fibre, fuel, and others)
3. Regulating services (climate regulation, flood regulation, disease regulation, water purification, and others)
4. Cultural services (aesthetic, spiritual, educational, recreational, and others)

The ecosystem services are assumed to affect important constituents of human well-being:
1. Security (personal safety, secure resource access, security from disasters)
2. Basic material for good life (adequate livelihood, sufficient nutritious food, shelter, access to goods)
3. Health (strength, feeling well, access to clean air and water)
4. Good social relations (social cohesion, mutual respect, ability to help others)
5. Freedom of choice and action (opportunity to achieve what an individual values doing and being)

Mediation and linkages between the natural/ecosystem components and the human/socio-economic components are assumed to differ between the types of services

Source: Millennium Ecosystem Assessment (2005)

elaboration and reflection includes the more principal question, whether the concept of ESS is useful or necessary in ecological research and for purposes of nature protection. This question is controversially discussed and answered since the terms are used, but it did not develop to the point to discuss services systematically in relation to theories of society, ecosystem theories, and theories of nature and society.

The classification from the Millennium Ecosystem Assessment shows efforts to connect ESSs to humans, their well-being, and to human society without investigating the interaction of social and ecological systems discussed above for the concept of SES. This classification avoids a social-scientific terminology, with ad hoc descriptors of social, economic, and health effects, based on some ethical assumptions. Such assumptions are explicitly formulated for cultural services and the constituents of human well-being, of good social relations, and freedom of choice. Ethical reflections and discourses are usually ascribed to social systems, societies, culture, or civilisations and as exclusive capacity of moral judgement to humans. In the construction of ESSs, important points are unclear and insufficiently reflected: the constituents of human well-being, the generation of services, and the interaction of social and ecological systems. Sometimes the services refer to political principles as visible in the description of human well-being, sometimes they refer to prior debates as that of basic human needs (well-being components one to three); only in some examples it is intuitively clear how they refer to ecosystem functions (services one to three). The connections between ESSs and human well-being, expected to specify and clarify the benefits for humans, are arbitrary in the attribution of services to ecosystems. Furthermore, ESSs have little explanatory value, but imply normative assumptions that require further justification in a critical debate of the ecosystem concept. Obviously, the description in the Millennium Ecosystem Assessment is for political use and communication of the idea of benefits from nature to many environmental actors in the policy process and the environmental discourse. The description does not clarify ESSs as scientific concept and the epistemological and methodological difficulties surging with the application of the term for the formulation of environmental policies. *To go beyond a value-based ascription of services to nature, the concepts of ESSs and well-being need to be connected, discussed, and assessed with scientific knowledge from ecological and social research: at this point, the concept of SES and the theory of society–nature interaction become important.*

The notion of ESSs has rapidly spread in the past decade, but does not develop from an epistemological "point zero"; it refers in manifold ways to prior analyses of ecosystems. Nevertheless, the notion of ESSs used in science and policy bears the burdens of unclear, diffuse, and contradicting attempts to reformulate ecosystem functions and processes for purposes of policy formulation and resource management, with epistemologically and methodologically doubtful assumptions. Descriptions of ESSs use the older terminology of functionalist systems theory, but this does not help to dissolve the controversy whether systems theory creates a unifying framework for the natural and social sciences and in doing that creates unifying theoretical explanations of processes of intersystemic interaction and social-ecological change. The nature of ESSs needs to be discussed further with theoretical concepts and perspectives that are part of the emerging theory of nature and society.

ESSs in Theoretical Perspectives

ESSs in the forms they are defined in recent ecological research and debates, paradigmatically in the Millennium Ecosystem Assessment, are designed for a new research programme in applied ecology, not for a theoretical clarification in a social-ecological theory. There may be implicit theoretical arguments and explanations built in the ESSs concept: how such services are produced, transferred between ecosystems and social systems, and used by humans. But these date back to ecological research and the established concepts in ecology, do not require a theory of interaction of nature and society. Similar as the concept and research on resilience was criticised as supporting the neoliberal mainstreaming of the sustainability discourse (see Chap. 3), ESSs can be criticised for the support of a monetarisation and commercialisation of ecosystems through payments for the services. As in the resilience debate, it needs to be asked whether payments are the unexpressed goal, or rather a political, power-based misuse of an ecological idea, pretending that monetary values and payments help decision-makers to use the idea. *In the social-ecological theory of nature and society, it seems important to distinguish between ecological and economic logics of ESSs to be able to separate ecological ends from economic means, and to find alternatives to use the term without economic valuation.*

A first step of clarifying the nature of ESSs theoretically is, as above with the concept of SES, the identification of ecological and interdisciplinary theories that can provide theoretical framing of ESSs (Table 5.5). These are partly the same theories used to clarify the SES concept.

Most of the theories and approaches in Table 5.5 are broader than ESSs that appear only as specific forms of interaction of society and nature, as part of the processes analysed with the broader concept of SES. In this broader perspective, the construction of ESSs needs to be clarified: whether the services are produced by nature or ecological systems and only consumed by humans, or whether human society, labour, and knowledge are involved in their production as productive forces that are throughout human history transforming nature through the appropriation and use of natural resources. The theories described above help to discuss ESSs in the broader context of interacting and coupled social

Table 5.5 Ecosystem services and theories related to that

Core themes and concepts	Theories and authors	Main ecological and social assumptions	Assessment
(1) Classification of ecosystem services	Millennium Ecosystem Assessment (2005) Ecological research (Wilson and Matthews 1970; Ehrlich and Mooney 1983; Daily 1997), sometimes referring to older biological and philosophical discussions of nature	Agency of nature Anthropocentric perspective of analysis (ecosystems services are for humans) although the concept is not inherently anthropocentric but develops in a broader bio- or eco-centric perspective	Evidently, the services described result for a summary of ecosystem functions and circular processes described in ecological research, only the cultural services seem an addition to take into account social aspects; no final classification of services can be complemented and specified
(2) Classification of ecosystems (structures, processes, functions, services) Classification of ecosystem types related to species and vegetation (bio-ecosystems) or areas and soils (geo-ecosystems)	Connecting ecosystem services to analyses of ecosystems	The core component of the heuristic framework, the idea that services can be derived from functions as concepts of "higher order" in ecosystem analysis	Specifying ecosystems services as generated by ecosystems, without clarifying how they are produced (involvement of humans in production)
(3) Theories of ecosystems: – Ecosystems ecology – Systems ecology – Systems theory	Theoretical conceptualisation of ecosystems (including humans) based on ecological research; the three approaches are overlapping	The idea of ecosystem functions and services develops from a functionalist perspective of ecosystem analysis	No clear criteria for functions or their theoretical analysis and explanation

(4) Intersystemic relations between social and ecological systems: – Societal relations with nature – Societal metabolism – Social-ecological systems	Critical theory of society Social-ecological theory as material and energy flows between society and nature Ecological framework/theory	Agency of society and co-production of society and nature in historically changing modes of production (historical materialism) Agency of society and co-production of society and nature in historically changing socio-metabolic regimes	Ecosystem services, in systematic classification on theoretical base, can be seen in theories of nature–society interaction as components of the "dialectical relationship" between social and ecosystems that develop through "coproduction" of nature and society in historically varying forms
(5) Sociology of flows	Theory in environmental sociology to analyse material and non-material flows between social and ecosystems (Mol et al.)	The intersystemic flows are so complex that they cannot be analysed exactly, in quantitative and causal terms	Theory aiming to reject ideas for the discourse of critical theory of society and social-ecological theory: assumption, that interactions and flows between social and ecosystems cannot be analysed quantitatively to show power asymmetry in the modern world system

(continued)

Table 5.5 (continued)

Core themes and concepts	Theories and authors	Main ecological and social assumptions	Assessment
(6) Philosophy of nature and nature–society interaction	Philosophical theories of nature: – Vernadsky (biosphere) – Bloch (ethics of hope) – Jonas (ethics of responsibility)	Creating the basic concepts for nature–society interaction, with various terms of nature, to develop the idea of services: nature as generating ("*natura naturans*") and generated ("*natura naturata*"), as super-organism (biosphere, noosphere, Vernadsky), "alliance" of society and nature (Bloch), nature as caring	Useful to clarify the implicit and controversial assumptions in the conceptualisations and classifications of ecosystem services: anthropocentric views of nature, capacities of nature ("to act", "to work for humans", attributing to ecosystems capacities to produce, act, regulate, care for humans)

Sources: Own compilation, sources quoted

and ecological systems. The notion ESSs came into wider use through the Millennium Ecosystem Assessment, although the idea originated earlier (see Table 5.5). The term bears connotations of older concepts and debates in- and outside ecology, for example, that of ecosystem functions, or in the philosophy of nature that of the concepts of generating and generated nature. The ascription of productive forces to nature may be the most general idea from which the specific meanings of ESSs can be derived.

Ascribing to nature capacities of generation or production implies various alternatives to conceptualise ESSs:

- The idea of ecosystem functions and services as specifying *productive or subject capacities of nature* dates back to older philosophical ideas of nature as supreme order.
- The idea of ESSs as being *part of the broader processes of interaction between social and ecological systems* requires explanations of the forms of co-production through humans, society, and nature.
- The interpretation of ESSs through the elements and processes of ecosystems and their life-maintaining processes connects the concepts of ecosystems and autopoietic or living systems in the sense of the biological theory of autopoiesis (Maturana and Varela), as the capacity of self-production of systems.
- The interpretation of ESSs with the concepts of biological and societal metabolism, described in social-ecological theory, is based on the analysis of material and energy flows and use of resources in ecosystems that are important for the understanding of ecosystems and of ESSs.

The theoretical interpretations and connections of ESSs to broader theoretical concepts show already: it is not a fixed concept, but open for different, sometimes complementary, sometimes competing interpretations. ESSs resembles the idea of sustainable development in several regards: being a controversial concept, catalysed through a political discourse and practical necessities of environmental policy and natural resource management, but insufficiently clarified through research and theoretical reflection. Also following the scientific history of the concept does not clarify all disputed interpretations. As for other interdisciplinary concepts in environmental research, the impression is that ESSs are created through analogies, concept transfer, and metaphorical use of notions, in empirical and model-dependent research. In the following discussion, ESSs are elaborated in the framework of the social-ecological theory with typologies of the forms of interaction and coupling between social and ecological systems.

Since the first publication by Ehrlich and Mooney (1983) explicitly using the term develops the field of ESSs-research rapidly. Costanza and Kubiszewski (2012) describe it vaguely as multi-, inter- and transdisciplinary crossing of knowledge boundaries (Costanza and Kubiszewski 2012: 21), echoing ideas from broader discourses as sustainability science, SES, and resilience research. Showing the interdisciplinary nature of ESSs does not create theoretical explanations, expresses more the expectation that interdisciplinary research may clarify the concept and the forms of ESSs.

Critical Discussion of ESSs

The following debate of ESSs starts from two critical reviews, that of Lele et al. (2013) arguing for a broadening of the knowledge base and that of Farley (2012) arguing for a shifting of the normative perspective.

1. *A critique of the concept of ESSs, regarding its conceptual inconsistencies, its instrumentalisation for specific political purposes, and the non-intended social consequences of its application refers to its economic use.* Lele et al. (2013) ask whether the concept is useful for framing research on the relationship between society and nature with its focus on biotic nature and the use of economic valuation for neoliberal policy programmes (Lele et al. 2013: 30). The authors do not seek for a social-ecological theory of nature and society to improve the concept, stop with the diagnosis of a narrow understanding of environmental problems and their causes given in ESSs. Beyond the normative and policy-related critique, their arguments can be summarised as follows:
    (a) *The progress with the term of ESSs in ecological research is threefold*: Firstly, it provides a better understanding of regulating services,

whereas provisioning services have since long been studied in applied research about forestry, fishery, land use, focusing on the goods harvested form ecosystems. Secondly, the scale of analysis expands from single ESSs to regional models including all ESSs. Thirdly, a better collaboration of ecologists and economists resulted in more careful elaboration of socially relevant variables. These advances seem somewhat vague and outweighed by the critical arguments.

(b) *The critique of the term refers to its economic bias:* ESSs become in practice monetary valuation of ESSs through the dominance of economists and policy-determined managerial interests in applying the concept. The economic valuation results in a simplified view of causes of degradation of ecosystems, based on the tautological reasoning, the benefits and their value are not known without pricing the services. Furthermore, an ecosystem process (e.g. soil erosion by streams) can generate dis-service (e.g. siltation of dams) or service (e.g. fertilisation of the floodplain). In this interpretation, the concept ignores that the benefits require investing human labour and capital and the flow of energy and material in the economy. Since the valuation of services cannot be separated from the social or socio-technical contexts of obtaining benefits, the mapping of ESSs and the transfer of benefits becomes doubtful with regard to the social consequences of ESSs. An incoherence of ESSs is seen in the anthropocentric utilitarian ethic connected with a biocentric perspective attributing services to nature, which implies a simplification of the analysis of human well-being through the choice of an economic valuation framework. These critical arguments converge to the point that ESSs are co-produced by nature, human capital, labour, and ecosystem processes, can only been valued within specific social contexts, including human agency in the production and consumption of ESSs (Lele et al. 2013: 7, 9ff, 24f).

(c) *In the economic valuation of ESSs, widespread misapplications can be found:* estimation of absolute value rather than marginal value changes, deriving global scale estimates from local studies, and double-counting of supporting and final services. More fundamental is, however, the critique of economic valuation of benefits with the argument: it becomes impossible and ethically doubtful to put monetary value on things that have intrinsic value. The conventional approach of estimating changes in economic welfare adds up benefits and costs across all individuals, regardless of the difference in their wealth. Carbon sequestration, for example, creates benefits from tree planting for the whole world, but the poor who need to collect firewood in the forests have to bear the negative consequences. Also the well-being of future generations is not sufficiently accounted for. The economic valuation of ESSs neglects the difference between choices made by individual consumers and public goods that have common-pool quality and specific ethical attributes. Altogether the authors see two major misapplications of ESSs: supporting human well-being through valuation of ESSs, but neglecting the maintenance of biodiversity; and insufficient analysis of the causes and consequences of ecosystem degradation that require other and more critical analyses of power relations and power differences. *With these arguments, economic valuation methods are discarded in favour of deliberative decision-making and critical approaches with more complex reasoning regarding the production, consumption, and unequal distribution of benefits and disadvantages from ESSs* (Lele et al. 2013: 25ff).

The critical review of the concept of ESSs by Lele et al. is devalued by their incoherent reasoning. They argue for a plurality of analytical perspectives to strengthen the concept and its coherence, but without a theoretical basis, only referring to the lack of engagement of more critical approaches as political economy, political ecology, environmental sociology, ecological anthropology, and human geography in the debate. With knowledge from these social-scientific subjects, it should become possible to show more clearly the deficits of the concept. These refer to the limits of ecological specialisation and application without epistemological, theoretical, and methodological considerations that show a much more complex reality. But the authors do not show how to reformulate ESSs or to correct the inconsistent applications resulting from the neoliberal mainstreaming of the idea. Their critique is limited to the basic points that the cooperation of conservation biologists and environmental economists is insufficient, not showing the co-production of ESSs through nature, human labour, technology, and capital. It can be assumed that broader interdisciplinary discussion will result rather in discarding of the concept than in its theoretical elaboration that would, according to the reasoning of Lele et al., become too complicated to achieve a theoretical consolidation of ESSs.

2. *An alternative interdisciplinary broadening is to connect the debate of ESSs with the sustainability discourse in search of a critical compass for its application and for shifting from efficiency- to justice-related assessment.* Although sustainability is a controversial concept and the discourse similarly disjointed as that of ESSs, it has advantages in comparison to the critique of Lele et al. The intensive critical debate of sustainable development helps to clarify and improve the concept of ESSs; and a justice-oriented perspective seems necessary to break the hegemony of environmental economics in nature conservation. The critical debate of ESSs begins with social inequalities generated and neglected with the application of ESSs in neoliberal policy programmes. Farley (2012) analyses critically the economic premises of ESSs, showing the inappropriateness of marginal analysis and monetary valuation for an ecologically oriented practice of managing ESSs. Such a practice requires similar things as in the sustainability discourse, difficult until today: to integrate social, economic, and ecological perspectives and contradicting goals.

   (a) Farley sees the advantage of an ecological debate by ecological economists and others in the goal of improved quality of life compatible with the conservation of ecosystems. For that purpose, economic institutions need to be adapted to deal with the physical qualities of ESSs and the qualities of ecosystems as adaptive and resilient systems. ESSs are not necessarily aiming at this goal which requires normative reasoning and choice among alternatives. But the logic of decision-making applied for ecosystem management opens possibilities of conscious choices between alternative objectives as monetary valuation, human quality of life, or conservation of nature because of its intrinsic values (Farley 2012: 40, 48). This view considers explicitly the relevance of ESSs for human welfare, taking into account the limits to economic growth and the limited possibilities to substitute natural resources through other natural or non-natural resources. Furthermore, the idea of resilience comes to the foreground, highlighting the possibility that ecosystems can change into alternative states which are less supporting human welfare than undisturbed ecosystems.

(b) In unfolding his critique, Farley argues that commodification of ESSs is not a consequence of the ecological concept, but of conventional economic ideas used in its application. A market-based monetary allocation fails in two decisive points: to achieve ecological sustainability and just distribution of natural resources between different social groups, countries and national economies, or generations. Farley is aware that this reasoning violates the conventional economic logic of allocation, assuming that markets always provide for efficient allocation of scarce goods between alternative forms of use and alternative user groups. Consequently, he asks whether an economy taking into account the physical and ecological qualities of resources can be market efficient (Farley 2012: 48). *With the unfolding of these ideas it becomes possible to connect several critical discourses, that of limits to economic growth, of degrowth, and of de-commodification of ESSs, for purposes of achieving sustainability in natural resource management.* Farley's arguments converge with the integration of analyses of vulnerability, resilience, and sustainability, discussed in Chap. 4, and it supports a more theoretical analysis of social and environmental limits to natural resource use that unfolds in social-ecological theory.

Similar as Lele et al., Farley discusses the economic perspective in the application of ESSs and its deficits, but goes a step further in connecting the discourses of ESSs and sustainability by showing the implications of ESSs for ecological sustainability. *Both critical reviews are compatible with the theoretical considerations in this chapter, although they do not elaborate a theoretical perspective of social-ecological analysis. This can also be said for the further debate of ESSs, where three questions are in focus*: which methodologies can be used for the valuation of services, how ESSs support conservation of nature and ecosystems, and how ESSs can be applied in practices of natural resource management?

3. *The further debates of ESSs are restricted to specialised ecological and applied research, less relevant for a theoretical elaboration of ESSs, but showing various methodological problems of managerial application of the term.*
   (a) *Methodologies for the analysis and valuation of ESSs*, especially the use of geographical information systems (GIS) and the mapping of services, show some potential strengths of the concept, supporting the use of ESSs in policies aiming at new forms of protecting nature and ecosystems. Methodological improvements of data collection and use would require theoretical discussion of ESSs and appropriate conceptual construction of services. But the methodological debate is limited to practices of conservation, assuming that ESSs is primarily for application in environmental policies and resource management. This purpose justifies methodological scrutiny, ignoring principal doubts about the use of the concept. For policies of conservation, no critical assessment of non-intended consequences is delivered with this methodological debate. Policies seem to be perceived in technocratic views of expert knowledge, supported by the terminology of policy analysis that focuses on formulating and implementing policy programmes and measures. These views ignore the controversial and power-related nature of politics and political decision-making, reducing it to the delivery of alternative options for policy decisions and finding adequate areas for protection of nature, biodiversity or ecosystems. The following arguments show the shortcomings in discussions of methodological questions of analyses of the value of ESSs.

   Schägner et al. (2013) develop a GIS-based matrix to classify studies according to the methodologies applied for mapping the value of ESSs. The advantage of GIS technology is seen in improving spatial valuation of ESSs, in improving information about trade-offs and synergies of alternative policy scenarios, and in identifying suitable locations for policy measures. However, practices of ESSs valuation differ widely in terms of spatial scope, purpose, disciplines, methodologies used to show how ESSs are supplied, and in their accounting for variation of values across space (Schägner et al. 2013: 44). For the different methods in value mapping, supply mapping of services, and spatial distribution of services, no consensus is available on which method is best for which purpose. Operationalisation and development of specific measurement procedures are discussed under the premise that the only relevant issue for the practice of biodiversity management is how to assess the value of ESSs. Still more limited is the discussion of possibilities of mapping ESSs as, for example, referred to in the EU biodiversity strategy 2020 (Hauck et al. 2013). There the critical question is that of the amount of information that can be used in the decision-making process, a reasoning that does not advance to the clarification of valuation methods. Hauck et al., being aware of the limits of valuation methods, conclude only that maps can be helpful but should be used carefully, whatever this implies. At the end, the discussion of policy instruments for measuring the value of ESSs falls back to scientists who need to clarify for the decision-makers which instruments to use. *More critical questions of methodology in the analysis and valuation of ESSs that are missing in both studies are that how methodologies connect to theoretical framing, reflection, and synthesis of knowledge on services.* With such questions, the discussion goes beyond the valuation of services which dominates in the methodological debate, reducing the use of scientific knowledge to that what can be measured with different methods of research and valuation.

(b) *The debate of ESSs and conservation of nature and ecosystems* comes closer to the core of the problem of constructing and classifying ESSs as water purification, recreational opportunities, erosion protection, soil regeneration, and so on. Simpson (2011: 16ff) discusses this point referring to the economic applications and the assumptions underpinning these, also mentioning dis-amenities that ecosystems generate for humans. Among the assumptions that enter in strategies to improve, conservation through economic valuation of ESSs and payments for these is one, which Simpson doubts: that about the incapability of local and poor communities to maintain these services. His point is that their welfare may not be improved when conservation is for the benefit of an international conservation community interested in the maintenance of biodiversity, loading the burdens of that on the shoulders of these communities. Also Simpson's critical assessment of the ESSs term does not show an alternative to the concept, nor a way out of the contested and doubtful assumptions in its application for improving conservation strategies. Rather this discussion allows for the conclusion that the perplexities of the concept in economic reasoning, and the simple solution mechanism of payments for ESSs, cause its problematic quality and practice. This has, with more detailed arguments, been formulated in the critique of Lele et al. and Farley discussed above.

A recent attempt to summarise the critical arguments against and for the use of the concept of ESSs by Schröter et al. (2014) evoked more attention. The authors show a series of contradicting interpretations of ESSs between which choices are required:

- Being anthropocentric or going beyond instrumental values
- Promoting an exploitative human–nature relationship or reconnecting society to ecosystems
- Conflicting with biodiversity conservation objectives or being complementary to these
- Focussing on economic valuation or accepting other values
- Promoting commodification of nature or rejecting market-based instruments
- Vagueness of definitions and classifications seen as deficit or as requiring transdisciplinary collaboration
- The controversy about the normative nature of the concept, whether all outcomes of ecosystem processes are desirable or not

Having driven the debate to the point of normative assumptions, a major weakness of the concept is laid open that cannot be removed with the help of valuation methods or the formulation of contrasting interpretations. The question emerges *whether the concept should be given up because no consensus can be found between incompatible values and interests guiding the interpretation. Alternatively, it may be helpful to transfer the normative debate into a theoretical one, showing the implicit or explicit views in connection to different theoretical arguments and lifting the debate to a theoretical one that can connect more theoretical and disciplinary perspectives.*

(c) *The concept of ESSs in the practice of political decision-making* is discussed by McAfee (2012) regarding the commercialisation and trading of services. These practices gave rise to controversies in the ecological discourse and in international policies, whether the idea of payments for ESSs or selling nature to save it has the intended effects. The author concludes that *ten years of experience with payments of ESSs shows, market-efficiency criteria are in conflict with poverty-reduction priorities, reinforcing a conflict between development and conservation goals in projects framed by the "asocial logic" of neoclassical economics.* Application of the market model in international conservation policy shows a redistribution of wealth from poorer to wealthier classes and from rural regions to centres of capital accumulation in the Global North (McAfee 2012: 105). This critique was implicit in several of the analyses and reviews discussed above, but it may not become effective unless it is elaborated in theoretical perspectives and arguments that show ESSs in the broader context of interacting social and ecosystems. Without unfolding such contextual perspectives and synthesising the arguments in the social-ecological theory of nature and society, the critique of commercialisation and commodification of ESSs by the World Bank and the United Nations may remain weak: an ethical reasoning against commercial misuse of payments for ESSs that refrains from a more consequent critique of the policy- and decision-making processes. Critical reasoning can be stronger when the implications of market-based strategies of conservation are shown empirically and through coherent theoretical arguments about the non-intended, power-based, socially and environmentally destructive consequences of commodification of nature.

To launch this critical perspective, arguments found in the critical debate need to be systematically discussed and assessed, as some of the reviews attempted. McAfee also refers to a variety of debates that showed attempts to operationalise ESSs. These attempts include elaboration of methods, construction of indicators, and quantification of ESSs, in knowledge practices of classifying, modelling, mapping, assessing, and valuing services, using GIS-based information. Such practices turn finally out to be attempts of methodological standardisation, of opening a field of applied research by a discourse community with vested interests, closed off against more critical interdisciplinary discussion, as McAfee described. After all discussion of ESSs, it remains still to be done what the social-ecological theory of nature and society requires: to reinterpret the technocratic concepts and to integrate them in an interdisciplinary theory, not only in explanatory functions, also for supporting the theory-based critique of environmental policies and managerial practices.

The main theoretical argument to be developed in the emerging social-ecological theory is beyond the critical review of manifold shortcomings and inconsistent applications of ESSs: a coherent ecological perspective requires a theoretical formulation of the concept and a classification of ESSs that is free of arbitrary ascriptions of services to ecosystems and shows how ESSs can be maintained without economic distortions and privileging of powerful actors, for the benefit of all, including disadvantaged groups. This would require a description of benefits of ESSs in ecological and social terms, renouncing to monetary valuation and resisting commodification of nature. Further knowledge from different disciplinary and interdisciplinary sources can help to achieve the ecologically justified targets of conservation by using ESSs, targets that are missed with "quick and dirty" methods, and strategies that support political and economic misuse of the idea.

Conclusion: Theoretical Elaboration of SES and ESSs

The discussion of SES and ESSs shows that empirical research has not brought a clarification and validation of new concepts, rather continuing disputes about their application and how to achieve the widely accepted goal of protection of nature or maintaining of functions of ecosystems. In the practice of environmental policy and resource management, it is more the power of institutional actors than the quality of scientific arguments that counts. The more complex goal of transformation to sustainability needs to be rethought with the help of more elaborate concepts and theories. For this transformation, the knowledge and research on risks, vulnerability, resilience, and governance of SES or ESSs, delivers only preliminary, often confusing, and contradicting results. The knowledge dilemmas cannot be dissolved through more and new empirical research which implies more research in the same forms, disciplinary specialisations, established institutions, and academic power relations that demonstrated their inefficiency already in earlier environmental research and policy making. Also in the new research areas, as research on resilience and ESSs, such scientific power structures are recreated rather than changed. An alternative seems more systematic use, critical review, combination, and synthesis of available knowledge and the development of inter- and transdisciplinary knowledge cultures. Some possibilities to advance in that direction became visible with the discussion of SES and ESSs.

1. The establishment of a new term of SES in ecological research was not thought as a traditional form of concept development: it signalled in ecology the effort to develop a more interdisciplinary research culture, which did not happen so far. New schools and epistemic communities with practices of specialisation-based knowledge enclosure and citation networks developed in competition with others. Attempts to integrate SES analysis in the theory of nature and society showed at least what would be necessary to transform the notion from a heuristic to a theoretical concept that can help to integrate knowledge from different disciplines and research areas and reflect critically empirical research and knowledge production. *The contours of the SES concept develop with the combination of social-scientific and ecological concepts and theories, and their reinterpretation in interdisciplinary knowledge syntheses:*

    – In social-ecological theory, SES include not only geo-biophysical systems with actors and institutions connected, but varying forms of coupled social and ecological systems that show how in modern industrial society nature and society interact and create successful forms of sustainable development or maladaptive change.

- SES cannot be conceived of in general and standardised forms, but as historically and spatially specific systems that develop with inter- and transdisciplinary research and knowledge integration and strategies that support in practice the networking of institutions and policies across several spatial and temporal scales.
- SES can, with the help of classifications discussed in this chapter, be integrated in the framework of a social-ecological theory of interacting nature and society which includes further knowledge than that produced in ecological research.
- The relevance of theoretically based SES analysis for practices of natural resource use and conservation is to develop alternatives to commercialisation of nature or ESSs, strategies supporting fair sharing of benefits from ESSs. This requires complicated and conflict-rich debates, negotiations, and deliberations to transform powerful vested interests of privileged groups and actors, nationally and internationally, combatting of unequal appropriation of benefits from ESSs and conservation, criticising powerful international conservation communities that presume to act in the global common interest but reinforce established power structures in the economic world system.

2. *The integration of the term of ESSs in the social-ecological theory does not develop through a new theory emerging from empirical research. Its theoretical contours develop through the elaboration and critique of the concepts and assumptions with which ESSs are constructed.* The classification of services and their connection to human well-being show incoherent, contradicting, and insufficiently reflected knowledge use that is hardly corrected through empirical research and require explanatory purposes to specify the forms of interaction and coupling between social and ecological systems. The new fields of applied research on SES and ESSs did not develop towards interdisciplinary and theoretical knowledge integration, rather as new specialised fields of research. Alternative concepts, improved classifications of services, and changed practices of resource management remain to be done in the further elaboration of the theory of nature–society interaction discussed here, or other forms of such theories. Main requirements in the clarification of ESSs are

- to show the co-production of services through "cooperation" of nature and society, and the transformation of ESSs through their use by humans;
- to deal with the disjointed practices of empirical research and postponed critical review of ESSs strategies that work with economic valuation and pricing of services;
- to develop alternatives to the commercialisation of nature by strengthening ecological perspectives of conservation and ecosystem protection.

3. The social-ecological theory to develop offers possibilities to construct ESSs in other theoretical perspectives than that of self-regulation of ecosystems that is a weak theoretical basis to understand the co-production of ESSs through nature, human labour, knowledge, and technologies in coupled and interacting SES. In the analysis of ESSs, economic and ecological forms of theorising are prevailing that resulted in contrasting interpretations. Further perspectives and theoretically reflected concepts can help to change the blocked discussion and the practice of mainstreaming and commercialising the services through payments for them. When the critique of economic valuation of ESSs is connected with theoretically developed critiques of growth and limits of resource use, and ideas for de-commodification of nature—it directs towards new strategies for social-ecological transformation that go beyond public policies and environmental governance. Critically reflecting the analyses of nature–society interaction in various social and ecological theories is a way to identify counterproductive ideas appearing in the ecological discourse and in practices of natural resource use and management, thus supporting the creation of capacities and agency for the global transition to sustainability.

References

Brondizio, E. S., Ostrom, E., & Young, O. R. (2009). Connectivity and the governance of multilevel social-ecological systems: The role of social capital. *Annual Review of Environment and Resources, 34*, 253–278.

Bruckmeier, K. (2014). Problems of cross-scale coastal management in Scandinavia. *Regional Environmental Change.* doi:10.1007/s10113-012-0378-2.

Costanza, R., & Kubiszewski, I. (2012). The authorship structure of 'ecosystem services' as a transdisciplinary field of scholarship. *Ecosystem Services, 1,* 16–25.

Daily, G. (1997). *Nature's services: Societal dependence on nature.* Washington, DC: Island Press.

Ehrlich, P., & Mooney, H. (1983). Extinction, substitution, and ecosystem services. *BioScience, 33*(4), 248–254.

Farley, J. (2012). Ecosystem services: The economics debate. *Ecosystem Services, 1,* 40–49.

Fischer Kowalski, M., & Haberl, H. (Eds.). (2007). *Socioecological transitions and global change. Trajectories of social metabolism and land use.* Cheltenham: Edward Elgar.

Fischer-Kowalski, M., Haberl, H., Hüttler, W., Payer, H., Schandl, H., Winiwarter, V., et al. (1997). *Gesellschaftlicher Stoffwechsel und Kolonisierung von Natur: Ein Versuch in Sozialer Ökologie.* Amsterdam: G+B Verlag Fakultas.

Folke, C. (2006). Resilience: The emergence of a perspective for social-ecological systems analyses. *Global Environmental Change, 16,* 253–267.

Gerst, M. D., Raskin, P. D., & Rockström, J. (2014). Contours of a resilient global future. *Sustainability, 6*(1), 123–135. doi:10.3390/su6010123.

Gómez-Baggethun, E., de Groot, R., Lomas, P., Montes, C. (2010), 'The history of ecosystem services in economic theory and practice: from early notions to markets and payment schemes', *Ecological Economics,* 69 (6): 1209–1218.

Gómez-Baggethun, E., de Groot, R., Lomas, P., Montes, C. (2010), 'The history of ecosystem services in economic theory and practice: from early notions to markets and payment schemes',*Ecological Economics,* 69 (6): 1209–1218.

Hauck, J., Görg, C., Varjopuro, R., Ratamäki, O., Maes, J., Wittmer, H., et al. (2013). 'Maps have an air of authority': Potential benefits and challenges of ecosystem service maps at different levels of decision making. *Ecosystem Services, 4,* 25–32.

Herout, P., & Schmid, E. (2015). Case study. Doing, knowing, learning: Systematization of experiences based on the knowledge management of HORIZONT3000. *Knowledge Management for Development Journal, 11*(1), 64–76.

Hornborg, A., & Crumley, C. L. (Eds.). (2006). *The World System and The Earth System: Global socioenvironmental change and sustainability since the neolithic.* Walnut Creek, CA: Left Coast Press.

Kulig, A., Kolfoort, H., & Hoekstra, R. (2010). The case for the hybrid capital approach for the measurement of the welfare and sustainability. *Ecological Indicators, 10,* 118–128.

Lele, S., Springate-Baginski, O., Lakerveld, R., Deb, D., & Dash, P. (2013). Ecosystem services: Origins, contributions, conditions, pitfalls, and alternatives. *Conservation and Society, 11*(4), 343–358.

Lenski, G. E. (2005). *Ecological-evolutionary theory: Principles and applications.* London: Pluto Press.

McAfee, K. (2012). The contradictory logic of global ecosystem services markets. *Development and Change, 43*(1), 105–131. doi:10.1111/j.1467-7660.2011.01745.x.

Millennium Ecosystem Assessment. (2005). *Current state and trends.* Global Assessment Reports, Vol. 1. www.millenniumassessment.org

Olsson, P., Gunderson, L. H., Carpenter, S. R., Ryan, P., Lebel, L., Folke, C., et al. (2006). Shooting the rapids: Navigating transitions to adaptive governance of social-ecological systems. *Ecology and Society, 11*(1), 18. http://www.ecologyandsociety.org/vol11/iss1/art18/.

Ostrom, E. (2007, February 15–19). *Sustainable social-ecological systems: An impossibility?* 2007 Annual Meeting of the American Association for the Advancement of Science, Science and Technology for Sustainable Well-Being, San Francisco.

Rockström, J., Steffen, W., Noone, K., Persson, Å., Stuart Chapin, F., III, Lambin, E. F., et al. (2009). A safe operating space for humanity. *Nature, 461*, 472–475.

Schägner, J.-P., Brander, L., Maes, J., & Hartje, V. (2013). Mapping ecosystem services' values: Current practice and future prospects. *Ecosystem Services, 4*, 33–46.

Schneider, N. (1998). Erkenntnistheorie im 20. Jahrhundert: Klassische Positionen. Stuttgart: Reclam

Schröter, M., van der Zanden, E. H., van Oudenhoven, A. P. E., Remme, R. P., Serna-Chavez, H. M., de Groot, R. S., et al. (2014). Ecosystem services as a contested concept: A synthesis of critique and counter-arguments. *Conservation Letters.* doi:10.1111/conl.12091.

Simpson, R. D. (2011). *The "Ecosystem Service Framework": A critical assessment.* The United Nations Environment Programme: Division of Environmental Policy Implementation, Working Paper Series Ecosystem Services Economics (ESE), paper No. 5. Nairobi.

Wilson, C. M., & Matthews, W. H. (Eds.). (1970). *Man's impact on the global environment. Study of Critical Environmental Problems (SCEP).* Cambridge, MA: MIT Press.

Young, O. R., Berkhout, F., Gallopin, G., Janssen, M., Ostrom, E., & van der Leeuw, S. (2006). The globalization of socio-ecological systems: An agenda for scientific research. *Global Environmental Change, 16*, 304–316.

6
Knowledge Transfer Through Adaptive Management and Environmental Governance

This chapter discusses two managerial approaches, adaptive management and environmental governance. The forms of exchange and transfer of knowledge between ecology, policy, and practices of natural resource management are analysed and reformulated in the context of the interdisciplinary theory of nature–society interaction. Both frameworks are critically reviewed in similar perspective as the concepts of SES and ESSs in the preceding chapter:

- To show how they can be modified and used in the social-ecological theory that works with various knowledge practices for synthesis, assessment, transfer, and application of scientific knowledge
- To show how transfer and sharing of knowledge between different knowledge spheres and practices become possible

Adaptive management develops from ecosystem research whereas environmental governance is a more diffuse set of ideas discussed in the social sciences, in political ecology, and in environmental policy. The knowledge practices found in both approaches show the changes of conventional forms of knowledge transfer in environmental research. The practices include interdisciplinary frontier research, blending of basic and applied research, and the governance of ignorance, insecurity, and risk, connecting with the research on vulnerability, resilience, and sustainability. In the perspective of the social-ecological theory of nature and society, both frameworks are interesting for their use in integration and transfer of knowledge for sustainable resource management: which forms of societal transformation to sustainability through local and multi-scale governance approaches do they support?

Adaptive management exemplifies the difficulties with interdisciplinary strategies of knowledge transfer. The approach is formulated with few and simple principles for ecosystem development in situations of ignorance, where new knowledge cannot be provided through research. Practical experiments in resource management and variation of management rules, trial-and-error-based procedures, steps of joint learning, and continuous improvement of governance systems are seen as ways to navigate towards sustainable resource management.

Environmental governance has some vague principles: focusing all policy, management, and action processes on the environment; organising economic processes according to ecological principles of a circular economy; and connecting people to ecosystems. Environmental governance aims at embedding communities and societies in the environment or in nature (Speth and Haas 2006; Delmas and Young 2009; Evans 2012). Global environmental governance is oriented to global resource flows and possibilities of managing global environmental change.

The following discussion shows that adaptive management and environmental governance are not standardised approaches and develop slowly through application, reflection, and methodological and theoretical refinement for which social-ecological theory provides critical reviews and some framing and bridging concepts. For both practices, empirical research on management and governance helped to formulate them, but simultaneously the limits of research as knowledge practice for governance became visible.

Knowledge Integration and Application in Public Policy and Resource Management

Duit et al. (2010: 363) describe a new dilemma of resource management as increasing interconnection of the world through information, trade, and technology and fragmentation and lack of coordination of decision-making and institution building. Decentralisation, growth of public–private partnerships, and growing influence of NGOs are seen as signs of fragmentation of policies and decision-making. It is, however, unclear, whether the consequences of the processes seen as fragmentation are only such of reduced coordination and integration of policies and resource management. Decentralisation, networking, and nesting of local management systems can also be seen as broadening of the repertoire of collective action, creating new possibilities for knowledge integration that support the transformation to sustainability. In both interpretations, knowledge problems come to the foreground that are so far neglected in the management and governance debates. These problems are discussed in the following sections to prepare the more detailed analysis of adaptive management and environmental governance in the perspective of social-ecological theory. The following five arguments describe *major problems of knowledge integration for purposes of natural resource management:*

1. *Methodological difficulties of knowledge integration and synthesis:* A renewal of resource management requires methodologies of inter- and transdisciplinary knowledge integration. Methods for combining, integrating, and synthesising knowledge from empirical research and theoretical analyses across disciplinary boundaries are poorly developed. The communication and knowledge transfer in applied research and resource management practices happen often in oral forms of discussion and synthesis, working with heuristic principles from management science. Although interdisciplinary knowledge use is practised in environmental research since many decades, rules and criteria for combination, integration, and transfer of social- and natural-scientific knowledge are not systematically investigated and discussed. As a consequence of these methodological deficits, the knowledge synthesis

and application for environmental management and governance suffer from the selectivity, the casualty, and the limits of knowledge exchange in the cooperation between scientists and practitioners. The practice of knowledge synthesis in environmental research focusses on knowledge from empirical studies; in the practices of natural resource management, different forms and sources of knowledge are used, but methodological guidance and structuring of knowledge application is low.

Tengö et al. (2014: 1) describe for the governance of ecosystems the knowledge integration and synthesis as creating synergies across knowledge systems and assume that the use of different knowledge forms can create new insights through complementarities, enriched understanding, and joint assessment of knowledge. Synergies between knowledge systems are discussed in terms of integration, cross-fertilisation, and co-production of knowledge. The knowledge processes studied are limited in scope and purpose: integration of local and indigenous knowledge in local case studies. The purpose of knowledge integration is seen as creating a richer picture and understanding of problems. With these vague ideas, the principles and problems of inter- and transdisciplinary knowledge syntheses cannot be described in methodological or epistemological terms. The example shows a form of knowledge integration discussed with insufficient epistemological concepts, more as a side aspect of empirical research and not with regard to interdisciplinary theory as part of knowledge integration.

2. *The practices of "muddling through" in natural resource management:* In applied research, the progressing specialisation of disciplinary research results in preliminary and incomplete syntheses and knowledge practices in forms of enclosures when scientific and epistemic communities protect their competing knowledge claims and domains. McAfee (2012) describes such practices of selective and competitive knowledge use with the proliferation of preliminary models, heuristics and inexact methods, insufficient analysis of intersystemic relations in SES, and ignoring of theoretical knowledge. Ways towards interdisciplinary concepts and frameworks and to improved conceptual models and knowledge integration in natural resource management are not

opened from the natural- and social-scientific disciplinary knowledge practices. A vague idea of scientific management strategies for conservation and natural resource use guides the integration of empirical knowledge in adaptive management. This idea is more misleading than developing the approach through interdisciplinary knowledge synthesis. The social-scientific knowledge that is used to reformulate adaptive management under the name adaptive co-management and as governance is limited to some forms of cooperation, power sharing, and adaptation to the institutional context.

The insufficiently developed concepts of SES and ESS and the limited approaches to knowledge transfer indicate the necessity to develop methods along the *knowledge chain of generation, integration, diffusion,* and *transfer and application of knowledge.* In all these phases of knowledge generation and application, the complexity of systems and processes in SES creates main epistemological and methodological problems. The development of theory-based knowledge synthesis requires more investigation and reflection than in interdisciplinary research and practice, where synthesis is often intuitively done in heuristic practices without epistemological guidance. Multi- and interdisciplinary knowledge integration is especially important in environmental research. New areas of research such as climate change and new approaches and frameworks in ecological research and natural resource management as adaptive management and environmental governance are confronted with many forms of interaction and coupling between social and ecological systems.

3. *The institutional crisis of natural resource management:* Knowledge strategies to deal with insecurity, risks, and future resource use problems are not advanced in management research that needs to deal continually with the cognitive problem that knowledge is about the past whereas the future can only be dealt with in the form of decisions that influence possible paths of development. Approaches discussed in this chapter and further ones as that of Ostrom's (2009) multi-tier framework show that complexity and insecurity in resource management research and practice are approached in simple methods of knowledge integration, mainly using knowledge from empirical research, case studies, and conceptual frameworks. Such methods may be sufficient

for limited purposes, integrating data from local case studies or in local management. Further forms develop in inter- and transdisciplinary knowledge integration: more systematic knowledge syntheses for research, and for the practice of resource management forms of networked, nested, or multi-scale management systems.

Insufficient knowledge and data, insecurity, and ignorance are known problems in environmental research. The approaches discussed here get rid of the methodological problems through the shifting of knowledge synthesis from science to practice, from research to resource management and policies. There the syntheses are often done in managerial deliberations and in attempts to negotiate and match the interests and knowledge of the actors pragmatically in management processes that are continuously under time and decision pressure, work without formalised methods, in oral forms, insufficiently documented, and evaluated. Such syntheses reduce the aspirations of knowledge integration, dictated by the constraints of the management situation and the practical purposes of knowledge application. The situation is similar to that described as postnormal science (Funtowicz and Ravetz 1993). The practice can be justified with the necessities of achieving consensus and joint decisions, therefore reducing methodological validation and quality control. In the cooperative and participatory resource management developed approaches, mainly in agricultural policies and international development cooperation, such imperfection is methodologically justified, for example, in the approaches of the rapid and participatory rural appraisal (Chambers and Conway 1991; Chambers 1994). The risks of managerial decisions are, in these practices, minimised through the local limits of management, the limited complexity of SES to manage, and the relatively transparent situation, that make corrections of management practices easier. It is, however, doubtful whether the toolbox of participatory methodologies offers sufficient methods for the complexity of problems to deal within multi-scale and sustainable resource management with its complex webs of research and management.

4. *The complexity crisis in social-ecological research—limits of knowledge synthesis from local case studies:* In the research of Ostrom et al., only preliminary forms of knowledge synthesis are developed, based on the

building of databases from local studies and the comparison of cases. Integration of knowledge from comparative empirical research, cognitive bridging strategies in form of heuristic frameworks, and some principles for guiding resource management with scientifically backed ideas are the main ideas guiding knowledge synthesis (see Chap. 4, the Appendix). More and other knowledge than from local studies is required for dealing with complex systems, resource use dilemmas, and wicked problems. The empirical and case-study-based research supports inductive generalisation, rather codification of knowledge, and helps to build gradually large databases for empirical research on resource management. But it is much less discussed how to codify this knowledge in overarching theoretical frameworks and to apply it in the practices of resource management. Ostrom (2007), asking for a science of complexity to deal with the much more complex multi-scale resource use problems, formulates only the necessity, with arguments that much more knowledge is required to deal with the complexity of global problems. She does not yet show how this complexity science can be realised by epistemologically and theoretically elaborated interdisciplinary perspectives of knowledge generation and application. Adaptive management and environmental governance are approaches that require for their further development such enriched and systematic empirical and theoretical synthesis. They are constructed as overarching approaches, which develop through bridging concepts mediating between research and policy or governance, insofar similar to the interdisciplinary and applied research on vulnerability, resilience, and sustainability (see Chap. 4) that is connected to these approaches; they can be applied in many fields of environmental policy and natural resource management, for example, for conservation of nature and natural resources, for ecosystem-based management, for renewal of energy regimes, and for searching local ways to adapt to global climate change.

5. *The challenges of transdisciplinary knowledge integration:* Knowledge integration becomes more difficult when scientific knowledge is connected with local, practical, and indigenous knowledge for which methodological and epistemological criteria differ from established scientific methodologies. Knowledge is usually classified in broad and

inexact forms as scientific, managerial, local, or practical and tacit knowledge. When non-scientific knowledge is adopted in environmental research, it is in exceptional cases and for "privileged" forms of knowledge that are directly relevant for processes of natural resource use. A paradigmatic example is local ecological knowledge, for example, in fisheries management Maes 2008. The utility of local ecological knowledge as complementary to scientific knowledge is evident, but its use is limited, complementing the dominant forms of scientific knowledge; it seems practically impossible to establish new knowledge cultures outside the dominant scientific knowledge and without control scientists. To be applied in research and resource management, non-scientific forms of knowledge need to be renewed as part of scientific knowledge production, not in their original forms as local or practical knowledge. The practices of new knowledge production called "mode 2" and "transdisciplinarity" are diffuse with regard to formulating criteria for knowledge integration. It seems that the development of these and other forms of knowledge integration need to work with more differentiated concepts of knowledge than the abstract forms mentioned above.

These reflections of difficulties in inter- and transdisciplinary syntheses of knowledge for purposes of natural resource management indicate a cognitive crisis in environmental research and management. The difficulties to deal with complex intersystemic processes in SES management evoked a crisis to which adaptive governance is a first reaction, similar to the approach of frontier research with combinations of interdisciplinary, basic, and applied research (European Commission 2005). Ideas of frontier and participatory research, stakeholder involvement, and transdisciplinarity do not yet show solutions where different forms of knowledge can be integrated with efficient and scientifically validated methods. Rather, they institutionalise the cognitive crisis of environmental research and resource management as a permanent crisis by way of continued use of preliminary concepts, models, frameworks, and inexact methods of knowledge search in situations where sufficient and sufficiently exact knowledge cannot be found.

Adaptive Management and Governance

The scientific knowledge and principles developed from management and planning research have since long supported lower ambitions in management because of the limited availability of knowledge or the limited capacity of humans to process knowledge in decision-making processes. This is articulated in different forms, for example, by Simon (1991) in the theorem of bounded rationality, in the insights that complexity of problems, time pressure for decision-making, and limits of applicable knowledge set limits to the practices of knowledge use. This was known long before ideas of postnormal science or adaptive management were formulated. For natural resource management, the limits and constraints were reflected again by Ludwig (2001) in the conclusion that adaptive management is management for the era when management is over. The formulation describes a situation where limits of knowledge, uncertainty, and ignorance enlarge the difficulties in natural resource and ecosystem management.

Principles of adaptive management and the later version of adaptive governance are derived from ecological knowledge and research. With adaptive management begins a critical reflection of the application of ecological knowledge, although not explicitly discussed in epistemological and methodological terms. The debate indicates growing awareness of the necessity of social-scientific and practical knowledge in natural resource management. It results in the support of cooperation of scientists, resource managers, and resource user groups. The awareness of knowledge limits and of the necessity of cooperation of different actor groups can be interpreted differently

- as a high ambition of integrating and synthesising different interests and knowledge forms to meet the difficulties of sustainable resource management, or, to the contrary,
- as a low ambition, describing an emergency solution in the sense that participation, cooperation, negotiation, and knowledge integration arrangements are required, because no form of knowledge, including scientific knowledge, is sufficient to deal with the complexity and uncertainty prevailing.

Arguments for both interpretations can be found; further clarification of the new management strategies and their application is necessary.

In the following discussion, adaptive management is described in its development and as a preliminary solution of to the continuous management problems of dealing with lack of knowledge, risks, and unforeseeable events in decision-making about the use of natural resources. With the critical discussion of the approach in the context of social-ecological theory, the intention is not to adopt or reject the ideas of adaptive management but to find out *how adaptive management can be developed, modified, and complemented through ideas from other knowledge sources and theories* that are relevant for interdisciplinary knowledge syntheses in environmental research and natural resource management.

Development and Critique of Adaptive Management and Co-management

Adaptive management starts from a conventional understanding of management requirements with the assumption that natural resource management is a form of scientific management. This assumption implies that knowledge can be transferred from ecological research to natural resource management. Resource managers are defined in similar ways to researchers in their practices of seeking information, reviewing knowledge, and dealing with questions of insecurity, risk, and ignorance. Adaptive management, seen in the context of sustainable development, reflects the social and environmental change in past decades and the complexity of problems that limits the formulation of "sound" methodological principles for resource management. Resource or ecosystem management cannot be improved further through scientific knowledge or research, but through controlled policy and management experiments, variation of management rules, iterative decision-making, monitoring and improving management systems through learning, and opening the management process for involvement of practitioners and other stakeholders in decision-making. These forms of seeking new knowledge sources and knowledge generation processes differ from the ideas discussed in cybernetics about dissolving uncertainty, Ashby's "law of requisite variety" or good regulator theorem, and Beer's "variety absorbs variety" (Beer 1979: 286). Adaptive management develops from ecological research and adopts more recent ideas of augmentation of managerial knowledge and improving management practice through social processes of cooperation.

In deriving management ideas from knowledge about ecosystem development, adaptive management faces difficulties when dealing with dynamics of social systems and social-scientific knowledge. This shows the discussion by Stringer et al. (2006: 1) with the discussion of dissolving uncertainty by use of knowledge from stakeholders, approaching sustainable resource management through cooperation, multi-directional information flows, collective learning, and flexible ways of managing the environment and natural resources—methodologically vague ideas of knowledge generation and integration. Adaptive management is stuck in difficulties of dealing with the complexity of interacting SES and system processes, reacting with reductive knowledge strategies and simplifications, leaving science. Other forms of seeking new knowledge for adaptive management would be possible: for example, ideas formulated with Ostrom's (2009) multi-tier framework, or ideas developing with the social-ecological theory of nature–society interaction that show other forms of dealing with complexity. Looking for new ways to develop adaptive management may be necessary: the knowledge from ecological research is limited and insufficient for dealing with complexity of social systems that influence natural resource management.

The stakeholders in natural resource management—political decision-makers, natural resource managers, governmental organisations, resource uses groups, enterprises, and environmental movements or NGOs—are directed in their practices more through their interests, goals, political and social attitudes, trust and mistrust, world views and relations of competition, conflict, and power asymmetries, not primarily through knowledge search and use. It can be asked, what is expected from new relationships, multi-directional information flows, joint learning, and new knowledge creation in complex situations of resource use? The answer by Stringer et al. (2006) is, to develop flexible ways of managing the environment and reformulate the dilemma without advancing the generation of solutions. In the debate of adaptive management, although opening the management process socially, not much is said about a major problem in collective resource management: the social and conflict dynamics that appear with the involvement of stakeholders. It is assumed that these dynamics can be managed with ecological knowledge and some heuristics and rules of communication and cooperation. The rules that Ostrom formulated for sustainable resource management (see Chap. 4, the Appendix) are more targeted and concrete; they include collective learning in difficult situations, situations described in the debate of postnormal science by a continuous lack of knowledge, existence of conflicting values and interests, and time pressure and deficits of coordinating and integrating processes and institutions. Continuous disputes, controversies and conflicts, and distrust and power struggles are consequences of these constraints. Moreover, the role of scientists in the management process is unclear when they can no longer take their conventional role as researchers who produce knowledge but refrain from decision-making. Through involvement in knowledge transfer, negotiation, and cooperation, scientists become stakeholders in resource management; this situation is not sufficiently reflected in the new resource management strategies.

The development and improvement of adaptive management shows three overlapping variants aiming at resource management in complex ecosystems under conditions of insecurity and lack of knowledge:

Adaptive management includes two schools, differentiated by McFadden et al. (2011) as ecological resilience-experimentalist school (discussed in the following) and decision-theoretic school with simpler models aiming at identification of management objectives through stakeholder involvement.

Adaptive co-management is a variant of adaptive management where the procedures of power sharing and cooperation are more explicit and dealt with in the management process.

Adaptive governance is a renewal of the ideas of adaptive management, according to Folke et al. (2005), addressing the broader social contexts of ecosystem-based management, including rules, conditions for collective action, and coordination.

The debate of adaptive management brought advances in ecological management thinking in several aspects:

1. *The basic ideas of adaptive management* as summarised by Allen et al. (2011: 1340) originated in a variety of disciplinary and interdisciplinary sources that are merged in an approach to natural resource management. The sources include practical experiences from business management, experimental science, systems theory, industrial ecology, and fisheries management. Holling (1978) created the new idea of adaptive management in the context of his resilience thinking. Adaptive management as part of strategies for building resilience requires from resource managers to observe thresholds that can change the state of the managed system. It seems difficult to maintain ecological systems in a favourable state for all potential resource users—but the exclusion of certain users creates problems as well. *Dynamic management of resilience through management experiments, enhancing collective learning and reducing uncertainty, became the guiding ideas of adaptive management.* Walters developed the approach further through mathematical modelling; he tried to bridge the gap between science and practice with designed experiments to reduce uncertainty. Thus, adaptive management appeared as the process of defining and bounding the resource management problems, using models of system dynamics, supporting joint learning and identifying possible sources of uncertainty. Policies should be designed to continue resource management while improving it through experience and learning. These ideas show a combination of abstract principles and heuristics to create pragmatic solutions and to dissolve uncertainty in natural resource management.
2. *The development of adaptive management and its changing ideas* described by Allen et al. gives the impression of a broadening of the interdisciplinary perspective with knowledge from several disciplines and research areas. This is not always confirmed through the debates in adaptive management. The ecological knowledge and thinking that characterises the approach shows that adaptive management cannot deal with all complexity problems (see below, Table 6.1) with its limited and selective set of management principles. Folke et al. (2005) show that adaptive management and governance emerged in the context of ecosystem-based management that brought also the concept of resilience; *adaptive management appears as a way to manage resilience.*

Table 6.1 Success and failure of adaptive management

1. *Strengths of adaptive management*
   Allen and Gunderson (2011: 1384) describe the situations where adaptive management is applicable: scientific uncertainty, availability of resources for experimentation with multiple treatments, competing hypotheses that are finite and testable, and leadership being able to overcome vested interests of self-serving stakeholders blocking experimentation.
   Allen et al. (2011: 1344) summarise the strength of adaptive management as recognition and confrontation of uncertainty while continuing, not precluding managerial actions, thus fostering resilience and flexibility to cope with an uncertain future, developing management approaches to deal with inevitable changes and surprises.

2. *Factors causing failure of adaptive management*
   (a) Allen and Gunderson (2011: 1381 f) describe the lack of stakeholder engagement, difficulty of experiments, inadequate reactions to surprises, too complex situations where adaptive management may fall back to following the original process rules, obstruction of action so that learning and discussion remain the only ingredients, learning not being used to modify policy and management, avoiding hard truths by conducting small experiments circumventing critical but controversial management challenges, lack of leadership and direction in the process, and focus on planning, not action.
   (b) Allen et al. (2011: 1341f) summarise additional causes of failure that converge to imperfect practices of adaptive management for natural resource management decisions: a lack of clarity in definition and approach; a paucity of success stories on which to build; management, policy, and funding that favour reactive rather than proactive approaches to natural resource management; failure to recognise the potential for shifting objectives; and failure to acknowledge the social source of uncertainty.

3. *Lack of conceptual and theoretical clarity*
   With the points above, Allen et al. (2011: 1341) touch theoretical deficits of the approach that oscillates between learning from the failure of prior management approaches and learning through feedback mechanisms (for continuous improvement of management practices with scientific principles). Adaptive management needs to address simultaneously the needs, interests, and the knowledge of scientists, managers, and other stakeholders that require further theoretical elaboration through interdisciplinary knowledge syntheses.

(continued)

Table 6.1 (continued)

4. *Deficits of application and implementation* Allen et al. (2011: 1344): The approach is not a panacea for dealing with many "wicked problems" that are on the agenda of global environmental policies now. Adaptive management does not produce easy answers, and is only appropriate for specific natural resource management problems where both uncertainty and controllability are high. In situations of high uncertainty and low controllability scenarios seem more appropriate. Furthermore, adaptive management is not applicable when natural resources cannot be sufficiently controlled through management or the system to manage is too complex as, for example, the climate system and global climate change: here, adaptive management can only help to mitigate some impacts such as shifting distributions of plants and animals, or changes in local availability of resources.

Sources: Quoted in the text

The social dimensions of adaptive management, co-management, and governance come more to the forefront in later discussions. Tyre and Michaels (2012) discuss forms of social and natural uncertainty that affect the management situation. In this critical analysis, the methodological profile of adaptive management and the view of complexity are still limited: reduced to illustrative examples from some case studies, showing that addressing uncertainty associated with the natural world is necessary but not sufficient to avoid surprise. Case studies of wildlife management show the indeterminism of management systems as non-stationary systems, systems that cannot be reduced to single and simple principles, where no probabilities of change can be formulated. The authors underline that uncertainty originates in ecological systems *and* in the perceptions, interaction, and decisions of human actors. Social complexity regarding the resource users; their ideas and interests; and their social, community, and group structures seem, however, difficult to apply in adaptive management.

3. *Reviews of the literature on adaptive management* show the limits of the uncertainty-based management philosophy. The question, how to deal with uncertainty, is answered through the management principle of gradually removing uncertainty by experiments and variation of rules.

Although it became obvious in the discourse that more knowledge about the interconnecting social and ecological systems is required to deal with uncertainty, social dynamics are widely neglected. Recent debates take up social aspects, reducing them to factors affecting cooperation and the management process; nothing from systems analyses of society, economy and politics appears in the discourse. *The main change of resource management principles that seems to be initiated through adaptive management is that "from maximum economic yield through maximum sustainable yield to management of ecosystem processes".* This change of principles implies a strengthening ecological sustainability in decision-making by way of learning from success and failure in ecosystem management. The later development of adaptive management can be described as broadening decision-making by taking into account more processes in ecosystems, still neglecting many processes in social systems that can affect resource management. This practice matches with the reductionist definition of SES as a geo-biophysical system with connected actors and institutions which excludes large parts of social-scientific knowledge from research and, as a consequence, from resource management. Folke et al. (2005: 443) describe this knowledge practice in their reconstruction of the development of ecosystem research that advances from single species management to management of ecosystems with different species in multiple-scale perspectives. With this broadening of the perspective, certain problems of biodiversity maintenance and of adaptation to climate change can be taken up in the analysis of resilience of ecosystems as complex adaptive systems. The human dimensions in shaping ecosystem processes and dynamics, the diversity of institutions and resource use practices, local interactions between actors, and selective knowledge use that shapes future social structures and processes are seen by the authors, but does not seem to be relevant for ecosystem-based or natural resource management. A more systematic use and integration of knowledge from social research is not recommended.

4. *In recent years, the critical discussion about success, failure, and limits of adaptive management intensified.* All improvements can be understood as applying the principles of adaptive management for the reformulation of the approach and its conceptual framework. However, such

reflexive knowledge practice has its limits: some deficits are connected to institutional structures of resource management, based on social norms, laws, and policy programmes that cannot be understood through reflection of ecological principles. At this point, it would be necessary to discuss policy reforms and management practices of organisational development (Bradford and Burke 2005). The institutional rules cannot be ignored in adaptive management, but they are not sufficiently reflected in the approach and its basic concepts rooted in ecological research. Institutional change and development requires social-scientific knowledge and changes of rules that cannot always be done by the cooperating scientists, managers, and resource users. With the multi-scale connections of resource management systems that develop in global environmental governance, these institutional constraints become more important in the management practice where institutional and governmental actors are dominant. The long history and development of co-management in European fisheries (Linke and Bruckmeier 2015) gives an example of the problem of slow and tough institutional change that impedes the development of new managerial approaches as participatory or adaptive management. It seems that the lacking success of adaptive management projects that is shown in many cases (see Table 6.1) is caused through the underestimation of the complexity of social systems in which adaptive management has to operate.

The literature on adaptive management (reviewed by McFadden et al. 2011: 1354) shows that it falls apart in the two different schools of thought mentioned above, a resilience-experimentalist school with emphasis on stakeholder involvement, resilience, and highly complex models, and a decision-theoretic school with relatively simple models through emphasising stakeholder involvement for identifying management objectives. This shows another dilemma with the use of adaptive management. According to the authors, acknowledgement of the simpler decision theory approach is growing, not that of the more complex experimentalist approach: the complex approach that is under continuous improvement finds less practical acceptance. The review does not show possibilities of improving adaptive management further to deal with the knowledge about interacting social and ecological systems. The main deficits and problems of adaptive management (summarised in Table 6.1) are still to be solved.

Factors and examples of success in adaptive management are few, factors of failure many, as Table 6.1 shows. Adaptive management appears as an early ecological approach, based on limited and selective ecological knowledge, and insufficiently elaborated as interdisciplinary and theoretical approach. Its design shows the intention to produce quickly applicable managerial rules and simple heuristics in situations where scientific knowledge is insufficient. A rapid and easy implementation of the approach did, however, not happen, and the newer improvements do not support a quick spreading. According to its epistemological characteristics, adaptive management is, as resilience, a reaction to environmental and natural resource use problems that cannot be dealt with by planning and technology. During the twentieth century, ecology advanced to the leading science producing knowledge to reduce or solve increasing environmental and resource use problems. As in other environmental research, the limits of knowledge and of inter- and transdisciplinary knowledge integration have been experienced. Adaptive management is an example for the limited possibilities of science to solve complex environmental problems. The search for possibilities to lift knowledge barriers where further empirical research is not helpful shows a new situation where policy experiments and variations of management rules advanced to principles of a heuristic approach to deal with ignorance. But if the limits of knowledge experienced are still the ones of ecology as a specialised discipline with a limited repertoire of methods, and knowledge from other disciplines, other disciplinary perspectives and interdisciplinary knowledge synthesis could be considered as further possibilities and ways out of the cognitive crisis.

The emphasis on continuous learning and improvements of the approach, especially in the variant called active adaptive management that aims at the improvement of the learning process, give the impression of a flexible and adaptable approach that has chances to develop into successful practices. In contrast to this impression, the conclusions by Allen and Gunderson (2011): 1383) give the impression of clear and final limits of adaptive management: that it is impossible when stakeholders have different understandings of the approach, when the managed

6 Knowledge Transfer Through Adaptive Management

resources continue to degrade while there is only discussion, and when no management experiments can be made. These conditions are often and temporarily given in environmental policy, as also for sustainable resource management, but such situations can change. As important as the difficulties met in the implementation practice seems the lacking reflection of the scientific experts and proponents of adaptive management with regard to their own roles, knowledge practices, limits, and selective knowledge use when they design, discuss, and assess the approach.

Both reviews summarised in Table 6.1 support the impression of limited utility of adaptive management for wicked problems and overused natural resources; they show conditions under which the approach works, where the ecological context factors are rather well described. The authors differentiate between two types of factors causing failure of the approach:

- Factors resulting from a lack of conceptual clarity of the basic ideas
- Factors referring to a lack of possibilities to apply and implement the approach which result from the institutional unwillingness and lack of interest of practitioners

The insufficient theoretical reflection of adaptive management that could be corrected with the help of several theories is not explicitly mentioned among the lack of conceptual clarity. It may become more important in the future development of adaptive management, when it is no longer seen as an ecological approach with few and limited ideas. The development of adaptive management can happen through adopting systematic knowledge about the complexity of resource management in coupled social and ecological systems and through interdisciplinary knowledge integration. The second group of factors referring to the contexts of application can include, as the first group, knowledge about social and ecological factors that block the implementation. Ecological factors are reflected better; social and institutional factors that block adaptive management are sometimes summarised in the simple reasons of institutional inertia and unwillingness of actors, but more factors influence the success (as Table 6.1 shows) and should be considered for improving adaptive management.

In spite of attempts of the authors cited to describe relevant social factors and conditions of adaptive management, it is obvious that a main weakness of adaptive management is connected with the selective knowledge use of the—mainly ecological—authors. Social conditions are described to a limited degree, often in simplified forms, in non-sociological terminology and a kind of everyday language. A detailed analysis of the structures and processes in social systems that influence or restrict management systems is widely missing. Social factors seen as relevant for success or failure or adaptive management are mainly economic conditions directly influencing management or such that describe institutional conditions of the management process. The wider social contexts of the political and economic systems are considered less. It seems that the specialisation and expertise of the scientists investigating adaptive management is responsible for the selectivity of knowledge used. The dominance of ecologists can help to explain to some degree the limited success of the approach. Adaptive management seems mainly limited through the impossibility to create sufficient knowledge through empirical research about ecosystems. Assuming that ecosystems are coupled with social systems, knowledge and analyses of social systems, their structures and dynamics that keep SES and resource use processes in the self-destructive processes of economic growth, maximisation of yields, and overuse of the natural resource base at a global scale could also be considered as blocking adaptive management and similar attempts of resource management based on ecological knowledge. Such socio-economic constraints are not adequately addressed in the discussion of the approach, and simplified explanations as "institutional inertia" can also become inadequate and misleading when the institutional power relations are not misjudged or ignored. The approach is so far based on ecological knowledge and works with few management principles. What would be required more seems

- to improve the practices of joint learning, adaptation, and experimenting formulated in this approach (self-correcting mechanisms that are already discussed to some degree);
- to improve interdisciplinary cooperation and joint learning for the further elaboration and refinement of the approach with empirical and theoretical knowledge about system dynamics of SES in which adaptive management happens.

Adaptive co-management is a variant of adaptive management that does not add many new ideas to the approach. Plummer (2009: 1) summarised the debate with regard to conceptual models and influential variables, confirming that this debate is not advanced: most experience with adaptive co-management is from recent cases. Plummer's definition of co-management is simple, implying the basic idea that rights and responsibilities should be shared among those with a claim to the environment or a natural resource. This is not specific for adaptive management. Plummer refers to the definition of Armitage that highlights conditions of change, uncertainty, and complexity. For Folke et al. (2005: 448), adaptive co-management includes flexible community-based systems of resource management in approaches tailored for specific places and situations. These systems work with various organisations at different levels by way of learning and controlled change, in processes of testing and revising institutional arrangements and ecological knowledge in continuous self-organised processes of learning by doing. Adaptive co-management combines the learning components of adaptive management with that of collaboration and sharing of management power at various levels of decision-making, which is a step further in dealing with social factors in the implementation process. There is not much more to find out than that what the term says: adaptive co-management is a strengthening of principles of cooperation in adaptive management. Both variants are characterised by the search for controlled changes of resource management practices, and both can be assessed in similar ways: the main question becomes that of clarifying the limits of adaptive approaches in ecological and social, spatial and temporal perspectives.

In the reviews referred to above, adaptive management is not seen as an approach applicable in all situations where complexity and uncertainty prevail in natural resource management. Adaptive co-management shows similar limits; it aims at improved cooperation in managerial practices, but does not clarify the possibilities of inter- and transdisciplinarity, of different forms of empirical and theoretical knowledge in the management of SES. The conditions of application specified by Allen et al. (2011) restrict the application of the approach severely. Rist et al. (2013: 1) discuss this question further and critically, doubting whether many of the assumed limitations found in the discussion are adequately assessed, arguing that uncertainty prevails in all natural resource management. Adaptive management helps in identifying and reducing critical uncertainties in environmental management, but the conditions of application of the approach are disputed. A review of the arguments for the applicability of adaptive management that is only based on case studies can be criticised for ignoring important knowledge and context factors that influence the implementation processes. In the renewal of the ideas of adaptive management, it can turn out that there are no definite limits and the difficulties of application result rather from the specific views or the resources available to managers. Not all failures attributed to adaptive management are specific for this approach; failure can also result from the complex policy, social, and institutional environment that creates difficulties for all approaches in environmental policy.

The critical discussion of adaptive management and co-management in science and practice supports a broadening of the knowledge perspectives, but continues to argue in the tradition and within the limits of ecological knowledge from which the approach develops. Rist et al. see adaptive management as an ecological approach that has chances to become successful in resource management practices. The factors of failure are not seen as related to the management principles of the approach, but as external and independent factors in the social, political, and institutional environment of the approach. The framework constructed by the authors is thought to evaluate adaptive management and to clarify its applicability (Rist et al. 2013: 6f). In evaluating the implementation of the approach, there is a need to differentiate clearly between two causalities, one where the approach was not appropriate to the specific management goals and another where it was not feasible or unlikely to work given the wider management context (e.g. failure to improve management outcomes by reducing ecological uncertainty). The approach with its experimental strategy can also be applied for large and complex areas and problems. By defining the number or type of uncertainties to be reduced, it can be found how to deal with them, for example, more complex natural resource use problems can be managed by reducing the number of uncertainties targeted and by increasing the managerial resources for solving the problems.

The discussion by Rist et al. seems determined through the felt necessity to develop the approach further—or to renew it—because the authors see it as a rational, sound and scientifically backed approach. Their guiding ideas for renewing adaptive management are insufficient for dealing with the social complexity of natural resource management, unrealistically narrowing down principles of a management approach, separating from it decisive context factors, selecting factors for analysis that are thought to make it successful. This is done by drawing artificial boundaries between the management approach and its external context, ignoring the discourse of "situationally specific management" of change (Baumöl 2008) and other management theories that helped to deal with context factors. These theories learnt from various social-scientific disciplines, since the early research on administrative behaviour by Simon (1945), are that management is the "art" of dealing with factual and value premises and constraints of different kind in decision-making that need to be considered simultaneously. Management theory cannot ignore facts and factors in the wider social, political, economic, and ecological contexts that affect the achievement of managerial goals positively or negatively. The ascription of failure to external factors by Rist et al. seems a methodological trick with the—wanted or unwanted—consequence of ignoring knowledge about the contexts in the implementation of adaptive management.

Adaptive Governance: Theory-Guided Reformulation

Governance, a new term that emerged in recent years in the "grey zone" between science and policy, has been critically reviewed in the social sciences. Governance seems a conceptual candidate for an "empty signifier" (Offe 2009), void of meaning and interpretable in many different ways. In spite of the controversial scientific discussion, governance has quickly made its way into disciplinary and interdisciplinary fields of applied research in the policy and management sciences. Also the ecological authors in resilience research and adaptive management have taken up the term to update their terminology. The expectations connected with the new term are unclear, varying, and so does its interpretation.

In the discourse of adaptive management, the new term governance was understood in a simple way, bypassing a difficult theoretical clarification. This gives an example of the knowledge strategy already critically assessed above: of simplifying the adoption of social-scientific terms, defining them in ways that fit in the prior adaptive management debate, and ignoring further and theoretical knowledge from the social sciences. Application of the governance terminology helped address some neglected social context factors of adaptive management, as explicitly formulated by Folke et al. This can be seen as the cognitive gain with a new concept: to address questions of participatory and cooperative resource management and broadening the scope of collective action. Folke et al. restricted the accounting for social factors to help deal with ecosystem complexity: creating conditions for ordered rule, collective action, and institutions of social coordination. This is done with the broadening of the management perspectives in the terminology of adaptive co-management and adaptive governance. The few arguments which Folke et al. develop for this purpose are characterised by a parallelising of the ideas for managing social and ecological components of SES, as the following description shows.

Adapting and managing change are components of resilience of SES that require from the actors the capacity to reorganise the systems to maintain desired states in response to changing ecological conditions and disturbances. Approaches as adaptive management, more flexible and with more criteria than optimal use and control of resources, have been developed to deal with ecosystem complexity. Some authors used the concept of adaptive governance to broaden adaptive management, including the wider social contexts supporting ecosystem-based management. Governance in that sense means creating conditions for ordered rule and collective action and for institutions of social coordination in efforts to make collective decisions and to share power. Adaptive ecosystem-based approaches include activities of

- governance for resolving trade-offs and providing a vision and direction for sustainability;
- management as the operationalisation of this vision;
- monitoring, providing feedback, and synthesising the observations to a narrative of the situation and its potential future unfolding (Folke et al. 2005: 444).

Reconstructing the concept of adaptive governance in this ecological perspective continues with the shortcomings and selectivity of an ecological reductionism where social system components and knowledge from social-scientific research are filtered in a naturalistic perspective. In ecological research, the knowledge generated is reduced to managerially applicable knowledge and governance as the form of action to generate adaptation and change of SES. The management perspective is only broadened to a limited degree with the governance term: to catch the social factors and processes required for a better functioning of ecosystem management. To make this perspective of adaptive governance more coherent, the authors of the resilience alliance, including Folke, "ontologised" the view that social and ecological systems are always coupled and cannot be separated. That implies a statement of the dominance of ecological over social processes without having investigated further the social-scientific knowledge about the forms of coupling between social and ecosystems and their relevance for the management of ecosystem processes. In this limited interdisciplinary perspective of resource management, theory-guided analyses of nature-society interaction are not considered as relevant. It seems necessary to elaborate the governance concept further by connecting it to the social-scientific research on environmental governance and formulating its theoretical implications in terms of an interdisciplinary theory of nature–society interaction.

Adaptive governance is sometimes only used as another name for adaptive management, but the choice of the concept of governance indicates more systematic and more critical reflection of the ideas unfolding with adaptive management. The governance term may help to overcome some of the limits of the management concept, which was already criticised by Ludwig with the basic ideas of adaptive management as "management for the time when management is over". Furthermore, in the adaptive governance debate, the perspective is broadened to show its connections to analyses of resilience and sustainability and to develop a more systematic analysis of SES as the context of adaptive management. Ideas and results from Ostrom's late social-ecological research become effective. Folke et al. (2005: 44) consider these ideas in their concept of adaptive governance. This seems to show the progress towards an interdisciplinary approach for the analysis of SES in this managerial-, policy-, and practice-oriented perspective.

Adaptive governance in an ecosystem-based management perspective involves polycentric institutional arrangements of nested decision-making units operating at local and higher organisational and spatial levels. In this governance perspective, the aim is to find a balance between decentralised and centralised control through multi-scale governance. The vertical connections of local, regional, national, and global institutions can strengthen adaptive governance in creating a larger diversity of response options to deal with uncertainty and match local and global governance. But adaptation can also be blocked when, for instance, national land-use regulations contradict or undermine informal local systems of land tenure and limit the possibilities of practitioners to collaborate in interorganisational networks. The advantage of polycentric arrangements is, by the authors, only seen in responding to ecosystem dynamics at different scales, as scale matching or "institutional fit" appropriate to the ecological systems without asking how far ecological adaptations support improved functioning of social systems. In this management perspective, the possibilities of developing adaptive governance in a theoretically informed social-ecological perspective are not yet adopted. The four components of adaptive governance of complex SES from Folke et al. include

- developing knowledge of ecosystem dynamics;
- applying ecological knowledge for adaptive management practices;
- flexible institutions and multi-level governance;
- dealing with external perturbations and uncertainty.

All these principles are formulated as such by developing ecological knowledge and managing ecosystem dynamics:

- Ecological knowledge of ecosystem processes and functions strengthens resilience, whereas knowledge about social systems is important only to better manage ecosystem dynamics.
- Ecological knowledge in adaptive management practices supports continuous testing, monitoring, and evaluation, thus enhancing adaptive responses in managing complex systems; also for that purpose social capacities of learning and leadership and changes of

social norms in management organisations are only valuable insofar they allow for better responses to ecosystem dynamics.
- Adaptive co-management is a further step towards changing social institutions, developing flexible institutions, multi-level governance, and multi-scale linkages of management; the purpose of cross-level interactions in social networks is to generate and transfer knowledge and develop social capital and support for ecosystem management.
- Multi-level governance systems require the capacity to deal with disturbance in many forms, with climate change, diseases, hurricanes, global market demands, subsidies, and governmental policies; resilient SES may even use disturbance as opportunity to transform into more desired states.

The main critical argument to support adaptive governance—beyond the limits of ecosystem-based management—seems to break the dominance of economic maximisation strategies in natural resource management. This remains a weak component of the governance approach insofar as it is not supported by stronger arguments for transformation of societal systems, derived from social-scientific knowledge and social-ecological theory. For adaptive governance, the postulates for building adaptive capacity and resilience and responding to ecological system dynamics could be reformulated to strengthen the ecological perspective of resource management in coherence with principles of sustainability. The imperatives of adaptation, resilience, and disaster management are rather undermining than supporting social-ecological transformations of societal systems.

The critique of weaknesses and selective knowledge use in the formulation of adaptive management and governance directs to the search of alternatives for improving these strategies, in line with long-term perspectives of transformation to sustainability and sustainable resource management. The social conditions under which adaptive management needs to be applied can be formulated more systematically with the social-ecological theory as creating conditions for changes of modes of production and societal metabolism. It seems unrealistic to expect such knowledge improvements from the procedures that make the core of adaptive management and governance: that of policy experiments and joint learning through collaboration of resource users for purposes of dealing with uncertainties, management dilemmas, resource use conflicts, building experience-based knowledge bridges into the future. Necessary as these procedures are to become aware of the limits of scientific knowledge, they cannot replace the use of further theoretical and social-scientific knowledge and more critical analyses of the "systemic contexts" of natural resource management in modern society. Conditions of adaptation, institutional contexts, participatory approaches, multi-scale management, and sharing of power and responsibilities require further social-scientific and interdisciplinary knowledge and knowledge synthesis. *The following arguments from the broader discourse of adaptive management support theoretical syntheses and improvements of resource management through adaptive governance:*

1. *Conditions of adaptation in adaptive governance:* Adaptation of SES requires institutions that need to endure throughout processes of adjustment and change, coping simultaneously with changes in ecosystems and social systems. Asking for possibilities to improve environmental governance, combinations of traditional market-, state-, or civil society-based strategies have been discussed. Berkes concludes that neither purely local-level management nor purely higher-level management works well, but lower-level management and community-based self-organisation tend to support sustainable resource management more effectively. These local approaches combine knowledge improvements and the normative message from resilience research that shared rights and responsibility for resource management and decentralisation are best suited to achieve resilience (Nelson et al. 2007: 9). It should be added that the local approaches support—with their success—also breakthroughs in transitions towards sustainable resource management at higher levels, in networked, nested, and integrated multi-scale approaches.
2. *The institutional context of adaptive management:* Administrative and governmental institutions change incrementally, in small changes of policies. Independent from the availability of scientific knowledge, there is no practically available knowledge to make large organisational changes, and standard operating procedures contribute to

organisational inertia, slowing the bureaucratic processes of management (Allen et al. 2011: 1343). It should be added that the interaction of knowledge and power that are intertwined in this argument need to be more clearly reconstructed. The analysis needs to show the possibilities of changes achieved through organisational development and changes that require further policy reforms, governance, and transformation of national economic systems and their accumulation regimes.

3. *Management of scale and rescaling:* Institutions matched to several scales of managing natural resources and ecosystems are ever more important for adaptive governance that helps to match different scales of management through the collaboration of stakeholders (including formal institutions, informal groups and networks, and individuals). To support this extension bridging organisations, enabling legislation and government policies that allow for the creation of long-term perspectives and visions in resource management can be useful (Allen et al. 2011). It should be added: the long-term perspectives and visions have two purposes that tend to be undervalued in the adaptive governance discourse. At first, institutions need to be created that can break the monopolistic rule of economic growth with the principle of maximisation of yields that turned out to be disastrous for the environment, causing a large part of the environmental problems to be dealt with today. Thereafter, institutions need to be created that can support the continued transformation of resource use systems towards sustainability—for that purpose, multi-scale governance is only a first step, not the complete change required.

4. *Multifunctionality of sharing of management power and responsibilities:* Adaptive governance and its collaborative and participatory practices require further processes of creating change and transformation (beyond leadership, legislation, and funding). These change-creating processes include the institutionalised procedures of public policy and management; the monitoring of ecological systems; information flow to build cross-scale linkages of concerted action; utilisation of a variety of sources and forms of knowledge; and venues or platforms for collaboration. All these procedures and processes are building resilience and sustainability in SES, reducing failures of management decisions

under uncertainty and with imperfect information. Furthermore, adaptive governance is dependent on social networks with the capacity for innovation, communication, and flexibility to mediate the interaction of dynamic ecological systems and rigid institutions or social systems as the globalising economy and governmental policies. Social networks can generate political, financial, and legal support for new forms of environmental management, but depend upon collaboration, joint learning, leadership and leaders for environmental management with the capacities of integrating local and other forms of knowledge from different sources (Allen et al. 2011: 1343). It should be added: multifunctional institutions and power sharing need to be based on theoretical systems analyses to identify the blocking factors, the requirements, and the possible paths of transformation to sustainability.

5. *Supplementary perspectives of adaptive governance:* From a review of the literature on adaptive governance, network management, and institutional analysis, some blocking factors of adaptive governance can be identified. These factors include the inability of practitioners and policymakers and cooperating actors to cope with complexity and various uncertainties. Furthermore, the objectives that can be achieved through governance should be reviewed more critically. This requires a clarification of the contextual conditions of resource use systems in which governance operates. The uncertainty that can be dealt with and the effectiveness of different governance strategies should be compared and assessed. The concept of fit-for-purpose governance can be used to analyse the effectiveness of governance structures and processes at a certain point in time. Adaptive governance focuses on responses to potential change, whereas fit-for-purpose governance considers future functions that the SES have to fulfil—thus providing possibilities to evaluate the effectiveness of governance arrangements and to predict the likelihood of success of institutional change (Rijke et al. 2012: 73). It should be added that more than predicting success of adaptions to meet future challenges of sustainability, adaptive governance requires awareness of the non-managed and not manageable forms of social-ecological change and transformation to counteract the illusion that societal transformation can be achieved through policy and governance only.

With all the weaknesses and selectivity of adaptive management thinking formulated in the critical reviews, this approach has coherent principles and shows consequent efforts to improve its effectiveness and applicability as an ecological approach. With the reframing in terms of adaptive governance, the coherence and continuity can be seen in

- anchoring the approach in ecosystem analysis;
- dealing with uncertainty and surprise;
- specifying the cooperative nature of the management process;
- requirements of change of rules in the political and institutional context of resource management;
- showing new capacities required from resource managers and decision-makers to cope with complexity and uncertainty.

From these points, the questions can be taken up that direct further discussion of adaptive approaches in the theoretical debate of the regulation of interfaces of society and nature. All five critical arguments found in the discourse of adaptive governance need to be connected with systems analyses of SES that show the constraints of transformation management.

Questions to deal with include the following:

- How are the knowledge transfer and the cooperation of scientists and practitioners with different forms of knowledge developed in the context of resource management?
- How are the practices of knowledge use in resource management improved methodologically?
- How is theoretical knowledge about society and nature in resource management used?

Deriving principles of knowledge use directly from ecosystem research, in the ecological discourse already done before adaptive governance, appears as insufficient in a social-ecological perspective of SES-governance. It should be reflected as to how it is possible to follow "nature's lead" without naturalistic fallacies in the sense of deriving that what ought to be (with ethical principles that confirm that nature is good) from statements about that what is.

To shift the limits of adaptive management and governance requires the discussion of multi-scale environmental governance by asking whether new models for natural resource management replacing present models need to be sought, or complementary models added to conventional management practices. Adaptive governance is thought primarily for environmental and natural resource management and should be developed with this perspective, not as a new and generally applicable management approach. The discussion of the governance concept in the context of environmental governance is useful for the further development of adaptive governance: for both concepts, the conditions of interdisciplinary broadening and use of social-scientific knowledge should be clarified.

Environmental Governance: Problems of Multi-Scale Governance

Governance is discussed at first in a general way, in the perspective of political science to specify its scientific meanings. Thereafter, the variants of environmental governance are discussed to specify a perspective of social-ecological regulation of nature–society interaction.

The debates on environmental governance are broad and include manifold ideas and approaches from various disciplines. Although most forms of environmental governance refer to sustainability as the guiding idea or goal of all environmental action and governance, the connection of the two terms is not necessary. Sustainability is in the policy-related discourse reduced to the normative goal of political action. Usually, this goal is referred to in the international policy discourse as a problem of policy integration, taking into account simultaneously the social, economic, and environmental dimension of sustainable development in integrated forms of decision-making, in public policy, and in private-sector organisations (Kanie and Haas 2004: 1). Reduced to a question of political action and policy integration, the term sustainability does hardly show connections to the interdisciplinary debate of sustainability as societal transformation discussed here. An interdisciplinary and social-ecological debate seems, moreover, necessary in a situation where sustainability is more critically discussed in the search for alternative concepts that can replace it (see Chap. 4).

The development of forms and mechanisms of environmental governance at various policy levels and in various countries is not reviewed here. This is the work of political science and empirical research that can be followed in specialised policy research (Kanie and Haas 2004; Newig and Fritsch 2009, Biermann 2011; empirical research and reviews: Biermann and Pattberg 2012; Hogl et al. 2012; Martin et al. 2012; Wurzel et al. 2013). In exemplary form, Tacconi (2011: 234) describes the requirements of developing environmental governance research: the limited interdisciplinarity in relevant subjects as ecological economics, political ecology, sustainability science, and Earth system governance; the postulate to develop transdisciplinary approaches; the necessity to integrate economic, political, social, and environmental aspects in governance research; and the lack of an encompassing theory and the improbability that such a theory develops. This broadening of the perspectives of environmental governance shows the growing influence of interdisciplinary thinking in environmental research and in the management sciences. Environmental governance appears as part of various governance forms, interwoven with the debates about transition management—transitions to sustainability and social-ecological transformation at various levels of society.

The concepts of adaptive and environmental management or governance require beyond definitions of management principles further reflections of their forms of knowledge application and practice. This is, as the review above showed, easier for adaptive governance that is more coherently applied in ecological debates, whereas environmental governance falls apart in heterogeneous interpretations and approaches. Two criteria are used to limit the discussion of environmental governance and to connect it to social-ecological theory:

- To follow the use the governance term in social-scientific research (which is often not oriented to environmental themes)
- To specify the governance term through additional descriptors of environmental governance (in specific approaches of local or global environmental governance, climate change governance, or biodiversity governance)

The discussion of governance in political science brought the following results:

Governance research includes many specialised themes and fields, without common theories, frameworks, and results. To clarify the use of the amorphous term Offe (2009: 552 f) suggests that the following social spheres should be excluded from it: (a) the sphere of private and civil society of citizen where coordination happens in spontaneous forms, (b) the sphere of market transactions, and (c) the state and governmental institutions for which the concept of government should be retained. The remaining phenomena to be described as governance are still broad, but can be specified further as

- *regulation of publicly relevant issues through non-state actors* (corporations, employer associations, labour unions, chambers of commerce, churches, mass media, and others), and
- *cooperation of governmental and private actors in the implementation of state policy*: co-optation of private actors may increase efficiency and effectiveness of policies through coordination of responsibilities. This cooperation implies the forms of participation and coordination under which large parts of environmental governance, including adaptive management and governance, can be subsumed.

Co-optation is a multifaceted phenomenon, including informal cooperation that may not only result in intended improvements of efficiency and effectiveness of policies; it can create dependencies that are difficult to separate from lobbyism and clientelism. From these specifications of social-scientific meanings follows: governance develops in spheres of action where governments cannot act or their action is inefficient. That is the main problem addressed with the term, differing from that which has been discussed above as the limits of management in the forms of ecosystem-based management. *Global governance* is a paradigmatic example described by Weiss and Thakur as

- the formal and informal institutions, mechanisms, relationships, and processes between and among states, markets, citizens, and inter- and non-governmental organisations;

– through which interests at the global level are articulated, rights and responsibilities ascribed, and differences of interests are mediated (Offe 2009: 553).

The general meaning of the term is in the political and managerial practices of its application watered down through diffuse normative connotations that appear under the valuing notion of good governance.

Although not specified for environmental policy and natural resource management, the clarification of governance by Offe seems useful for the discussion of environmental governance. His analysis of governance in the context of political science brings governance back to its origins in spheres of public policy and action where hierarchy and control-based forms of governmental action do not work and broader forms of public and political are required. This includes action within and between countries and collective action for the common interest of citizens or in the public interest. To understand governance as a more general and abstract phenomenon than governmental action is the popular interpretation that ignores the historical situation and modern world order where the term emerged. Governance develops in specific forms of broadening the organisational form of policy processes beyond states and their political authority. Yet, governance is not a new form of organisation of public action that shows a future order independent from states. It implies different ephemeral and residual forms of action that can be understood as related to and complementing governmental action.

In the context of a social-ecological theory of nature and society, environmental governance is no more than a specific form of public political action dependent on states; environmental governance is not explainable without referring to the governmental action to which it is connected, and governmental action is the dominant form that will not vanish. *Governmental environmental policies* and e*nvironmental governance do not show new forms of social agency that could make transformation of societal systems as demanded in the sustainability discourse more easy or abbreviate them. They are only developing into more complex forms of collective action that can take up more knowledge, and deal with more complex problems as the climate changes. For that purpose, they need to take up knowledge from system analyses of society, economy, and interacting SES.* Governance has no greater explanatory function than to understand how public policies broaden, especially as multi-scale governance. Governmental action and governance are subordinate functions in the societal processes of socio-economic reproduction and in the social metabolism of modern society. Also for the societal transformation to sustainability, governance is not the only form of action and change. The relevant capacity of governance is that of power-dependent coordination of collective action in manifold forms, necessary in societal transformation towards sustainability, but not providing all capacity and knowledge required for that.

The discourse of environmental governance includes various components:

1. The publications of Kanie and Haas (2004) and Kanie et al. (2014) show the influence of ideas of multi-scale environmental governance. Improvements of institutional structures and processes in management are sought with ideas of network management and best practices. Governance deficits and problems managing environmental problems internationally and in the perspective of sustainable development guide the reflections towards policy reforms and alternatives of global governance similar to ideas articulated by Ostrom and in broader debates of civil society action: network-based management, inclusion of new actors and stakeholders, and multiple and overlapping functions of management. The discussion of the necessity and role of a global environmental organisation remains controversial.
2. Like the research on adaptive governance, that on environmental governance is insufficient to formulate the knowledge components of an interdisciplinary theory of nature and society. This conclusion is supported by an early review of environmental governance research since the second half of the twentieth century by Davidson and Frickel (2004). The governance research is structured in methodological terms and fragmented along substantive or topical lines. Methodologically, the research on environmental states is based on case studies in specific nation states, especially industrialised states in North America and Western Europe. The insights from these case studies are of limited utility in other contexts, neglecting differences in environmental policies. The preponderance of cases representing a few economically and politically powerful states does not allow for generalisation. Broadening

the knowledge base analyses of developing and newly industrialising countries is required. Such analyses provide a better understanding of the potentials of global environmental governance. The conditions for continued dependence of the countries in the Global South on extractive industries and their dependence on international institutions for financial support need to be analysed. Furthermore, levels of democratisation, variation in institutional capacity, and presence of alternative ideological conceptions of the society–nature relationship are important variables for the reconstruction of capacities for environmental governance. Research and cross-national comparisons of various political, cultural, and economic contexts can enhance theory building for global governance (Scott and Frickel 2004: 485).

According to the review of Scott and Frickel and other reviews discussed above, the theoretical reflection of governance studies ends before problems and question of a theory of society or nature–society interaction are taken up. Such a theory would require a connection of governance analyses with that of the complex system-maintaining structures and processes in societal systems of SES and investigation of the possibilities of transformation of these systems. Adaptive management and environmental governance have only limited significance for these theoretical components of social-ecological theory; they are more relevant for the knowledge transfer and application of the theory in policies and practices of environmental policies and natural resource management.

3. *Various frameworks for analyses of local and global environmental governance* are described by several authors as multi-scale governance (Görg and Rauschmayer 2009), global governance (Biermann 2011), and adaptive governance (Williams 2011). These frameworks specify the salient meanings of environmental governance in theoretical terms. Each framework exemplifies a dilemma of environmental governance that requires the reconstruction of the governance term in theory-guided perspectives: the difficulties of rescaling environmental governance as multi-scale governance, of matching and integration in global governance, and of dealing with ignorance in managing ecosystem dynamics.

4. *A green economy is a dominant neoliberal form of policy in the discourse of global environmental governance.* In the global policy process after

the Rio conference in 1992, three variants of bio-economy, green economy and eco-economy developed (see Chap. 4) varying between incompatible approaches of neoliberal globalisation and a great transformation towards global sustainability. The first two forms build on institutional reform within the established power structures; the third one is more critical, but lacks the theoretical perspective of transformation of the global economy. All three forms are, in spite of their differences, part of the practices of environmental governance supported by the United Nations Environmental Programme with the expectation of improved human well-being and social equity, of significant reduction of environmental risks and ecological scarcities through a low-carbon and resource-efficient economy (Never 2013: 4f). The debate of another great transformation to sustainability begins in the social-ecological and related discourses and is better connected to critical theories of modern society that show the limits of policy reforms.
5. *Mechanisms of "greening the economy" and transforming the national economies focus on the notion of "green power"* to create a potential for transforming established political and economic power structures towards multi-actor, multi-level, and global environmental governance. Never (2013) assumes that countries with the capacity to develop green power will be those who manage and shape change by combining ideas of sustainability, innovation capacities, and transforming power structures in a competitive process. This concept of green power is not theoretically rooted; it shows in the version of Never and her illustration for climate change and clean technology the weak theoretical and analytical capacity of the governance concept applied in the policy discourse of sustainable development. System transformation is reduced to policy processes that can be managed by governments and cooperating actors.

The debate of environmental governance shows similarities with the situation in other fields of the environmental discourse that are discussed in this book, in various chapters: the discourses about sustainable development, about climate change adaptation, and about transformation of the industrial energy systems to more environmentally sound energy use. In the more theoretical and in the more practical discourses, a plurality of differing and competing approaches and interpretations needs to be dealt with, in research and in the policy discourses about social-ecological transformation. In further research on environmental governance, it seems necessary to separate the analytical components of governance studies from diffuse normative notions of governance which are widespread in the ecological discourse. A prominent example is that of International Union for Conservation of Nature (IUCN 2014) where multi-level environmental governance appears in formal and informal interactions of state, market, civil society, and further institutions. The purpose of such concerted action is to formulate and implement policies in response to environment-related demands and inputs from the society. This process should be guided by widely accepted rules, procedures, processes, and behaviour, showing the characteristics of good governance and supporting the achievement of environmentally sustainable development. Such "wishing list" forms of defining environmental governance correspond to similar normative principles in defining sustainable development as only ethically reflected approaches (Chen 2012) that ignore the analyses of social and ecological systems. More than directly seeking explanations of societal transformation to sustainability through new or improved forms of governance, it seems promising

- to develop in social-ecological theory of nature and society a step-wise improved concept of governance through comparison and classification of the variants and forms of governance in different countries and cultures;
- to connect the debate of global environmental governance with that of sustainability;
- to develop further concepts to describe the complex processes of societal and systems transformation with a variety of action- and agency-related concepts.

With such theoretical concepts and their combination, it becomes possible to shift the limits of governance and to deal with non-manageable change of social and ecological systems:

1. *From the various concepts and approaches to governance, the following examples seem relevant as conceptual elements for the analysis of environmental governance in theoretical perspectives:*

    – *Multi-scale governance* (Görg and Rauschmayer 2009) and multi-actor governance can help to analyse horizontal and vertical cooperation and to improve concerted action in global policies.
    – *Global governance* tends to become a theoretical term to guide the integration of knowledge about international policies and earth system governance, as described by Biermann (2011). Critical reviews of global governance studies as that of Hakovirta et al. (2002) brought insights on types of rationality of global policy processes, but not on the connections between policy processes and transformations of SES.
    – *Adaptive governance* is one of the few theoretical concepts to formulate policies from an ecological knowledge base, for decision-making and policies in situations of uncertainty and lack of knowledge for management action, using knowledge from the monitoring of policy experiments and from iterative learning of the actors (Williams 2011: 1348). Adaptive governance connects the ecological research and debates of vulnerability, resilience, and sustainability to the environmental governance debate.

   Further theoretical reflection of environmental governance should be part of epistemologically and methodologically guided forms of knowledge synthesis. Comparison and combination of further concepts or typologies of governance, as that of Leach et al. (2007: 32), differentiating adaptive, dynamic, and deliberative governance, seem of limited value for a social-ecological theory; they show thin and formal descriptions of governance processes that can be relevant in empirical research.

2. *With regard to the transformation of societal or social and ecological systems, governance requires connections to further, system-maintaining processes.* These processes include reproduction, maintenance of stability, boundary maintenance of social and ecological systems, and growth processes—all important to find possibilities of influencing, changing,

and transforming complex SES. For social-ecological research, the building of knowledge bridges from theory to practice includes the development of *the concept of social-ecological regulation that broadens the environmental governance debate by connecting it to a theory of nature and society and supporting knowledge transfer from this theory.* This is not self-evident, rather an exceptional form of ecological research. Theories are, conventionally seen, not constructed for enabling knowledge transfer and practical application of scientific knowledge, but for guiding, systematising, and explaining scientific knowledge production. To illustrate the difference between theories of environmental governance and social-ecological theory, the analysis of regulation in social ecology can be used. Social-ecological analysis of regulation of nature–society interaction shows that frameworks of adaptive governance and environmental governance cannot be directly connected to the broader social-ecological theory. These frameworks need to be reformulated in other terms by taking into account the differences between regulation as managed processes of resource use and self-regulation of ecosystems. Self-regulation is, in another perspective and independent from the management and governance debate, discussed in the analysis of ecosystem services. How social-ecological analyses of regulation processes differ from the policy-centred studies of adaptive governance and other forms of environmental governance can be shown in the discussions about the development of interdisciplinary social-ecological research in Germany and Austria.

The methodological and theoretical considerations to conceptualise social-ecological regulations can be summarised as follows, following the analysis by Hummel and Kluge (2004: 45ff; further discussion: Brunnengräber et al. 2008; Brunnengräber 2009):

- Social-ecological analysis of regulation needs to connect regulation to a theoretically reflected conceptualisation of SES where societal, natural, and technical components are analysed in their interactions.
- To deal with the intersystemic interaction processes requires, methodologically seen, several steps of analysis: system analyses of social

and ecological systems, causal analyses of the interrelations in the networks of intersystemic processes (flows of information, energy, matter), and analysis of regulation processes. Such combined analyses can adopt a theoretical concept of regulation for the reconstruction of the complexity of interaction processes—for example, in a cybernetic model with positive and negative feedback loops and feedbacks between governance processes and self-regulation of ecosystems.
- Finally, a regulatory type for specific SES and interaction processes needs to be chosen: this requires knowledge about the specific system qualities of social and ecological system components and knowledge about the specific forms of coupling. At this level of analysis, it can be decided whether simpler types (e.g. hierarchical or linear), or more complex types of regulation (e.g. second-order cybernetics, negative feedback ,and mechanisms to prevent collapse of the system) are required for a specific resource use problem.

The social-ecological regulation as "regulation of regulations" in the sense of second-order cybernetics implies a further broadening of regulation processes beyond the scope of policies. The methodology is connected to the analyses of SES and the theory of nature and society that develops in social ecology. The theoretical methodology of regulation analysis implies the specification of the social forms of regulation of societal processes and of processes of interaction between nature and society—for example, as political, legal, economic, cultural regulation, and connections of these human forms of regulation and the self-regulation of ecosystems. For all forms of regulation, the knowledge components of scientific and other knowledge need to be specified. This form of cybernetic analysis of social-ecological regulation types for improving environmental agency requires further theoretical reflection of the concepts of self-regulation and context regulation, developed in the social-scientific discussion of the theory of autopoietic systems. Furthermore, according to the ideas of a critical theory of society to which the social-ecological discourse connects, the forms of social-ecological regulation cannot be applied in the political practice of governance with some simple criteria and rules as suggested in the debate of adaptive management. Social-ecological regulation implies knowledge from systems analyses, whereas adaptive management tries to get rid of the complications of using and transferring scientific knowledge into managerial action with the suggestion to train decision-makers, resource managers, and politicians in the use of ecological knowledge. In difference to such shortcut ideas of knowledge transfer from science to practice, *social-ecological regulation requires the formulation of specific forms of knowledge transfer (including knowledge sharing, collective learning, and learning to change learning principles) to anchor the regulation processes in society and in different contexts of social action*. These contexts include the social systems involved in natural resource use, the lifeworld contexts of social actors involved, and the different fields of collective action, political, and other ones that enable transformative agency. Such agency as the capacity to transform the systemic structures of complex systems as the modern society and the economic world system through multi-scale regulation is not possible without theoretical knowledge about the functioning and interaction of social and ecological systems. A part of the transformation of societal systems happens with managerial and regulatory approaches, but the whole process of adaptation and transformation is more complicated, and requires the conceptualisation of these processes in theoretical terms, as discussed in the following chapter for the adaptation to climate change.

Conclusions: Environmental Governance and Social-Ecological Regulation

In this chapter, the interconnecting ideas and approaches of adaptive management and environmental governance were discussed with regard to empirical research and reviews of these approaches. It is not necessary for the purposes of developing an interdisciplinary social-ecological theory and the integration of these approaches in that theory, to give a more detailed, country- or case-study-based account of the practices, successes, and failures of adaptive management and environmental governance. The reviews summarise the state of the debates sufficiently and provide more systematic arguments than a patchwork of case studies, supporting the following conclusions:

1. *With the ideas of adaptive management and environmental governance, earlier principles of environmental research and resource management are abandoned.* These principles to abandon include that

    – of economic maximisation of yields;
    – of research guided by single theories that should be verified or falsified through empirical research;
    – of working with linear models and single causes, with theoretically derived hypotheses, with quantitative data and with exact knowledge;
    – of knowledge transfer for resource use planning as the guiding idea of transfer.

    New principles of interdisciplinary research and management are discussed in preliminary forms, for example, in sustainability science and as postnormal science (Funtowićs and Ravetz)—in efforts to turn the lack of positive knowledge, of adequate concepts, of exact methods and epistemological foundations of knowledge synthesis into practices of working with limited knowledge. The debates on limits of knowledge support often inexact methods and creation of robust knowledge to deal with complex and wicked environmental problems. With the epistemological formulation of these approaches, conventional principles of disciplinary science and research are given up. Yet, the consequences remain unclear: how can further theories, theoretical concepts, methodological procedures and claims of verification, validity, and reliability be maintained in environmental research? The reduction of cognitive aspirations was seen as a consequence of the complexity of problems and processes in SES that exceed the capacity of available theories, methods, and explanations. Non-predictability of development and change in SES (with manifold interactions of system components and processes, feedback loops, insecurity, and surprise) create new problems of production, transfer, and application of scientific knowledge.

2. *For adaptive management, knowledge transfer and application is organised in a simple form:* It is assumed that the principles derived from ecological research can be directly applied as management principles: no translation or methodological transformation of scientific knowledge is required to formulate managerial knowledge. The formulation of some

normative principles or heuristics that characterise the management approach is sufficient; for specific situations, for example, in the use of ecological knowledge for managerial decisions, the advice of researchers may be required, but this is a practical question, not one of methodological quality. *For the knowledge on environmental governance* resulting from social-scientific research, the knowledge transfer is less clear; it can be assumed that it is mainly done in the conventional forms of scientific advice to decision-makers. In some cases, it may be done by training of decision-makers and in the newly emerging forms of participatory research. For both approaches, it seems self-evident that knowledge transfer comprises empirical knowledge; management principles; and information from monitoring, evaluation, and implementation research. Only in exceptional cases are theories used in knowledge transfer; although theoretical concepts are used in resource management, their application happens mainly through definitions of concepts.

3. *Adaptive management and environmental governance are not core components of the social-ecological theory; they are only useful for operationalising this theory and transferring knowledge from it for purposes of sustainable resource management.* This application does not require aggregation and synthesis of data from empirical studies; such synthesis is, however, required for theory construction, especially for the use of empirical knowledge in the overarching social-ecological theory. For this purpose, the reviews of adaptive management and environmental governance have shown the principal difficulties to make use of the frameworks for theoretical purposes of SES analysis. Most empirical studies use the concepts and frameworks in exemplary ways and without discussing the vagueness, inexactness, and the multiple meanings of the terms; the terms are applied in case-specific form. The systematisation and synthesis of knowledge from the empirical research end at the level of constructing conceptual models for their integration, or with the construction of different types of governance as by Leach et al. (2007: 32): adaptive, dynamic, and deliberative governance. Also Davidson and Frickel (2004), asking for sociological research and theories of governance, do not connect their analysis to the discourses of theory of society or nature–society interaction, but ask for specific governance theories in a more limited sense. In a social-ecological review and discussion of the governance debate, Scheer

(2006) describes the process of development of more decentralised, effect-, and result-oriented processes of policy and management, differing from older bureaucratised and centralised approaches that still prevail in managerial practice. His conclusion is that the involvement of always more actors in governance makes this process more complicated, with the unforeseeable results, including unexpected ones and surprise (Scheer 2006: 71). This confirms in another perspective than that of Davidson and Frickel that the theoretical reflection of governance does not include theories of society or nature and society, or reduces these to normative views, world views, or paradigms.

4. *The two main questions that remain for further discussion of adaptive management and governance in the context of social-ecological research and a theory of nature and society* are: How can adaptive management and governance be developed further and connected to such a theory? How can the process of reconnecting theory and practice of resource management be developed beyond the simple transfer in the form of reformulating ideas for ecological research as principles of resource management? These questions can be answered in preliminary forms with regard to cooperation in resource management and the emerging debate about social-ecological regulation.

The cooperation of researchers and practitioners in resource management is intensively discussed to make policies and resource management more effective and sustainable. This discussion gives possibilities to develop forms of social-ecological regulation that connect management of resource use with the self-regulation of ecosystems. But the approaches of resource management suffer from under-reflection of the cognitive and knowledge problems that need to be dealt with in connecting science and practice in SES management. The knowledge problems include knowledge synthesis and transfer, applied research, knowledge sharing, joint learning, systems analysis of social and ecological systems, and theoretical conceptualisation of forms of governance and regulation. Knowledge transfer practices require the epistemological analysis and reflection of the significance of theory and theoretical knowledge in governance processes.

The concept of social-ecological regulation requires further analysis and theoretical reflection to become a bridging concept between the theories and approaches of environmental governance and the theory of nature–

society interaction. For the debate summarised above, this implies the inclusion of cybernetic ideas in regulation processes. Interdisciplinary knowledge synthesis requires elaborate theoretical frameworks for analyses of SES, of functions, structures, and processes in social and ecological systems. Such analyses may require one or several theoretical languages and ontological and epistemological reflection (e.g. how to deal with different epistemologies as constructivism and realism, with normative assumptions in knowledge syntheses). Controversial is whether the synthesis process can be carried out as a synthesis "ex ante", where the integration is done by choice of an interdisciplinary terminology and frameworks as in systems theory, or whether synthesis requires more complex processes of knowledge integration that can be described as synthesis "ex post", which advances in iterative steps of integrating empirical and theoretical knowledge where the synthesis is achieved at the end of the processes of research and knowledge integration.

References

Allen, C. R., Fontaine, J. J., Pope, K. L., & Garmestani, A. S. (2011). Adaptive management for a turbulent future. *Journal of Environmental Management, 92*, 1339–1345.
Allen, C. R., & Gunderson, L. H. (2011). Pathology and failure in the design and implementation of adaptive management. *Journal of Environmental Management, 92*, 1379–1384.
Baumöl, U. (2008) Change Management in Organisationen: *Situative Methodenkonstruktion für flexible Veränderungsprozesse*. Wiesbaden: Gabler Edition Wissenschaft.
Beer, S. (1979). *The heart of enterprise*. London: John Wiley.
Biermann, F. (2011). *Reforming global environmental governance: The case for a United Nations Environment Organisation (UNEO)*. Amsterdam: Earth System Governance Project and VU University Amsterdam.
Biermann, F., & Pattberg, P. H. (Eds.). (2012). *Global environmental governance reconsidered (earth system governance)*. Cambridge, MA: MIT Press.
Bradford, D. L., & Burke, W. W. (Eds.) (2005). *Reinventing organization development*. San Francisco: Pfeiffer.
Brunnengräber, A. (2009). Die politische Ökonomie des Klimawandels Ergebnisse sozial-ökologischer Forschung, Band 11. München: Ökom-Verlag.

Brunnengräber, A., Dietz, K., Hirschl, B., Walk, H., & Weber, M. (2008). *Eine sozial-ökologische Perspektive auf die lokale, nationale und internationale Klimapolitik.* Münster: Westfälisches Dampfboot.

Chambers, R. (1994) 'The Origins and Practice of Participatory Rural Appraisal', *World Development,* 22 (7): 953-969.

Chambers, R., & Conway, G. (1991). *Sustainable rural livelihoods: Practical concepts for the 21st century.* IDS Discussion Paper 296. Institute of Development Studies, University of Sussex.

Chen, B. (2012). Moral and ethical foundations of sustainability: A multidisciplinary approach. *Journal of Global Citizenship & Equity Education, 2,* 2, 20 pp. journals.sfu.ca/jgcee

Davidson, D.J., Frickel, S. (2004) 'Understanding Environmental Governance: A Critical Review',*Organization & Environment,* 17, 4: 471-492.

Delmas, M. A., & Young, O. R. (Eds.). (2009). *Governance for the environment: New perspectives.* Cambridge: Cambridge University Press.

Duit, A., Galaz, V., & Eckerberg, K. (2010). Introduction: Governance, complexity, and resilience. *Global Environmental Governance, 20,* 363–368.

Evans, J.P. (2012) *Environmental Governance,* New York: Routledge.

European Commission (2005) Frontier Research: The European Challenge. High Level Expert Group Report, EUR 21619 (Brussels, Directorate General for Research). Luxembourg: Office for Official Publications of the European Communities.

Folke, C., Hahn, T., Olsson, P., & Norberg, J. (2005). Adaptive governance of social-ecological systems. *Annual Review of Environment and Resources, 30,* 441–473. doi:10.1146/annurev.energy.30.050504.144511.

Frickel, S. (2004). Scientist activism in environmental justice conflicts: An argument for synergy. *Society and Natural Resources, 17,* 1–8.

Funtowicz, S., & Ravetz, J. (1993). Science for the post-normal age. *Futures, 25*(7), 739–755.

Görg, C., & Rauschmayer, F. (2009). Multi-level-governance and the politics of scale—The challenge of the Millennium Ecosystem Assessment. In G. Kütting & R. Lipschutz (Eds.), *Environmental governance, power and knowledge in a local-global world* (pp. 81–99). London: Routledge.

Hakovirta, H., Herne, K., Jokela, M., Lähteenmäki-Smith, K., & Salmio, T. (2002). *Global problems and their governance: The contribution by the Figare/Safir project.* Report. Department of Political Science, University of Turku.

Hogl, K., Kvarda, E., Nordbeck, R., & Pregernig, M. (Eds.). (2012). *Environmental governance: The challenges of legitimacy and effectiveness.* Cheltenham: Edward Elgar.

Holling, C. S. (Ed.). (1978). *Adaptive environmental assessment and management.* Chichester: John Wiley.

Hummel, D., & Kluge, T. (2004). *Sozial-ökologische Regulationen*. Demons Working Paper 3. Frankfurt am Main: Institut für sozial-ökologische Forschung.

IUCN (2012) Position Paper on Institutional Framework for Sustainable Development. https://www.iucn.org/news_homepage/events/iucn_rio_20/iucn_position/environmental_governance/ (Accessed 27.02.2016).

Kanie, N., & Haas, P. M. (Eds.). (2004). *Emerging forces in environmental governance*. Tokyo: United Nations University Press.

Kanie, N., Haas, P.M., Andresen, S., eds. (2014). *Improving Global Environmental Governance: Best Practices for Architecture and Agency*. Cheltenham: New York and London: Routledge.

Leach, M., Bloom, G., Ely, A., Nightingale, A., Scoones, I., Shah, E., et al. (2007). *Understanding governance: Pathways to sustainability*. STEPS Working Paper 2. Brighton: STEPS Centre.

Linke, S., & Bruckmeier, K. (2015). Co-management in fisheries: Experiences and changing approaches in Europe. *Ocean and Coastal Management, 104*, 170–181.

Ludwig, D. (2001). The era of management is over. *Ecosystems, 4*, 758–764.

Martin, P., Ziping, L., Tianbao, Q., Du Plessis, A., Le Bouthillier, Y., & Williams, A. (Eds.). (2012). *Environmental governance and sustainability*. Cheltenham: Edward Elgar.

McAfee, K. (2012). The contradictory logic of global ecosystem services markets. *Development and Change, 43*(1), 105–131. doi:10.1111/j.1467-7660.2011.01745.x.

McFadden, J. E., Hiller, T. L., & Tyre, A. J. (2011). Evaluating the efficacy of adaptive management approaches: Is there a formula for success? *Journal of Environmental Management, 92*, 1354–1359.

Nelson, D. R., Adger, N., & Brown, K. (2007). Adaptation to environmental change: Contributions of a resilience framework. *Annual Review of Environment and Resources, 32*, 395–419.

Never, B. (2013). *Toward the green economy: Assessing countries' green power*. GIGA Working Paper 226. Hamburg: German Institute of Global and Area Studies.

Newig, J., & Fritsch, O. (2009). Environmental governance: Participatory, multi-level – and effective? *Environmental Policy and Governance, 19*(3), 197–214.

Offe, C. (2009). Governance: An 'empty signifier'? *Constellations, 16*(4), 550–562.

Ostrom, E. (2007, February 15–19). *Sustainable social-ecological systems: An impossibility?* 2007 Annual Meeting of the American Association for the Advancement of Science, Science and Technology for Sustainable Well-Being, San Francisco.

Ostrom, E. (2009). A general framework for analyzing sustainability of social-ecological systems. *Science, 325*, 419–422.

Plummer, R. (2009). The adaptive co-management process: An initial synthesis of representative models and influential variables. *Ecology and Society, 14*(2), 24. http://www.ecologyandsociety.org/vol14/iss2/art24/.

Rijke, J., Brown, R., & Zevenbergen, C. (2012). Fit-for-purpose governance: A framework to make adaptive governance operational. *Environmental Science and Policy, 22*, 73–84.

Rist, L., Felton, A., Samuelsson, L., Sandström, C., & Rosvall, O. (2013). A new paradigm for adaptive management. *Ecology and Society, 18*(4), 63. doi:10.5751/ES-06183-180463.

Scheer, H. (2006). *Energy autonomy: The economic, social and technological case for renewable energy*. London: Routledge.

Simon, H.A. (1947) *Administrative Behavior: a Study of Decision-Making Processes in Administrative Organization*. New York: Macmillan.

Simon, H. (1991). Bounded Rationality and organizational learning. Organization Science, 2, 1: 125–134.

Speth, J. G., & Haas, P. M. (2006). *Global environmental governance*. Washington, DC: Island Press.

Stringer, L. C., Dougill, A. J., Fraser, E., Hubacek, K., Prell, C., & Reed, M. S. (2006). Unpacking "participation" in the adaptive management of social–ecological systems: A critical review, *Ecology and Society, 11*(2), 39. http://www.ecologyandsociety.org/vol11/iss2/art39/

Tacconi, L. (2011). Developing Environmental Governance research: The example of forest cover change Studies. *Environmental Conservation, 38*(2), 234–246.

Tengö, M., Brondizio, E. S., Elmqvist, T., Malmer, P., & Spierenburg, M. (2014). Connecting diverse knowledge systems for enhanced ecosystem governance: The multiple evidence base approach. *Ambio, 43*, 579–591. doi:10.1007/s13280-014-0501-3.

Tyre, A. J., & Michaels, S. (2012). Confronting socially generated uncertainty in adaptive management. *Journal of Environmental Management, 92*, 1365–1370.

Williams, B. K. (2011). Adaptive management of natural resources – framework and issues. *Journal of Environmental Management, 92*, 1346–1353.

Wurzel, R., Zito, A. R., & Jordan, A. J. (2013). *Environmental governance in Europe: A comparative analysis of environmental policy instruments*. Cheltenham: Edward Elgar.

7

Climate Change and Development of Coastal Areas in Social-Ecological Perspective

Climate change and its influence on development of coastal areas are discussed in this chapter in the perspectives of adaptive and transformative governance. Seven competing constructions of climate change and adaptation strategies are compared and analysed in their consequences for coastal development strategies. The failure of earlier policies of integrated resource management in coastal zones is described and reflected using concepts and knowledge from the social-ecological theory described in preceding chapters for various purposes: to integrate various perspectives in coastal management, to support knowledge transfer between science and resource management, and to build transformative agency and governance under conditions of global social and environmental change.

European countries are not expected to suffer most from the negative consequences of climate change. In countries in the tropical climate zone effects of global warming, sea level rise, and extreme weather situations are more dramatic; the problems are aggravated through the bad socio-economic conditions of large parts of the populations. The poorest people in the tropical zone contribute least to emission of greenhouse gases, have to suffer most from climate change, and have least possibilities to protect; this is discussed more critically in the social-ecological climate discourse (Brunnengräber et al. 2008, see below). A paradigmatic case is the Bay of Bengal in South Asia, the largest bay of the world, where the densely populated coasts are the home of about 40 % of the global population, mainly poor. This coast and the inhabitants are highly vulnerable to climate change (Samarakoon 2004). As a consequence of climate change, the management of coastal resources need to be rethought; this requires the integration of the perspectives of resilience and sustainability. Weinstein et al. (2007: 43) describe the challenges of sustainable coastal management in the twenty-first century as sharing of space and resources between humans and other species, as conflict mitigation, consensus building, sacrifice, and compromise.

In some regards, the situation in coastal areas is similar in all countries. Coastal landscapes are the preferred human habitats; two-thirds of all megacities are located at the coast. The global economy, global transport, and communication are strongly dependent on coastal cities and ports. In all continents, the population in coastal areas increases. Duxbury and Dickinson (2007: 319) describe the situation demographically: approximately 41 % of the world's population is living not more than 100 kilometres away from the coast: in 1990, 2 billion people and in 2000, 2.3 billion people. In 2025, the number is expected to be 3.1 billion people (Duxbury and Dickinson 2007: 319). Because of the concentration of natural resources in coastal waters, the coastal ecosystems are important as life-supporting systems. Ecologically seen, coastal areas are vulnerable to climate change and sea level rise, storms, floods, earthquakes, and tsunamis.

People continue to move to the coasts, although disasters increase and the impacts of climate change and sea level rise can already be experienced. After the devastation of New Orleans through Hurricane Katrina in August 2005, people moved back in the city and continued to live as usual. Inhabitants of coastal areas have always lived with floods and storms, but the problems are increasing with global climate change, especially in cities at the coast. The flood protection technologies of dams and dykes situations are necessary more than ever, but they become less effective with continuing sea level rise and more extreme weather situations. Technically and economically, it is not possible to build always more, stronger, and higher dams. Flood protection walls as the Thames

Gateway in London and the storm gates in Rotterdam show more the limits of protection through physical engineering than the successful adaptation to climate change. In future, coastal areas may be confronted with necessities of resettlement and with conflicts because of that.

Adaptation to Climate Change: Problems of Adaptation Strategies

Adaptation to environmental change and social-ecological regulation require differentiated forms of reacting to the development of local social and ecological systems and their specificities. From the economic development, of coastal areas, whether they are industrialised or not, developing through tourism or revitalising local cultural and economic traditions, result different conditions for adaptation to global change. Local adaptation cannot avoid social conflicts, disputes, and dilemmas that come through reactions to global change and sustainable development. Four kinds of cognitive problems need to be dealt with in adaptation to climate change and other forms of environmental governance.

1. *The possibilities of governance of global environmental change cannot be sufficiently described with the abstract term of adaptation that includes limited and passive forms of reaction.* More proactive, change- and development-oriented, and differentiated forms of adaptation to climate change would require perspectives and strategies of sustainable development. The process of adaptation is generally described as an activity of a system that generates change in its environment with the consequence that this change feeds back to the system and causes a change in the system that matches with the change in the environment. In this form, adaptation can be conceived as a continuous interaction between system and environment, based on feedback mechanisms, and as a mechanism to maintain stability of a system. Adaptation does not require the human capacities of reflection, anticipation, and action; the term can also be applied for mechanical and physical systems.

Living systems, ecosystems, social systems, and SES as systems developing and changing over time have complex forms of adaptation, boundary maintenance, and reproduction; they maintain continuous exchange of energy, matter, or information with the environment. Human action as constituent of social systems is based on anticipation and reflection, language and learning: capacities that differentiate the forms of adaptation to the environment. Humans are the most adaptable biological species (Crosby 1986; Moran 2000); they colonised all continents and adapted to different natural and climatic conditions.

In the analysis of SES and their adaptation, different epistemologies, concepts, and terminologies of social- and natural-scientific origin need to be integrated in interdisciplinary knowledge syntheses. The conventional form of differentiating human action from other forms of reaction is to use the term "behaviour" to describe biological and ecosystem-based processes with simplified mechanisms of stimulus, response, and feedback. Human action, more complex, knowledge-, and reflection-based, generates further possibilities and improved adaptation, but also maladaptation. Maladaptation is part of explanations of environmental problems. Forms of adaptation that may have been successful in earlier phases of development can turn out as a maladaptive change when the environmental problems and risks change or become more severe. Social reactions of humans to climate change in coastal areas include technical or physical protection and different forms of mobility and resettlement. Dams and dykes give examples of adaptation that became maladaptation, keeping coastal cities in a trap at high levels of development and modernisation. These cities are no longer stable and safe habitats; they need to be protected against sea level rise and floods, mainly because of the high investments in urban land.

2. *Climate change governance exemplifies wicked problems* that are difficult to solve and require complicated forms of environmental action. For such problems, it is assumed that only clumsy solutions can be found. Through social-ecological transformation develop more possibilities of problems solving, in a series of interconnecting steps and processes of managed and non-managed change. Adaptation to climate change and development of new energy systems cannot be reduced to problems

that require technical solutions, engineering, and geoengineering. In coupled and interacting technical, social, and ecological systems, the processes of adaptation and transformation follow the complex dynamics of nature–society interaction. Climate change adaptation and transformation of energy systems, closely connected, need to be analysed as specific problems of modern society and economy. Historical knowledge about former societies that did not use the modern energy sources of coal, oil, gas, nuclear energy may be useful only to some degree. The insecurity and the risks in the governance of energy systems (see Chap. 8) are caused through intervention of humans in natural process and their modification.

3. *Integrating adaptation with ideas of sustainable development in environmental governance* brings a controversy about the interrelations between resilience and sustainability. This controversy can be formulated with the question whether adaptation in the sense of building resilience is all what can be done in environmental governance; or whether sustainable development can provide further changes in the sense of transformation of society and the mode of production and practices of natural resource use. In the practice of environmental policy, this becomes a dispute whether sustainable development should be given up (Benson and Craig 2014). In coastal governance, the race against time may be lost and large parts of coastal areas will be left to the flood before sustainable development becomes effective. This would imply that the future to anticipate for coasts and coastal cities is not transition and management for change but management of decline, coping with disasters, and retreat. This does not seem to cover the possibilities of adaptation, change, and transformation in environmental governance adequately. System dynamics can vary in the course of time; economic decline and transition are possible in varying degrees and in different phases. Adaptation to climate change is not a final solution, only meaningful as a step in the transition to sustainability and climate-neutral energy systems and economies that require reduction of CO_2 emissions in the atmosphere.

4. *Adaptation strategies in forms of adaptive management and environmental governance* are complex forms of collective action. This action is influenced and constrained more through the structures and processes

in modern society than through the limited possibilities of intervention in ecosystem processes. Regarding adaptation to climate change, other forms of reaction can be differentiated from technical forms of engineering climate change: a series of social and ecological processes of transformation of SES. Climate change adaptation in coastal areas has logically seen a limited repertoire of alternatives (protection, adaptation or accommodation, and retreat), but these can happen in many different forms of action, of socially differentiated practices.

- *Protection* is mainly done in technical forms as physical protection through dams, dykes, and flood protection measures.
- *Accommodation* includes more flexible forms of reaction, where the term *adaptation* changes its meaning: it becomes an interactive social process of attempts to live with the negative effects of climate change and adapt flexibly, for example, in temporary retreat or through mobility.
- *Retreat* as the final form of adaptation in the worst case of permanent flooding implies resettlement. But also in that case the process can be more socially and temporally differentiated, with several options and phases.

There are no historical examples where cities of the size of modern megacities have been left and drowned in the sea. Cities in earthquake zones have rarely been left, but rebuilt after each catastrophe. How long this can be done in coastal cities under influence of climate change is unclear. What will be done when cities like New Orleans can no longer be effectively protected against floods? Will the buildings and the technical infrastructures drown or be left, with all consequences of pollution and loss of material for human use? Or will there be planned and phased retreat and resettlement, removing buildings and infrastructures? What are the personal, social, cultural, and economic consequences for the people moving? How will they be prepared for the changes? Improving adaptation strategies beyond technical and engineering solutions implies analysing the dialectics of adaptation and maladaptation, the manifold interactions between processes in social, technical, and ecological systems, and the consequences of climate change and climate change adaptation, especially social consequences.

The following section discusses differences of understanding and assessing adaptation to climate change in seven scientific approaches and theories. Comparing these approaches provides ideas for discussing adaptation and transformation problems more concretely—in the context of strategies for integrated and sustainable development of coastal areas.

Adaptation to Climate Change in Different Scientific Perspectives

Research on climate change developed in international cooperation and coordination. The IPCC is the institution for reviewing research and assessing policy options for climate change adaptation and mitigation. The IPCC reports are detailed assessment reports that synthesise new results from climate research and make the knowledge available for decision-makers. The process is transparent and the material published, but scientific as well as political controversies continue: about the reasons and causes, consequences and impacts of global climate change, the knowledge transfer, the formulation of climate policies, and the elaboration of policy instruments to adapt to global climate change. Critical debates about climate research were initiated by several environmental movements and scientific groups (see Table 7.1). Sceptical environmentalists see climate policy as useless, environmental technocrats see it in terms of engineering, management, and investments, and environmentalists argue for better protection of coastal habitats and their inhabitants. These disputes show more than the simple fact of sea level rise: climate change can be constructed differently and potential governance forms differ significantly. In international climate research and policy developed a climate technocracy with a managerial style of coordination and control of climate policy that creates expertise and powerful knowledge, but allows to bypass controversies and to neglect the seeking of socially differentiated solutions. In Great Britain where larger parts of the population are threatened by climate change, sea level rise, and flooding, began debates and experiments with different forms of adaptive governance (Landström et al. 2013; Lane et al. 2011). The disputes can have consequences for climate policy; the forms of interdisciplinary climate

Table 7.1 Constructions of climate change

Mainstream variants
1. IPCC (2013)—adaptation and mitigation of climate change: climate change as a global environmental change that touches humankind in total and requires global solutions
2. Economics of Climate Adaptation Working Group (2009)—advantages of climate change adaptation in economic terms
Critical variants
3. Social topology of climate change (Blok 2010b)—climate change in the perspective of actor-network theory: multiple globalities appearing in the construction of climate change
4. Critique of ideology (Swyngedouw 2009)—the post-political neoliberal consensus as neglecting the social differences and differences in vulnerability to climate change
5. Social movement theory and climate change (Jamison 2010)—different forms of knowledge influencing the construction of climate change
6. Environmental sociology (York et al. 2003; Urry 2011)—controversy between the theories of ecological modernisation and critical political-economic theories in describing climate change
7. Climate change in social-ecological research (Brunnengräber 2009)—social dilemmas of climate change

Sources: Own compilation; described in the text

research, methods of knowledge syntheses, the mechanisms of knowledge sharing, transfer, policy advice, consensus building, policy formulation, and implementation can change.

International climate policy works with an economic logic of cost–benefit thinking that is most explicitly articulated in the dominant positions of the IPCC and the described below. These two scientific institutions represent the interests of influential scientific actors and powerful global players in climate politics. The international climate policy connects with neoliberal "green economy" strategies through which countries and social groups may be affected differently by climate adaptation measures (Brunnengräber 2009). This gives rise to conflicts and disputes that can be seen from the contrasting constructions of climate change (Table 7.1).

The global climate policy based on the IPCC reports works with the construction of climate change as a phenomenon of global environmental change and as a problem that touches humankind in total, requiring global strategies and solutions. These premises structure the knowledge transfer and the formulation of policies and strategies for climate change.

The arguments from the Economics of Climate Adaptation Working Group are supporting and complementing the IPCC reports that present the scientific results systematically. A joint argument in several critical variants of climate change analysis is that the construction of a common global climate problem ignores manifold social differences in the threatening of different countries and social groups. The burdens of climate change adaptation are distributed unequally; they add up to already existing social, economic, and environmental problems and conflicts. The resulting new conflicts about unequal access to and distribution of natural and social resources are programmed through inadequate policies. The ignored interests and problems reappear as non-anticipated consequences in the further discourse and in policy implementation processes and need to be dealt with attempts to renegotiate, modify, 'repair', or correct earlier decisions. Two new approaches developed in the discussion of adaptation to climate change:

- *Analyses of social vulnerability to climate change* (Adger et al.). In these studies, the questions asked are which social groups in which societies are affected, who wins and who loses economically through climate change.
- *Analyses of contextual vulnerability to climate change* (O'Brien et al.). These studies broaden the analysis of climate change and its effects for society in multi-dimensional and multi-scalar perspectives; they ask for the dynamics of interaction of climate change and other forms of change in society.

Both approaches can produce additional knowledge for climate change adaptation, in reaction to the inflexibility of mainstream climate policy; they support more open and critical discussion of alternative approaches and differentiation and combination of several approaches.

The controversies in climate policy can be described with the variants of constructing climate change and possibilities of reacting to it (Table 7.1).

1. *IPCC—adaptation and mitigation of climate change:* The IPCC approach includes a global strategy for adaptation to and mitigation of climate change to avoid harmful climate impact (IPCC 2013) in which the perspectives of resilience and sustainability are blended in

unclear forms. Mitigation and adaptation are seen as complementary; increasing levels of mitigation to reduce global climate change imply less future need for adaptation. To stabilise greenhouse gas concentrations, the literature reviewed points to a wide range of mitigation pathways. Choices between these pathways are required, for example, the possibilities to bring atmospheric CO_2 concentrations to a particular level; the technologies that are deployed to reduce emissions; the degree to which mitigation is coordinated across countries; the policy approaches used to achieve mitigation within and across countries; the choice between alternative forms of land use; and the manner in which mitigation is connected with the policy objectives of sustainable development. Decisions about mitigation pathways can be made by weighing the requirements of different pathways against each other. Mitigation pathways involve a range of synergies and trade-offs connected with other policy objectives such as energy and food security, the distribution of economic impacts, local air quality, other environmental factors associated with different technological solutions, and economic competitiveness (IPCC 2013: 22). These issues require further clarification of the interrelations between adaptation and transformation, resilience and sustainability.

Economic effects play a dominant role in the internationally coordinated climate change and adaptation research. The economic perspectives in the IPCC and in the following example of the Economics of Climate Adaptation Working Group are strongly influenced from mainstream economics and the neoliberal discourse. A main form of arguing is in terms of income losses and gains through adaptation measures. Although in both approaches the situation of the economically poor, especially in the tropical countries, is included in the discussion, the abstract economic reasoning does not show concrete social consequences of adaptation measures. The question which parts of coastal lands to protect against climate change and which to give up can be answered differently. With the economic logic of decision-making and thinking in terms of return of investment, it is easy to argue for protecting land with high economic investments, as in cities, and to give up land with low investments, for example, agricultural land in rural areas. Taking into account all difficulties with multiple-criteria deci-

sion-making, decisions about land use under the influence of climate change become much more difficult.
2. *Economics of Climate Adaptation Working Group:* The working group formulated four key arguments:

   – Knowledge for climate adaptation is available. Despite uncertainty about the possible effects of global warming on local weather patterns, society knows enough to build plausible scenarios on which to base decision-making, even in developing countries, where historical longitudinal climate data may be limited. Using scenarios helps decision-makers to find decisions that take into account a series of consequences of climate change.
   – Significant economic value is at risk. If current development trends continue to 2030, the locations studied will lose between 1 and 12% of gross domestic product as a result of existing climate patterns, with low-income populations such as small-scale farmers in India and Mali losing an even greater proportion of their income. Within the next twenty years, climate change could worsen this picture of losses significantly.
   – Adaptation measures can reduce the expected economic losses and economic benefits can outweigh the costs. These measures include infrastructure improvements, such as strengthening buildings against storms or constructing reservoirs and wells to combat drought; technological measures, such as improved fertilizer use; systemic or behavioural initiatives, such as awareness campaigns; and disaster relief and emergency response programmes.
   – Adaptation measures can strengthen economic development: Measures with demonstrated net economic benefit are also more likely to attract investment—and trigger valuable new innovations and partnerships. Early investment to improve climate resilience is likely to be cheaper and more effective for the world community than disaster relief efforts after the event (Economics of Climate Adaptation Working Group 2009: 10f)

   The message from this economic analysis of climate change adaptation is that the risks quantified in income losses require adaptation measures to reduce income losses. Adaptation can strengthen economic development. It can be assumed that adaptation measures are cheaper than disaster relief. In economic risk assessment, some differences in the exposure and affectedness by climate change are taken into account. For adaptation measures, income losses and investment protection are dominant arguments; social consequences of climate change are seen in this economic perspective.

3. *Social topology of climate change—climate change in the perspective of actor-network theory:* Actor-network theory (Blok 2010a, b) shows climate change adaptation and its consequences in a social constructivist perspective. The main theoretical argument unfolds in the construction of the topography of networked locales connected through sociotechnical relations. The theory breaks with the idea of hierarchically nested scales from the local to the global, arguing that everything happens on the same topographical plane of networked locales. The scales are no natural phenomena but products of scaling of the actors. The notion of one global environment is abandoned with the notion of "oligoptica" as multiple coexisting globalities (Blok 2010a: 900). The theoretical reasoning is difficult to translate into political options for climate change adaptation. The main value of the theory seems to be critique of the social consequences of dominant global climate policy that can be summarised in the following arguments (Blok 2010a: 909):

– The construction of global climate change through the scientific networks of the IPCC resulted in contested forms of action in global climate policy.
– Alternative constructions of climate change take into account political, economic, and ethical concerns beyond technoscience, where shifting topologies of knowledges, symbols, commitments, and practices are found.
– In the global climate policy, neoliberal forms of carbon accounting, different forms of moral reasoning and moral ambiguities, and socio-natural concerns articulated on behalf of marginal communities in the Global South contrast with each other.

Although the climate discourse is described in differentiated forms, the "locked" theoretical reasoning with a hybrid terminology does not provide concrete arguments for differentiating adaptive and transformative governance of climate change. The theoretical core argument comes close to the reasoning about circulation and flows in actor-network theory, formulated by Latour (1999: 17ff): macrostructures and microinteractions, the local and the global are components of circulating entities. The constructivist reasoning in actor-network theory has two major disadvantages:

- *This theory uses a complicated language for analyses of "relationality" and the reconstruction of different views of the global with an often metaphoric vocabulary and new, hybrid terms that mask the arguments more than clarifying them.* Much of the reasoning is dependent on the use of a terminology that is difficult to reproduce in other theories, approaches, and in the policy discourse. In Bloks' description, the consequences of the constructions are unclear with regard to alternative approaches that could help to break the powerful neoliberal climate policy.
- *The theory and its "relational-scalar topology" is not underbuilt with stronger theoretical arguments from societal and economic systems analysis.* The analysis of society–nature interaction is reduced to variants of communication about this interaction, showing the contingency of interpretation and knowledge creation: it is always possible in different ways. The theory identifies types and forms of networks and their spatial dynamics on the "communication surface" of SES, neglecting the analysis of systemic interaction. With the incompatible constructions of climate change, it is no longer possible to achieve consensus and concerted action. Climate policy is becoming politics, the continuous power- and knowledge-based struggle between heterogeneous interpretations and interests of the actors.

Blok mentions two points that are to be discussed methodologically for all attempts to inter- and transdisciplinary knowledge integration in environmental research:

- *Actor-network theory has a normative message for social scientists:* Through the power of their descriptions, they become part of scientific and political projects of constructing the socio-spatial worlds of global environmental change (Blok 2010a: 910). This implies a transdisciplinary perspective for analysing knowledge practices,

where scientists appear no longer as neutral knowledge producers but to be participating in social and political practices—willingly or not, researchers become stakeholders with specific interests.
- *It is difficult to integrate global environmental governance processes with the heterogeneous and competing knowledge of different actors.* Global scientific knowledge and local ecological knowledge build on different ontologies, not on differences in scale but on phenomenological differences of understanding (Blok 2010a: 901). In transdisciplinary knowledge combination, more complicated methods are necessary to communicate and connect different constructions of the world. Knowledge practices need to be transferred from epistemically different forms of reasoning to reasoning in forms of social action, negotiation, and decision-making.

4. *Critique of the ideology of post-political environmental consensus:* Actor-network theory in its attempt to critically deconstruct the knowledge regimes for climate change is in its cognitive aims similar to critique of ideology in the older forms of critical theory, although, as post-modern theory, it does not use the ideology term. Swyngedouw gives an example for an approach connecting to older critique of ideology. He unfolds the argument that climate change was one of the discourses that brought significant changes in science and politics: the emergence of technocratic-managerial forms of policies, where political controversies and forms of democratic decision-making are replaced by post-political and post-democratic populism. This is described by Swyngedouw (2009: 11) in two points:

Post-political environmental consensus includes a chain of arguments:

- Social and ecological problems caused by modern capitalism are external side effects, not integral part of the relations of liberal politics and capitalist economies.
- A populist reasoning with the interest of an imaginary "the people", nature, or "the environment" is lifted to the level of the universal rather than opening spaces that permit to universalise the claims of particular socio-natures, environments, or social groups or classes.
- The side effects are constituted as global, universal, and threatening.

- The "enemy" or the target of concern becomes socially disembodied is always vague, ambiguous, unnamed and uncounted, and ultimately empty.
- The target of concern is managed through a consensual dialogical politics whereby demands become depoliticised and politics naturalised.

The harmonious view of nature in the post-political environmental consensus is, according to Swyngedouw, "radically reactionary" and rejecting the articulation of divergent, conflicting, and alternative trajectories of future socio-environmental possibilities. Much of the climate change and sustainability thinking seems to build on such harmonious views that eliminate alternatives. Such thinking silences the antagonisms and conflicts that are constitutive of socio-natural orders in modern society. The current post-political condition constitutes a particular fiction that forecloses dissent and the possibility of a different future. There is a need for different stories and fictions, for formulation of different socio-environmental futures, for recognising conflicts and differences, and for struggle over the naming and trajectories of these futures.

Swyngedouw's arguments seem clearer than that of Blok at the point of reclaiming democratic politics to fight for alternative views and futures. For that purpose, his main argument is that socio-environmental conflict should be legitimised as constitutive of a democratic order, turning the climate question into a question of democracy and its meaning. Democracy requires the expression of conflicts, agonistic debate and disagreement, and naming of different possible socio-environmental futures. The reasoning is similar to the post-Marxist reasoning of Mouffe who uses Schmitt's concept of politics in critical analyses to argue for a radicalisation of democracy as "agonistic pluralism" (Mouffe 2013). How the critique of post-political consensus and harmonious views of nature, formulated in abstract terms, can be transformed into new practices of climate change action, or in politics of resistance to technocratic-managerial climate policies, is rather unclear. This transformation would require more than critique of ideologies and different climate stories: a reconstruction of political knowledge practices, formulation of strategies for transformative governance, identification of political subjects that can change the economic and political power structures in the present world system, and formulation of strategies of collective action.

Other critical research in the social and political sciences supported the critique of the policy-oriented climate research where coalitions of governmental, economic, and scientific power elites transformed the culture of research. This is described in the "triple-helix" hypothesis or in the theory of new knowledge production and "mode 2" (see below, Jamison).

5. *Social movement theory and climate change:* Jamison (2010) reconstructs in a critical sociology of knowledge several contrasting discourses on climate change with empirical and theoretical knowledge. His theory differs from the reasoning in actor-network theory and critique of ideology through its concrete analyses of knowledge practices connected with social movements. Jamison identifies the producers of scientific and other knowledge, among these also social movements as knowledge producers in their own right (for further discussion, see Chesters 2012). This description of knowledge producers is useful in discussing the contrasting and competing forms of knowledge production about climate change according to

- *types of reasoning:* oppositional (climate scepticism), dominant (green business), and emerging (climate justice);
- *types of movements*: neoconservative, neoliberal, and global justice;
- *scientific perspectives*: academic/disciplinary, entrepreneurial/non-disciplinary, and cross-disciplinary;
- *knowledge forms:* traditional/personal, contextual/proprietary, and hybrid/public.

For Jamison, the dominant approach in knowledge production for climate change corresponds to the new "mode 2" of knowledge production, where the traditional boundaries between science and politics and the borders between the academic and commercial spheres are transgressed. In the newly emerging knowledge practices, science is not carried out in a disinterested and impartial fashion; research is funded by non-scientific sources and interests in order to contribute to policymaking and technological development. Such policy-relevant research dominates in many climate research centres and in the IPCC. The research is often carried out in networks of academic, governmental, and business organisations, in projects to provide policy advice as well as profitable "solutions" to climate change. "Mode 2" implies epistemic criteria differing from academic science, with different methods of investigation, different forms of interpretation, and different rationales of justification and verification of knowledge claims and practices (Jamison 2010: 819f).

This approach of social movement theory applied for climate research is concrete in its arguments about knowledge production and application. In difference to the two approaches of actor-network theory and critique of the ideology of a post-political environmental consensus, Jamison has no difficulties to translate arguments about knowledge production and changing knowledge practices in descriptions of social and political action, as in the analysis of climate policy. In parallel to a broader typology of changing forms of knowledge production that Jamison developed in earlier writings, three kinds of knowledge practices can be observed in the climate discourse:

– A variant of traditional academic science (also called "mode 1") that is connected with the movement of environmental scepticism, questioning the existence of climate change.
– A variant of "mode 2" or transdisciplinary knowledge production done in cooperation of academic researchers and private business, described in the "triple-helix" theory of new knowledge production as cooperation of science, government, and business. This is the mainstream of climate research that includes the IPCC, although it does not mean that research is done in direct cooperation of science and business. It is enough that the research takes into account economic interests of certain actors as the ones that opt for the neoliberal "green economy".
– A third variant, one of critical "mode 2"-knowledge production, is connected with the newly emerging movement of environmental or climate justice as a critical movement. This seems to be a concrete form and example of a new critical practice in climate policy that takes into account the social inequalities that appeared in the

critical reasoning of Blok and Swyngedouw, although not concretised in terms of strategies for collective action.

The three knowledge regimes appear in a temporal sequence: the traditional academic "mode 1" is the one developing in the first half of the twentieth century as a dominant form, but less important although not vanishing with the new forms of knowledge production. The newly emerging movement for climate justice described by Jamison includes different actors and interests. The question is whether environmental justice can become the guiding idea for formulating political strategies and knowledge practices and is able to generate new climate politics to deal with social inequalities in climate change mitigation and adaptation. Political strategies require more than normative ideas of social and environmental justice and fair sharing of natural resources. The environmental justice movement seems to be locked in ethical thinking that needs to be reinforced through political or social strategies and finding of partners in the power struggles in environmental politics.

6. *Environmental sociology—the controversy between the theories of ecological modernisation and critical political-economic theories:* Analysis of climate change requires use of sociological knowledge and theories for interpreting and reconstructing the natural-scientific knowledge on man-made global climate change. In the theories of ecological modernisation (Mol) and critical political-economic theories ("treadmill of production", Schnaiberg; "ecological rift", Marxist theory), climate change is interpreted on the basis of systems analyses of society and economy. York et al. (2003, 2004) attempted to test the arguments from these theories with data available about greenhouse gas emissions, methane, and carbon dioxide that cause climate change. They use a modified variant of the ecological IPAT model where the total environmental impact (I) is a multiplicative function of population (P), per capita consumption or affluence (A), and impact per unit of consumption or technology (T), transforming it in stochastic form (York et al. 2003: 280f). The results of their global analysis include that population size, affluence, industrialisation, and urbanisation increase emissions, and that tropical countries have lower emission than non-tropical countries. The largest part of emission is from industrialised countries in the moderate climate zone, a result that is not

falsified through the data of growing emissions from the newly industrialising countries since the last decade. The authors interpret the results as not supporting

- the theory of ecological modernisation (Mol), where technological change is assumed to result in lower emissions, or
- the hypothesis of a Kuznets curve, where it is assumed that in the course of economic modernisation emissions are reduced.
- Both approaches are reasoning with technological improvements and institutional change. Arguments from other theories are supported by the data:
- Those of the "treadmill of production" theory, arguing that the industrial-capitalist mode of production causes the environmental disturbance, including climate change.
- The hypothesis of the "metabolic rift" dating back to Marx, arguing that modern capitalist production, agricultural and industrial, interrupted the natural cycles of nutrients and materials that are necessary for maintaining the functions of ecosystems.

York and Rosa argue that institutional and technological change, the factors on which ecological modernisation and the Kuznets curve hypothesis are built, cannot be neglected, but more fundamental changes are required for transitions to sustainability. Technological and institutional change cannot transform the systemic mechanisms of the capitalist mode of production in the industrial society as the main causes for emissions of greenhouse gases. Transitions to sustainability require forms of production and consumption with lower environmental impact, taking into account the variation in emissions at a given population size and level of development. More complex strategies to reduce negative environmental impact can build on interdisciplinary research of material and energy flows in industrial ecology (York et al. 2004).

Although the interpretation, the methodology, and the data used by York and Rosa are criticised by Mol, their main argument of system transformation is not dependent on these data or their interpretation alone; it is supported from further research, including the analyses of global resource flows in social ecology. The authors reject ideas and strategies of technological and institutional adaptation that are part of continuing economic modernisation, but do not support social-ecological transformation and significant reduction of negative environmental impacts. Some critical sociological analyses of climate change, neoliberal capitalism, and the dominance of economics in climate change adaptation (Urry 2011: 119f) do not sufficiently analyse requirements of social-ecological transformation of society. Urry seeks a solution through a renewed capitalism called "resource capitalism", where nature would not be separated from economy—but he concedes doubts, whether a solution of climate change problems is possible within capitalism. More critical analyses of social-ecological transformation include political-economic theory, world system theory, and theories of societal metabolism in social ecology or their synthesis.

7. *Climate change in social-ecological research:* Brunnengräber (2009: 62 f, see also Brunnengräber et al. 2008) analyses the social dilemmas of climate change. This research argues that the present ecological crisis results from contradictory interactions between nature and society and contradictory forms of their regulation. The actors on the markets use natural resources without taking into account externalities, social inequality, and environmental degradation. Governmental regulation is adapted to neoliberal globalisation and global exchange of commodities. Many citizens demand nature protection and contribute simultaneously as consumers, through their forms of life to destruction of nature. Social inequality shows not only within countries but also between industrial and developing countries The poor countries in the tropical zone contribute least to emission of greenhouse gases and have to suffer most from climate change; the industrial countries can protect against negative consequences of climate change. But also in the Global North, the social differences between rich and poor imply differences in vulnerability.

The arguments from social-ecological research sum up to a differentiated reconstruction of the climate change discourse:

– Climate change effects are not global in the sense of globally equal or common problems; they differ strongly in the social classes of society and are influenced by "multiple inequalities". The heterogeneous

consequences and vulnerabilities prevent a common diagnosis of the problem of climate change.
- Climate change as part of complex structures, of socio-cultural and socio-ecological relations, is simplified and subjected to socially selective (ideological) reasoning when it is seen as a global and common problem of humankind.
- The natural-scientific reasoning of a "crisis of nature", the discussion of global catastrophes in the media, and the moral reasoning in terms of a global problem of humankind do not adequately describe the phenomena of crisis.
- The question that needs to be asked is which interests and interpretations of the problems and the political perspectives guide the political and economic reactions to climate change. Climate politics is a conflicting field in which not necessarily protection of the climate is at stake, but the control of societal and economic crises and their regulation.

The social-ecological research about climate change and adaptation to it can be complemented through the elements of a new social-ecological theory of nature–society interaction. With this theory develops a framework for the analysis of social-ecological regulation of societal interaction with nature that includes the themes of climate change and sustainability. This operational framework derives from the theoretical concept of a social-ecological regime (Fischer-Kowalski et al.; see the detailed description in Chap. 4).

From the comparison of the seven approaches to climate change analysis, two conclusions can be drawn:

1. *The discourse of climate change and adaptation to it is trapped in a vicious circle of constructing competing scientific narratives and models for political action.* In the transfer of knowledge less the quality of scientific arguments counts, more the power relations and vested interests in science and politics.

    - The first two approaches, the global strategy formulated by the IPCC and the adaptation strategy of the Economics of Climate Adaptation Working Group, represent the powerful positions that

inform the neoliberal, market-based climate policies, connecting natural-scientific with economic arguments. All other approaches broaden the knowledge perspectives to include more social-scientific knowledge and draw attention to the inconsistency and the ignored or non-anticipated social consequences in the dominant climate narratives.
- The climate discourse is an arena where the dispute about adapting to climate change or transforming industrial society and economy is carried out. In the dominant positions, the transformation debate is avoided with the dominant terms of adaptation and mitigation of climate change. In the critical approaches, insofar they are not stuck in the knowledge construction debate; the connections between climate change adaptation, mitigation, and transformation to sustainability are more clearly articulated. But also with—and between—them the controversies about possibilities, pathways, and methods of societal transformation to sustainability continue. Minimal consensus about a sustainable society is that it becomes climate-neutral through a decarbonised economy. The technical change in the form of "green technologies" and renewable energy sources to reduce CO_2 emissions does not cover the whole transformation process and its forms as societal change.
- The description of the programme of industrial ecology by York and Rosa can help to analyse societal system dynamics with analyses of material and energy flows through industrial systems. With such analyses, the quantitative and qualitative dimensions of natural resource flows and consequences of their appropriation and transformation in economic production can be assessed. Results include, for example, socially unequal exchange and "hidden flows" that make a large part of materials extracted that do not enter in the production process appear in the economic system as waste.
- Industrial ecology has a limited knowledge perspective with the focus on technological innovation to improve the eco-efficiency of production and resource use. This limitation has been criticised, for example, regarding non-intended negative effects of technologies. In social-ecological analyses, it is necessary to describe and analyse resource flows not only empirically as flows of materials, energy, and

information, but the flows are constituents of the metabolism of industrial society and its economic structures that block the transformation to sustainability and require further empirical and theoretical knowledge to identify possibilities of transformative governance.

2. *Social-ecological and other critical analyses of global environmental problems made, with the analysis of climate change, visible the difficulties and the necessity of including in environmental governance conflicting interests, contrasting arguments, and different scientific narratives.*

   – A critical point of scientific knowledge construction became visible in the discussion of climate change analyses and policies of adaptation: the dominant neoliberal climate policy and its scientific underpinning through economic analyses of climate change adaptation ignore contrasting forms of knowledge use and contrasting interests. Large parts of social-scientific knowledge from societal systems analyses are neglected; concrete and contrasting interests of social groups are suppressed with the construction of "common interests". The challenge of climate change analysis and climate policy has been formulated by Sayre (2012: 67) as thinking, studying, and acting across different spatial and temporal scales of climate change and its complex web of causes and consequences. This is not yet concretised in knowledge practices that can support the cooperation of different groups of knowledge bearers with different and conflicting interests and values. To find such knowledge practices cannot be left to the political process where they need to be found somehow. The political process itself needs to be broadened and changed.
   – Climate research is influenced by different worldviews or paradigms for the production of scientific knowledge and by different interests articulated in the climate discourse. The separation of scientific knowledge from power relations in science, politics, and economy supports an idealist view of interest-neutral science. The difficulties of theoretical, epistemological, and methodological kind in knowledge synthesis cannot justify such idealism that works with constructions of "our common future" and "common interests of

humankind". The processes of transferring knowledge into practices of resource management and policy formulation seem to be contingent, depending on the selective knowledge use of epistemic communities. It seems necessary to find additional and new forms of knowledge communication, integration, and cooperation between different groups of knowledge producers, users, and appliers. Conflicts and disputes need to be dealt with in other, more elaborate and practical forms than ethical reasoning.

– In the debate of *transformative learning, literacy, and agency*, the understanding and shaping of societal transformations is at stake (WBGU 2011; Scholz 2011; Schneidewind 2013), in efforts to connect transformative research, transformative education, and transformative action. Other, more specific forms of building action capacity and transformative literacy are developing in the British research and practice of climate change adaptation (Landström et al. 2011; Lane et al. 2011), where futures are imagined and constituted through knowledge practices that connect policy, management, and science. In both forms of creating transformative agency, one can find three components: involvement of the target groups (for example as resource users or as citizens who have to live with specific consequences of climate change such as flooding); inter- and transdisciplinary forms of knowledge integration; and knowledge generation practices for anticipating and constructing the futures of a society that learns to change, adapt, and transform. Transformative learning and literacy is a broad research theme in pedagogics and other disciplines which can also be applied in the environmental governance debate. However, as the similar debate about transformative agency, it is often limited to individual learning and behaviour changes or to learning within a single organisation (Haapasaari et al. 2014). In ecological research, a theory of transformative agency in SES has been discussed (Westley et al. 2013); it suffers from its limitation to ecological resilience research and the adoption of the adaptive cycle model. Application and combination of the concepts and theories of transformative learning, literacy, and agency in transdisciplinary environmental research can help to create new forms of transformative governance. For the

analysis and discussion of social-ecological transformation, these concepts are not yet sufficiently developed and synthesised.
- At the end, the question of creating new knowledge practices becomes again one of social-ecological theory. How can the social-ecological transformation discourse deal with incompatible forms of knowledge production and construction, different epistemologies and methodologies, heterogeneous interests of social actors, and contradicting worldviews and paradigms supporting a plurality of climate narratives? The differences cannot be dissolved through the construction of one theory or by way of explanation, but only through a theoretical discourse where cooperation between different knowledge forms and practices is discussed, beyond the analyses and explanations created through the theory. The social-ecological theory is in continuous development, it reconstructs global environmental change, interconnections of social and environmental problems, and possibilities of societal transformation. A theory of this kind requires combinations of different theories, interdisciplinary knowledge syntheses, and a mapping of knowledge practices for social-ecological change and transformation. The aim of such a theory is not to integrate all competing theoretical variants of climate change analyses but to show their differences, their justification, and relative validity.

Adaptation to Climate Change and Sustainable Development in Coastal Areas

The paradigm of a coastal society in European history was the Roman Empire. The coast was the nexus of the economy in this historical world system of political nature (empire). In the Roman society, provision of food resources, maintenance of the economic reproduction, and the development of the society were dependent on maritime transport routes along the Mediterranean coasts. The continuous expansion of the societal system from a local society to a political world system became possible through coast-bound expansion. Roman society was studied in several scientific disciplines, also, by human ecologists (Tengström) and environmental historians (Sieferle). The Empire seemed to show the archetype of the modern economic world system, also, with regard to its socially, economically, and environmentally unsustainable forms of development. At the end, the Roman Empire collapsed for different reasons, among these ecological ones (the overuse of natural resources and destruction of ecosystem in the colonies) and political ones (the Empire expanded to a size which exceeded the political and military possibilities to maintain and defend it).

In the Roman Empire, all important flows of material resources were dependent on the maritime transport system, the only form of mass transport of goods in human history before the technical invention of the railway in the nineteenth century. To protect the continuous resource flows through the Mediterranean Sea with a big commercial fleet required a military fleet for protection (mainly against pirates). The Roman Empire overcame the limited, land-based expansion of earlier empires through expansion across the sea, which increased the possibilities to grow and compensate the limited resource base in the centre through imports from the colonies in the periphery. The similarities with the modern capitalist world system do not help to explain the functions and problems of coasts in the modern society with its capitalist economy and mode of industrial production. Coastal development problems in modern society can be described in three main aspects:

1. *Coasts in the modern society are still important for maritime transport and natural resource use, but their dynamics as SES have changed* with the social-ecological change of coastal SES in later history, the high population density in coastal areas and their cultural and economic importance for recreation and tourism. As a concomitant of socio-cultural modernisation in Europe these changes are, for example, studied in the social sciences by Corbin (1990). He analysed the discovery of the coast and the sea by tourists in European countries since the mid-eighteenth century, more or less parallel with the development of industrial society. The changes described at the beginning of the chapter show the social, economic, and ecological importance of coasts in modern society: the concentration of population, settlements and cities, infrastructure systems including transport and communication systems, and the abundance of natural resources in terrestrial and maritime ecosystems.

2. *The ecological functions and processes in coastal areas are important for the material and symbolic reproduction of modern society, but this society is not a coastal society in the strict sense of the term.* Certain countries are located more or less completely in the coastal zone, but policies cannot only concentrate on these. The globally networked modern society requires more complex, multi-scale perspectives in the analyses of climate change. The modern economic world system as reference system for the analysis of society encompasses the total global space, land, and sea. There are only few unexplored spaces that have not yet colonised nature, especially the deep sea, which becomes more and more important for natural resource extraction and mining. The system-maintaining interactions in modern society are, since the beginning of modernity with the building of the modern world system in the early sixteenth century, global relations and processes, flows of goods, exchange and communication that are part of the building of the system of modern capitalism. Topological metaphors, for example, of coastal cities as nodes in global networks, do not show the systemic nature of the global processes and their causality.
3. *In the twenty-first century, coasts as the territorial parts of modern society that are most vulnerable to climate change gain new importance in the ecological discourse and in environmental governance.* Climate change cannot be reduced to natural processes and its description as environmental change is inexact. It is a change in which society and nature change simultaneously, in complicated interaction, structured through the modern economic world system. This system followed in its history the logic of expansion through conquering of new territories and colonisation. Today, expansion has become economic growth, capital accumulation, and neoliberal globalisation. The social and economic structures and deficits of this system and the maladaptation of modern capitalist society to its natural environment cause societal crises in which social and natural disasters are blended.

What appears for the people at the coast as natural disasters through climate change appears in the perspective of the modern capitalist system as an economic disaster of investment, as disinvestment and blocked growth. Climate change enforces ever larger protective activities, defensive costs of protection, and the reorganisation of the accumulation process.

Defensive costs for the repair of environmental damages and restoration of ecosystems are since longer seen as indicators that economic growth happens in a vicious circle of resource use and pollution and is slowing down in the long run, when the global economic system reaches the global limits to growth. The socio-economic crisis is not a consequence of the normal, periodical crises of accumulation and economic reproduction of modern capitalism. It shows a dysfunctionality of modern society at higher levels of organisation, including malfunctioning interaction between SES; it can be the beginning of involuntary transformation of the modern society. This transformation can include the biggest migration and resettlement processes in history, processes of disinvestment, devaluation of capital, destruction of physical capital and infrastructure, and violent processes with wars and civil wars.

The empirical research on coastal and marine management and development in Europe shows that many problems of environmental and transformative governance are unsolved and wait for solutions in future, under higher problem- and time-pressure.

1. *Coastal development:*
    - *Integrated coastal zone management* (ICZM; see Appendix) has been practised in Europe since 1996 and in other continents still longer. This approach with few and vague principles was not effective to deal with the consequences of social-ecological change. The EU policy of ICZM has failed, to a large degree, because of neglect of new scientific research and failed attempts of learning ways towards transitions to sustainability (McFadden 2007; McKenna et al. 2008). With the new initiative of marine spatial planning no significant change and renewal happens, more a broadening of conventional management and planning practices from the land to the sea.
    - *The Common Fisheries Policy of the EU* achieved little success in converting top-down governance and implementing fisheries co-management in efforts to control overfishing. After more than twenty years of experimenting with cooperation and participation (McCay and Jentoft 1996; Symes 2006; Coers et al. 2012), the further development of co-management is unclear, rather expected

through influence of the ideas of adaptive governance and sustainable resource management (Linke and Bruckmeier 2015). Participatory management in marine protected areas is studied in few case studies only (for example Vasconcelos et al. 2013); the debate about different understandings of the purposes of protection continues (see below).
– *Adaptive management in coastal fisheries* has, in Europe and elsewhere, brought few successful examples (Walters 2007). In coastal areas, adaptive management could be adopted for the management of water- and land-based resources. But the experience with its introduction showed that it did not spread rapidly and easily, also not with newer improvements (see Chap. 6).
– *Coastal conflict management* is badly supported from governmental institutions and shows big deficits, as well as in the framework of ICZM, as in other approaches (Bruckmeier 2014; Stepanova 2015). The spreading of newer approaches, for example, conflict mediation, is a slow and difficult process, the necessity of conflict mitigation itself disputed in coastal research.
– *Ecosystem-based management* in coastal areas and in fisheries management, intensively discussed in the USA, is not widespread in Europe. The majority of European fisheries is still based on single-species assessment and ignores, according to Mollmann et al. (2013), the wider ecosystem context. These authors studied the situation in the Baltic Sea; they identify as a reason for the slow progress of ecosystem-based management the lack of a coherent strategy, although integrated ecosystem assessments offer such a strategy.

2. *Marine governance*:
In the broader discussion of protection and management of marine areas and resources, four models of management have been found to describe the practices: government protected areas, private protected areas, co-managed protected areas, and community conserved areas (Borrini-Feyerabend et al. 2006). The question of effectively combining protection and (sustainable) use of resources is, after repeated discussion (for example Noel and Weigel 2007), still controversial. Its

clarification would be decisive for advances in the processes of ecosystem-based and sustainable resource management. Additionally, protected areas are disputed whether they help to resolve conflicts and to integrate the interests of stakeholders. Recent debates confirm that designation of protected areas evokes new conflicts when trying to solve other ones. Gaines et al. (2010: 18251) summarise the trends observed in research on marine reserves: the ideas of reserves, ecosystem-based management, and marine spatial planning spread, but the consequences remain unclear. There is not yet a breakthrough to social-ecological transformation, although the protection processes become more complex; they are broadening beyond single reserves and ecological components to reserve networks and taking into account socio-economic factors and impacts. The networking of protected areas results in large areas for marine management (LAMM), for example, seascapes, marine ecoregions, large marine ecosystems, regional seas, integrated coastal management, all of which relate to coast-adjacent areas (Bensted-Smith and Kirkman 2010). Social complexity implies to take into account in marine protected areas what is often neglected in the design and management of protected terrestrial areas: the role of people, of local users of the area, and potential conflicts between conservation and resource use (Andrew-Essien and Bisong 2009).

3. *The development of climate policy for coastal areas in Europe:*
The policy and legislative documents of the EU follow to a large degree the principles and ideas found in the IPCC reports, highlighting the economic importance of European coasts, the potential damages and income losses through climate change, and the high population density in coastal areas where 30 % of the European population lives at the coast: not more than 50 kilometres away from the coastline. In all policies and strategies in coastal areas, the EU follows similar principles of integration and coordination: with ICZM that is still applied although it was not successful, and the Water Framework Directive, the Floods Directive, or the Marine Strategy Framework Directive that aims to achieve good environmental status of European marine waters by 2020. The policy programmes, the legal instruments, and the research and development projects in coastal areas supported by the EU follow the overarching principles and goals of sustainable development.

Nevertheless, the European coastal areas are not on the way towards sustainability, which can also be seen from the data published by the EU that confirm growing risk and vulnerability, threatening of coast through erosion and flooding.

The empirical research and the policy discourse show that the ways to transform coastal and marine SES to sustainable systems need to be found in the future, with knowledge practices that include research and others. More important become the combinations of different knowledge practices in governance processes—transdisciplinary research, experimenting, simulating, modelling, envisioning, and others. The problems to deal with in transitions to sustainability—reducing the environmental impact per unit of economic activity, lowering the worldwide rate of economic growth, and addressing global income inequality (Stutz 2009, 49)—are not yet specified in governance forms. These goals are "heroic abstractions" and nearly utopian formulations, derived from an inexact diagnosis of present trends of global social change. To deal with these problems requires significant changes in coastal, marine, and climate change management.

Rebuilding coastal management: Renewing coastal management in the perspective of transitions to sustainability requires first of all better connection and integration between different territorial and sectoral approaches in environmental governance. Terrestrial, coastal, and marine management need to be spatially connected, the management of climate change and sustainability policies thematically. Climate change is a major challenge in coastal areas that cannot be left to the particularism of national environmental policies. The policy and management institutions at national and international levels, necessary for integrated policies and global governance, have not yet developed mechanisms supporting social-ecological transformation. Governance of climate change implies adaptation in the sense of resilience and, beyond that, governance of transitions to sustainability. The connection of the two processes requires learning and correcting failures of past policies and a change in governance beyond these forms inbuilt in the policy process as routines: a struggle for new political orders and hegemonic constellations that are not programmed from the economic power relations in the modern world system.

- Climate change, resilience, and sustainability should be seen as conflicting issues that require transformation of political institutions and cannot be left to specialised governmental organisations coordinated in top-down management. The new organisational model for integrated administrative systems for coastal management has not yet been found with ICZM. Much more than temporary, task force and project organisation will be required. It seems time to leave the older ideas of ICZM and find more knowledge-based approaches that support transformative governance.
- Governance of climate change and sustainability cannot be based on some normative policy principles in the sense of panaceas, trying to address all environmental problems with one simplified approach. Social-ecological research as that of Ostrom et al., discussed in preceding chapters, showed the problems with panaceas, opting for the networking and nesting of many and different local management approaches. This is still thought in a narrow perspective of management and needs to be elaborated for higher-order governance processes.

In climate and sustainability, governance developed only same management principles during the past three decades that can become elements of future strategies:

- Environmental governance in multi-scales policies
- Mitigation of conflicts between resource users about unequal access to and distribution of resources, conflicts within and between countries ("ecological distribution conflicts": Martinez-Alier)
- Opening and broadening of strategies for resource management, beyond policy instruments, to connect managed and non-managed processes of social-ecological change.

Various difficulties and hindrances in the processes of integrating and broadening policies and governance can be envisaged:

1. *Climate change is not the only problem that needs to be dealt with in environmental governance.* Reducing global governance to climate governance is another dead end that implies reducing the transition to sustainability to managing climate-related resilience and disasters, as

argued in some forms of resilience research and practice. Ideas of creating resilience as a form of or instead of sustainability are misleading and neglect the challenges of societal transformation and social-ecological regulation that sustainability implies. Climate change adaptation should not be used to level down environmental governance to resilience. Not merging, but connecting analyses of resilience, climate change, and sustainability as specific fields of knowledge production for interdisciplinary knowledge syntheses opens possibilities to develop workable forms of sustainable resource management. How far this reformulation has advanced can be seen in the debates about global environmental governance with regard to climate change adaptation. Deere-Birkbeck (2009) summarises the debates as requiring strategies to deal with interlocking environmental risks (environmental disruptions and disasters that exacerbate social vulnerabilities). Solutions to the risks need to be politically and legally feasible, ethically acceptable, publically discussed, and democratically legitimised. Similarly, Bernstein formulated some requirements of new legitimacy as social acceptance and justification of shared rules for the global community to achieve new policies for knowledge production and application, for financing, development, and security.

2. *The ideas of global environmental governance to adapt to climate change* in the global governance discourse (Biermann 2004; Bernstein 2005; Deere-Birkbeck 2009) show that important requirements for earth system governance are not yet realised. Improvements need to happen in future, under deteriorating conditions and with more or less resistance to be expected. Among the improvements that are insufficiently discussed and underestimated is that of interdisciplinary knowledge synthesis, of knowledge and burden-sharing in the regulation of global change. Global governance develops under the auspices of economic globalisation, which explains to some degree the distorting influence of neoliberal policies of deregulation and market-oriented policy reforms on the climate and sustainability discourses. New and improved forms of participation that can be described as "multilevel democracy", where citizen are actively included in decision-making at local community level, in national and international politics, are difficult to develop under conditions of de-politicising and de-democratising policy

reforms under neoliberalism. At least in the initial phase of changing global governance, it is necessary to work with weak, temporary, and informal forms of participation and local power sharing, without strong, democratically legitimised legislative and regulatory powers at international levels.

3. *New forms of environmental governance* need to deal with multi-scale problems and conflicts, with insecurity, risks, and disasters, with redistribution and sharing of resources. Creating democratic legitimation for institutional changes and long-term transformative agency are the power-, knowledge- and resource-related problems to deal with. The accumulation of scientific knowledge will not be sufficient to transform societal and economic systems. Assumptions of the kind "global problems require global management institutions" are misleading the sustainability process when solutions are sought through the building of centralised institutions. The reduction of governance to normative principles of the kind of ecological citizenship, environmental justice, environmental democracy, and new normative orders is also misleading when these principles are not connected with mechanisms to transform the policy processes and the processes that maintain the economic world system. Sustainable resource management in multi-scale processes and transitions to sustainability requires more and other forms of interdisciplinary knowledge synthesis, knowledge practices, power sharing, and changing power asymmetries in the political and economic systems. In the sociological debates of actor-network theory, risk society, and reflexive modernisation, such concepts as "sub-politics" (Beck), "existential politics" (Giddens), and "parliament of things" (Latour) show more lack of clarity than principles of new environmental governance that can be realised through institutional reforms. Offe (2009) argued in the review of the governance concept that governance develops in areas, where governments alone cannot act, or not act successfully, where cooperation of governments, non-governmental actors, private corporations, and civil society actors is required. This is the situation in environmental policy where the communication barriers between science and policy are discussed more intensively in the last years in ecological and social-scientific environmental research.

For coastal development, climate change adaptation, natural resource management and sustainable development, and the overarching concepts of governance of transformation or transformative agency are discussed in ecological and social-ecological research. Theoretical debates of social-ecological transformation and regulation (Becker and Jahn 2006; Fischer-Kowalski and Rotmans 2008; Brunnengräber 2009; Chapin et al. 2011; Brand 2015b) are the basis for the development of ideas of transformative governance. The ideas of social-ecological transformation and transformative agency reached the agendas of governmental policies in several countries (for Germany, see WBGU 2011) and are discussed everywhere in science and policy (Pereira et al. 2015). New ideas for reformulating strategies of sustainable development come with the transfer of transformation research into approaches of "governance for social-ecological change" (Siebenhüner et al. 2013). In addition to more conventional ecological debates of transformative agency and adaptive governance (Westley et al. 2013, see Chap. 6), the social-ecological debate generates ideas for the analysis of long-term policy problems and for use in participatory policy, knowledge integration, and joint learning.

Governance of transformation, to achieve a future sustainable society, is confronted with a series of difficulties. A debate about ways of achieving global transformative governance and agency unfolds with the following points:

1. *Sustainability paradoxes:* Sustainable development seems a contradiction in itself, a non-growing economy, impossible in the modern world system and the industrial society. Sustainability policies exist mainly in national programmes for sustainable development, but how to achieve global sustainability from the heterogeneous and often badly working programmes is unclear. One does not know what can happen in the future, but acts as if one would know what can happen. The management of complex SES is seen as impossible, but continuously programmes and strategies for ecosystem-based management are discussed and newer ecological research is working with the paradigm of complex adaptive systems; in ecological resilience research, the future-oriented action is described as planning for that which cannot be planned. The future society that will develop through social-ecological

transformation is unknown, but transformation needs to be initiated with some ideas about this future society. Research cannot improve much the "informed guesses" about future development and change. The near future of the coming decades seems rather clear, overshadowed by the rapidly increasing global environmental change and social conflicts. The distant future is not foreseeable, not with the construction of global scenarios and other methods. Whether the new governance processes discussed and constructed with the ideas of cooperation, experimenting, joint learning, preparing for disturbance, adaptive governance, with elements of anticipation, reflection, planning, and replanning, are fulfilling the expectations to construct the future is insecure. The ways into the future are paved through trial-and-error processes and many of the new governance ideas are not yet practised.

2. *Global transformative governance develops with knowledge from ecological and social-scientific research,* in attempts to cope with complex environmental problems and to proceed towards global sustainability through multi-scale policies. These policies include combined local, national, and global action strategies for which multiple interests and different views and constructions of the environmental problems or their solutions need to be discussed, negotiated, and matched. The rebuilding of environmental governance works better in bottom-up processes of networking and nesting local policies and management practices than in top-down approaches of coordination and governance that tend to fall back in default options for standardisation, centralised coordination, and abbreviation of necessary knowledge integration in such forms as found in climate adaptation research and policy. There is no historical experience with such problems as how to achieve global sustainability. Global transformative governance needs to develop from knowledge and experience gained in the sustainability process, from experimenting, and learning how to build more complex, multi-scale governance approaches; it requires indirect management approaches for social-ecological processes that cannot be managed and changed directly through political decisions. Transformative governance is higher-order governance. Ideas for the management of change described in organisational research (Anderson

and Ackerman 2010) do not cover the necessities of change in transformations of SES, and are useful only at the level of individual organisations participating in the process. Further, principles of transformative agency described in ecological research on resilience and adaptive management are of limited use as well. This research, discussed in preceding chapters, tends to bypass the complicated translation of scientific knowledge about SES-dynamics to knowledge for governance. It works with quickly formulated ideas that can easily be rejected through further discussion. Some ideas emerge from empirical research, some from reacting to broader governance debates, but they do not foster a consolidation of research and new forms of governance, rather support governance practices of muddling through. What unfolds with the critical social-ecological debates of transformation is another view of governance processes and practices in the meaning of higher-order governance or regulation of governance (Brunnengräber et al. 2008; Brand 2015a, b). Some preconditions for such governance are formulated in research on environmental governance:

- *Political framing strategies to develop new institutions:* Not coordinating existing institutions but building transformative capacity at different levels of policies is the main requirement of social-ecological transformation. To initiate such systemic changes in politics and in the economic system, possibilities to control and regulate the disembedded markets need to be explored. Neoliberal environmental governance, also with its negative social consequences that reinforce inequality, tends to develop from governance through deregulated markets to more authoritarian political forms of governance that reinforce the power asymmetries in global economy and politics.
- A *long-time perspective of several generations* is required for transitions to sustainability, to establish phased processes of transformation with the constitutive components of *managed change* (transformative governance, including changes of political power structures), of *non-manageable change* ("indirect management" of autonomous social and ecological processes), and of *social-ecological regulation* (higher-order governance or regulation of governance) in temporally structured processes with many phases and knowledge

feedback where progress in change can be monitored and measured.
- *Multi-scale approaches for the networking and nesting of institutions and for connecting the governance processes at various levels.* The levels and scales of processes in social and ecological systems are variable and changing and cannot be derived from institutional characteristics of established political systems, and not from physical characteristics of ecosystems; this has been discussed in geography (Lefebvre 1991; Brenner 2000) and ecology (Reid et al. 2005), with the consequences that politics of scale in transformative governance cannot be simple processes of integrating or coordinating processes and institutions at several—local, regional national, international, global—levels in political systems. Multi-scale governance is more complex governance of *continued rescaling and reconnecting scales.* Coordination of multi-scale management systems requires global coordination of policies, without hierarchical and linear top-down processes; such coordination can be effective to the degree that successful strategies at lower levels can be networked and several integrative mechanisms reinforce and control each other.
- *Combining and synthesising of knowledge from different fields of research* requires criteria for the knowledge use and critical assessment of the ecological and economic knowledge that is so far prioritised in global climate policy. Knowledge syntheses for transformative governance imply several components, knowledge about coupling and interaction of SES, analyses of vulnerability, resilience, and sustainability, and knowledge about mitigation of resource use conflicts. The syntheses of scientific knowledge support the creation of new knowledge practices and the use of theoretical knowledge from social-ecological theory to reconstruct the systemic structures of SES and compensate the limits of empirical research.

3. *Higher-order governance, multi-scale governance, and knowledge-based strategies* working with interdisciplinary knowledge syntheses are basic principles of transformative agency and governance. With them appear the knowledge- and power-related problems so far ignored in the debates about coordinating, integrating, networking, and nesting

environmental governance across several scales of societal and ecological systems. Transformative governance is a combination of many activities and political programmes, of technological, social, and ecological strategies, of capacity building and knowledge integration, of policy reforms, and restructuring of power relations. All that cannot be organised in one globally coordinated top-down approach of governance or in a hierarchy of command-and-control, but requires a variety of nationally and locally adapted and differentiated policies. Transformative governance, where different perspectives and contradicting views and interests need to be integrated, develops through networking and "soft coupling" of processes that include the following components:

- *Creating redundancy and anchoring governance mechanisms in many social and ecological processes:* This implies the development of multifunctional management mechanisms, overlapping management systems, polycentric, nested, and networked governance systems that are described in research on environmental management and governance (some elements developed in Ostrom's research: 2007, 2009). In ecological research as in governance research, simple ideas about integrating and managing complex and interacting systems need to be criticised and developed further. Ecosystem-based management is, for example, based on the assumption that social and political scales of action and decision-making can be adapted to ecological scales, without further discussion of the problems of scaling and rescaling social and political action. Other forms of reductionism can be observed in the global governance debate when rescaling of governance is reduced to political problems of power sharing and global coordination, neglecting other aspects of integration: knowledge exchange, deliberation and syntheses, restructuring of power relations through institutional reforms, and connecting governance to the self-regulation of ecosystems.
- *Matching of social, economic, and ecological requirements of sustainability and sharing of knowledge, power, and responsibility* at several levels, between different managerial and political institutions, and

between many actors are necessary—but do not necessarily generate capacities of transformative governance. The process can temporarily fail because of the many difficulties involved that cannot be solved with old management principles. The scaling and rescaling of governance processes implies the identification of actors in many different political arenas and spheres of collective action. Actors in transformative governance include far more than the politically and formally legitimised political actors—governmental organisations, international institutions, political parties, and co-opted non-governmental actors in the policy processes. With the broadening of management perspectives, with participatory management and co-optation, it becomes necessary to decide which actors should participate in which governance processes, and who should be excluded. But who can decide about that, with which criteria?

Since Selznick's (1949) classical study of water- and land-related policies of the "Tennessee Valley Authority and the grassroots", the forms and processes of power sharing, participation, and co-optation are continually discussed and disputed, broadened, and modified. What Selznick (1949) and Gamson (1968) observed, power asymmetries, emergence of risks and conflicts at community levels, may be core phenomena to motivate participation and co-optation, but do hardly explain management practices, their success and failure. What transformative governance has to add to earlier experiences and learning about the processes of institutional change can be described as second-order governance beyond top-down command-and-control management (changes of indicators and settings in the governance process: Spangenberg 2008; Noteboom and Marks 2010), and third-order governance (changes of policy and governance paradigms: Hall 2011).

Conclusions: The Future of Coastal Governance in a Sustainability Perspective

The discussion of coastal development and climate change policies is an exemplary theme in the work with and the application of social-ecological theory for analyses of transformation to sustainability. Further themes and fields of research could be discussed in similar ways: rural development and food production (for Europe, see, for example Dwyer 2013; Rutz et al. 2014), global land use change (for example Klein Goldewijk 2001), and global urbanisation (Birch and Wachter 2011).

The integration of local and regional governance systems in networks of managed areas tends to become difficult with the integration of large areas, countries, continents, and oceans. But the focus on the managerial problems of coordination, integration, and control directs the efforts of coastal governance away from the search for transformation paths and transition to global sustainability. Strategies for adaptation to global environmental change tend to become managed, decline and retreat when no further development perspective is created for coastal areas through sustainable development and social-ecological transformation. Social-ecological transformation is not a single strategy but a continuous process of developing transformative literacy, agency, and governance at many places and in many forms. Interdisciplinary knowledge syntheses in social ecology and the social-ecological theory can only be used as knowledge compasses to approach the goal of transformation to sustainability.

The complicated situation and the lack of strategies for social-ecological regulation and transformation at the beginning of the long process of transformation to sustainability do not justify giving up the idea of sustainability in the policy process. Concentrating instead on adaptation, resilience, and disaster management does not touch the core and the cause of global environmental change, economic growth, and the capitalist accumulation process that undermine the long-term development of SES. The way towards new interdisciplinary knowledge production and synthesis begins with the reflection of the deficits of present management practices and attempts to deal with uncertainty and lack of knowledge. The ideas of joint, social learning and cooperation of scientists and practitioners discussed in adaptive management and governance do not say all what is required to achieve sustainability; they can become misleading when the aim of sustainability is not clarified and remains a vague principle. Joint learning and cooperation of actors in participatory management as components of governance suffer from the idealistic premises of earlier participation debates. Transformation to sustainability implies difficult forms of learning—learning to use knowledge from different sources to solve conflicts, share power and cooperate, and to deal with vested interests and more complicated forms of reflexive learning, for example, double-loop learning (Argyris) as an experience-based change of methods and goals. Such forms of learning have been discussed for single organisations, but hardly in complex systems and contexts of multi-scale governance and transformation of SES.

Main ideas about further learning and capacity building for transformative agency and sustainability that can be learned from the experiences with climate change policy and coastal management are connected with further conditions of improving governance:

(1) *Clarifying the requirements of transformation in the sustainability process:* A precondition for that is that the interaction between social and ecological systems to be governed in the sustainability process can be clarified with the help of knowledge, concepts, and models from interdisciplinary research on society and nature, especially from recent research in social and political ecology (Haberl et al. 2011; Brand 2015b).

(2) *Learning about the possibilities of transformation of global systems:* Such learning happens presently with insufficient knowledge practices: in global scenarios for sustainable development (Raskin et al. 2010) that aim at a "great transition" but conceptualise this societal transformation in unrealistically simplified system models and with insufficient knowledge. Changes towards global sustainability may include catastrophic forms—collapse of the economic and political systems in different parts of the world, degradation of ecosystems, and dramatic forms of poverty, hunger, diseases, and resource scarcity—but the future development is not only disaster management. Developing adaptive and transformative governance requires ideas described above (last section) as higher-order governance.

(3) *Converting the undermining of democratic institutions* in public policy processes and regaining capacities of political action, decision-making, and control of economic processes in transformative governance. Constitutional changes in national political systems are necessary to restructure power relations in national and international systems and to regain political control of markets that are presently deregulated. Institutions at international levels that have the power

and legitimacy to redirect and control economic development and resource use cannot be the institutions that were created to maintain the global economic order. Strategies for international redistribution of resources, for limitation of economic growth, and for changing the unequal flow of materials and energy in the global economy do not yet exist. They require complicated dialogues in science, policy, and society to provide knowledge and ideas about systemic changes and the building of a new world order.

(4) *The significance of theoretical knowledge about change and transformation of SES:* The development of a social-ecological theory of nature–society interaction is a means to gain further clarity about transformation to sustainability. The theory can be seen as abbreviating the search for sustainability strategies and the erratic learning processes that happen in specialised research when the knowledge and the results are not critically assessed and synthesised, as, for example, the research on ecosystem services or the research on governance falling apart in different areas of governance and empirical policy research.

Appendix: The Practice of Integrated Coastal Zone Management in Europe

The short history and failure of ICZM in Europe: The integrated policies of coastal development spreading since the 1980s mainly in countries of the Global South as reactions to environmental and resource use problems of coastal areas were too simple in the construction of policy instruments to deal with global environmental change. The policies of ICZM in Europe had, ten years after their introduction through an experimental demonstration programme in 1996, failed in many countries. ICZM as a first step towards sustainable management of natural resources in the coastal zone was conceptually badly developed and politically insufficiently supported. The strategy did not help to prepare coastal areas for the future problems and conflicts (for further details, see: McFadden 2007; Bruckmeier 2008, 2012). Also regarding climate change adaptation the approach needs to be developed further. The diffuse goals of ICZM did not initiate the intended process of sustainable development in coastal areas (White et al. 1997: 335). "Integration" referred to management activities for land resources, coastal waters, and living marine resources (OECD 1993: 50). ICZM programmes created spatial connections between

- landward management of water resources (for which the EU enacted the Water Framework Directive), following an ecosystem approach, including river basin management and coastal waters, and
- the management of the open ocean, for which effective international rules and management programmes are not advanced, develop gradually through marine spatial planning.

ICZM developed through the integration of rules and resource management systems in the coastal zone, with a vague idea of integration as transsectoral integration, without clear requirements of scientific knowledge and knowledge syntheses. After the Global Environmental Conference in Rio de Janeiro from 1992, resulting in the global policy programme "Agenda 21" (where ICZM strategies were demanded in chapter 17 from coastal countries), the EU started a European demonstration programme from 1996 to 1999. After that, ICZM was introduced in EU countries with a recommendation by the European Council and Parliament (2002/413/EC). The European demonstration programme adopted an open approach that was never scrutinised through clear criteria and evaluation, referring to practical examples and "lessons from experience". The subsequent policy continued with a soft approach and the weak legal instrument of a recommendation, leaving considerable freedom to the member countries on how to use the principles and build their national strategies for ICZM. The following basic principles of ICZM are formulated in the Recommendation 2002/413/EC from May 30, 2002:

- A broad thematic and geographical perspective
- A long-term perspective
- Adaptive management (as a gradual process of adjustment of management rules when problems or knowledge change)
- Taking into account the local specificity and the great diversity of European coasts (specific and flexible measures)
- Working with natural processes and respecting the carrying capacity of ecosystems
- Involving all parties concerned

- Involving all relevant administrative bodies at national, regional, and local levels
- Combining instruments to facilitate coherence between policies and between planning and management

The vague principles echo the debates in environmental research and policy with the concepts of adaptive management and the carrying capacity of ecosystems as ecological components, and participatory management, involving stakeholders and administrative bodies as governance components of ICZM. The principles can be interpreted differently by the decision-makers and administrators applying them. The EU Recommendation should be implemented through national programmes to be prepared by 2006. At that time the policy was already in crisis, as the evaluation showed. Not all member countries with coasts had finished such strategies and the implementation slowed down, in spite of attempts to revitalise the process. The EU and still many more national governments did not sufficiently support ICZM. In March 2013, the EU Commission adopted a new and modified proposal for a Directive to establish a framework for maritime spatial planning and integrated coastal management. At that time the idea of ICZM had already lost scientific and practical support, for several reasons. Throughout the practice of ICZM it remained unclear how it connects with research, and how much and what kind of research is required to support, implement, and improve the approach. McFadden (2007) saw science as disappearing from ICZM. With intensifying coastal research, complex problems like climate change, and new conflicts in coastal resource use for which the approach was not constructed and the decision-makers not prepared, it appears today as outdated and as a failure.

The twenty years from the Rio conference 1992 to the "Rio+20" conference mark in European coastal policy the delayed adoption, weak support, and practical failure of a common coastal policy. Improved strategies, a second generation of ICZM to strengthen sustainable development and resource management in coastal areas and support social-ecological transformation, are not yet available. ICZM helped to initiate institutional changes, reacting to deficits and failures in coastal management, for example, bureaucratic segmentation of administrative sectors, lack of coordination, lack of efficient control of pollution and overuse of resources. Coastal managers are confronted today with more difficult and multi-scale problems, more stakeholder groups with changing interests, more conflicts, and recurrent disasters for which ICZM was not prepared. The approach was insufficiently supported by governmental institutions, implemented without accompanying institutional reforms; how it should help to meet the challenges of sustainability and of climate change remained unclear. In the practice of coastal management did not develop efficient forms of joint learning of the actors and continuous improvement, although adaptive management was among the principles of the approach. Cooperation remained often limited to some core groups in political institutions, resource managers, municipal administrations, specialised coastal and marine scientists; efficient participation of resource users did not develop. The reform of fisheries policy in the EU and attempts to establish co-management showed similar difficulties and deficits of participatory management (Linke and Bruckmeier 2015).

The short list of ICZM principles had the quality of a bureaucratic and legal document where the principles were not concretised and operationalised to support compliance and rule enforcement. In the review and evaluation processes, the real problems of practising ICZM were not sufficiently clarified. Rather, the vague principles allowed the member states and the practitioners to interpret the rules as they found it sufficient to follow the EU policy, sometimes also to bypass the strategy with arguments of the kind that it has been done in other ways than a national ICZM strategy. Coastal research, management, and implementation of policies need to be reviewed, regarding the practices of knowledge generation, transfer, and application in research, policy formulation, and the routines of implementation. This implies to rebuild coastal management in the perspective of adaptive governance, for climate change, and for transitions to sustainability.

References

Anderson, D., & Ackerman, L. (2010). *Beyond change management: How to achieve breakthrough results through conscious change leadership*. San Francisco: Pfeiffer.

Andrew-Essien, E., & Bisong, F. (2009). Conflicts, conservation and natural resources use in protected area systems: An analysis of recurrent issues. *European Journal of Scientific Research, 25*, 118–129.

Becker, E., & Jahn, T. (Eds.). (2006). *Soziale Ökologie. Grundzüge einer Wissenschaft von den gesellschaftlichen Naturverhältnissen*. Frankfurt am Main: Campus.
Benson, M. H., & Craig, R. K. (2014). The end of sustainability. *Society and Natural Resources, 27*(7), 777–782.
Bensted-Smith, R., & Kirkman, H. (2010). *Comparison of approaches to management of large marine areas*. Cambridge/Washington, DC: Fauna & Flora International/Conservation International.
Bernstein, S. (2005). Legitimacy in global environmental governance. *Journal of International Law & International Relations, 1*(1–2), 139–166.
Biermann, F. (2004). The global governance project. *IDGEC News, 8*, 10–11.
Birch, L., & Wachter, S. M. (Eds.). (2011). *Global urbanisation*. Philadelphia, PA: University of Pennsylvania Press.
Blok, A. (2010a). Topologies of climate change: Actor-network theory, relational-scalar analytics, and carbon-market overflows. *Environment and Planning D: Society and Space, 28*, 896–912.
Blok, A. (2010b). Divided socio-natures: Essays on the co-construction of science, society, and the global environment. PhD thesis, Department of Sociology, University of Copenhagen.
Borrini-Feyerabend, G., Johnston, J., & Pansky, D. (2006). Governance of protected areas. In M. Lockwood, G. L. Worboys, & A. Kothari (Eds.), *Managing protected areas – A global guide* (p. 116). London: Earthscan.
Brand, U. (2015a, June 10–12). *How to get out of the multiple crisis? Contours of a critical theory of social-ecological transformation*. Paper presented at the conference The Theory of Regulation in Times of Crises, Paris. Accessed November 10, 2015, from https://www.eiseverywhere.com/retrieveupload.php?
Brand, U. (2015b). Sozial-ökologische Transformation als Horizont praktischer Kritik: Befreiung in Zeiten sich vertiefender imperialer Lebensweise. In D. Martin, S. Martin, & J. Wissel (Eds.), *Perspektiven und Konstellationen kritischer Theorie*. Münster: Westfälisches Dampfboot.
Brenner, N. (2000). The urban question as a scale question: Reflections on Henri Lefebvre, urban theory and the politics of scale. *International Journal of Urban and Regional Research, 24*(2), 361–378.
Bruckmeier, K. (2008). Integrated regimes for natural resource management in a transdisciplinary perspective – The Swedish example of sustainable coastal zone management. In A. Dehnhard & U. Petschow (Eds.), *River basin management*. Berlin/München: IOEW/Ökom Verlag.
Bruckmeier, K. (2014). Problems of cross-scale coastal management in Scandinavia. *Regional Environmental Change*. doi:10.1007/s10113-012-0378-2.

Brunnengräber, A. (2009). Die politische Ökonomie des Klimawandels Ergebnisse sozial-ökologischer Forschung, Band 11. München: Ökom-Verlag.

Brunnengräber, A., Dietz, K., Hirschl, B., Walk, H., & Weber, M. (2008). *Eine sozial-ökologische Perspektive auf die lokale, nationale und internationale Klimapolitik*. Münster: Westfälisches Dampfboot.

Chapin, F.S., III, Kofinas, G.P., Folke, C., eds. (2009) *Principles of Ecosystem Stewardship: Principles of Ecosystem Stewardship: Resilience-Based Natural Resource Management in a Changing World*. New York: Springer.

Chesters, G. (2012). Social movements and the ethics of knowledge production. *Social Movement Studies, 11*(2), 145–160.

Coers, A., Raakjær, J., & Olesen, C. (2012). Stakeholder participation in the management of North East Atlantic pelagic fish stocks: The future role of the Pelagic Regional Advisory Council in a reformed CFP. *Marine Policy, 36*(3), 689–695.

Corbin, A. (1990 (1988)) *Meereslust: Das Abendland und die Entdeckung der Küste*. Berlin: Wagenbach.

Crosby, A. (1986). *Ecological imperialism: The biological expansion of Europe, 900–1900*. Cambridge: Cambridge University Press.

Deere Birkbeck, C. (2009). Global governance in the context of climate change: The challenges of increasingly complex risk parameters. *International Affairs, 85*(6), 1173–1194.

Duxbury, J., & Dickinson, S. W. (2007). Principles for sustainable governance of the coastal zone: In the context of coastal disasters. *Ecological Economics, 63*, 319–330. doi:10.1016/j.ecolecon.2007.01.016.

Dwyer, J. (2013). Innovation for sustainable agriculture: What role for the second pillar of CAP? *Bio-based and Applied Economics, 2*(1), 29–47.

Economics of Climate Adaptation Working group (2009). Shaping climate-resilient development: a framework for decision-making. Retrieved from: *mckinseyonsociety.com/shaping-climate-resilient-development/*

Fischer-Kowalski, M., Rotmans, J. (2009) 'Conceptualizing, Observing, and Influencing Social-Ecological Transitions', *Ecology and Society* 14(2): 3. (online) URL: *http://www.ecologyandsociety.org/vol4/iss2/art3/*.

Gaines, S. D., Lester, S. E., Grorud-Colvert, K., Costello, C., & Pollnac, R. (2010). Evolving science of marine reserves: New developments and merging research frontiers. *PNAS, 107*(43), 18251–18255.

Haapasaari, A., Engeström, Y., & Kerosuo, H. (2014). The emergence of learners' transformative agency in a Change Laboratory intervention. *Journal of Work and Education*. doi:10.1080/13639080.2014.900168.

Haberl, H., Fischer-Kowalski, M., Krausmann, F., & Martinez-Alier, J. (2011). A socio-metabolic transition towards sustainability? Challenges for another great transformation. *Sustainable Development, 19*, 1–14.

Hall, C. M. (2011). Policy learning and policy failure in sustainable tourism governance: Form first and second order to third order change? *Journal of Sustainable Tourism, 19*(4–5), 649–671.

IPCC (Intergovernmental Panel on Climate Change). (2013). Technical summary, Working group III – Mitigation of climate change. Accessed February 27, 2016, from http://www.ipcc-wg3.de/

Jamison, A. (2010). Climate change knowledge and social movement theory. *Wiley Interdisciplinary Reviews: Climate Change, 1*(1), 811–823.

Klein Goldewijk, K. (2001). Estimating global land use change over the past 300 years: The HYDE database. *Global Biochemical Cycles, 15*(2), 417–433.

Lane, S. N., Odoni, N., Landström, C., Whatmore, S. J., Ward, N., & Bradley, S. (2011). Doing flood risk science differently: An experiment in radical scientific method. *Transactions of the Institute of British Geographers, 36*(1), 15–36.

Landström, C., Whatmore, S.J., Lane, S.N. (2013) http://dx.doi.org/10.1177/0162243913485450 Learning through Computer Model Improvisations', *Science, Technology and Human Values*, 38 (5): 678-700.

Latour, S. (1999). On recalling ANT. In K. Law & J. Hassard (Eds.), *Actor network theory and after* (pp. 15–25). Oxford: Blackwell.

Lefebvre, H. (1991 [1974]). *The production of space*. Oxford: Basil Blackwell.

Linke, S., & Bruckmeier, K. (2015). Co-management in fisheries: Experiences and changing approaches in Europe. *Ocean and Coastal Management, 104*, 170–181.

McCay, B., & Jentoft, S. (1996). From the bottom up: Participatory issues in fisheries management. *Society & Natural Resources: An International Journal, 9*(3), 237–250.

McFadden, L. (2007). Governing coastal spaces. The case of disappearing science in integrated coastal zone management. *Coastal Management, 35*(4), 419–443.

McKenna, J., Cooper, J. A. G., & O'Hagan, A. M. (2008). Managing by principle: A critical analysis of the European principles of Integrated Coastal Zone Management (ICZM). *Marine Policy, 21*, 941–955.

Mollmann, C., Lindegren, M., Blenckner, T., Bergstrom, L., Casini, M., Diekmann, R., et al. (2013). Implementing ecosystem-based fisheries management: From single-species to integrated ecosystem assessment and advice for Baltic Sea fish stocks. *ICES Journal of Marine Science*. doi:10.1093/icesjms/fst123.

Moran, E. F. (2000 [1979]). *Human adaptability: An introduction to ecological anthropology* (2nd ed.). Boulder: Westview Press.

Mouffe, C. (2013). *Agonistics: Thinking the world politically*. London: Verso.

Noel, J.-F., & Weigel, J.-Y. (2007). Marine protected areas: From conservation to sustainable development. *International Journal of Sustainable Development, 10*(3), 233–250.

Noteboom, S., & Marks, P. (2010). Adaptive networks as second order governance systems. *Systems Research and Behavioral Science, 27*(1), 61–69.

OECD (Ed.). (1993). *Coastal zone management – Integrated policies*. Paris: OECD.

Offe, C. (2009). Governance: An 'empty signifier'? *Constellations, 16*(4), 550–562.

Ostrom, E. (2007, February 15–19). *Sustainable social-ecological systems: An impossibility?* 2007 Annual Meeting of the American Association for the Advancement of Science, Science and Technology for Sustainable Well-Being, San Francisco.

Ostrom, E. (2009). A general framework for analyzing sustainability of social-ecological systems. *Science, 325*, 419–422.

Pereira, L., Karpouzoglu, T., Doshi, S., & Frantzeskaki, N. (2015). Organising a safe space for navigating social-ecological transformations to sustainability. *International Journal of Environmental Research and Public Health, 12*, 6027–6044.

Raskin, P.D., Electris, C., Rosen, R.A. (2010) The Century Ahead: Searching for Sustainability, *Sustainability*, 2, 2626–2651 (doi:10.3390/su2082626).

Reid, W. V., Berkes, F., Wilbanks, T., & Capistrano, D. (Eds.). (2005). *Bridging scales and knowledge systems: Concepts and applications in ecosystem assessment*. Washington, DC: Island Press.

Rutz, C., Dwyer, J., & Schramek, J. (2014). More new wine in the same old bottles? The evolving nature of the CAP reform debate in Europe, and prospects for the future. *Sociologia Ruralis, 54*(3), 266–284.

Sayre, N. (2012). The politics of the anthropogenic. *Annual Review of Anthropology, 41*, 57–70.

Samarakoon, J. (2004) 'Issues of livelihood, sustainable development, and governance: Bay of Bengal', *Ambio*, 33 (1–2): 34–44.

Schneidewind, U. (2013). Transformative Literacy: Gesellschaftliche Veränderungsprozesse verstehen und gestalten. *GAIA, 22*(2), 82–86.

Scholz, R. (2011). *Transformative literacy in science and society: From knowledge to decisions*. Cambridge: Cambridge University Press.

Selznick, P. (1949). *TVA and the grass roots: A study in the sociology of formal organization*. Berkeley, CA: University of California Press.

Siebenhüner, B., Arnold, M., Eisenack, K., & Jacob, K. (Eds.). (2013). *Long-term governance for social-ecological change*. New York: Routledge.

Spangenberg, J. (2008). Second order governance: Learning processes to identify indicators. *Corporate Social Responsibility and Environmental Management, 15*(3), 125–139.

Stepanova, O. (2015). *Conflict resolution in coastal management: Interdisciplinary analyses of resource use conflicts from the Swedish coast*. PhD dissertation in Human Ecology, School of Global Studies, University of Gothenburg, Sweden.

Stutz, J. (2009) 'The three-front war: pursuing sustainability in a world shaped by explosive growth', *Sustainability: Science, Practice*, & Policy 6 (2): 49–59.

Swyngedouw, E. (2009, September 22–25). *Climate change as post-political and post-democratic populism*. Paper presented at DVPW conference, Kiel, Germany.

Symes, D. (2006). Fisheries management and institutional reform: A European perspective. *ICES Journal of Marine Science, 64*(4), 779–785.

Urry, J. (2011). *Climate change and society*. Cambridge: Polity Press.

Vasconcelos, L., Pereira, M. J. R., Caser, U., Goncalves, G., Silva, F., & Sá, R. (2013). MARGov – Setting the ground for the governance of marine protected areas. *Ocean and Coastal Management, 72*, 46–53.

Walters, C. J. (2007). Is adaptive management helping to solve fisheries problems? *AMBIO, 36*(4), 304–307.

WBGU. (2011). *Welt im Wandel – Gesellschaftsvertrag für eine Grosse Transformation. Hauptgutachten*. Berlin: WBGU.

Weinstein, M. P., Baird, R. C., Conover, D. O., Gross, M., Keulartz, J., Loomis, D. K., et al. (2007). Managing coastal resources in the 21st century. *Frontiers in Ecology and Environment, 5*(1), 43–48.

Westley, F. R., Tjornbo, O., Schultz, L., Olsson, P., Folke, C., Crona, B., et al. (2013). A theory of transformative agency in linked social-ecological systems. *Ecology and Society, 18*(3), 27. doi:10.5751/ES-05072-180327.

White, A., Barker, V., & Tantrigama, G. (1997). Using integrated coastal management and economics to conserve coastal tourism resources in Sri Lanka. *AMBIO, 26*(6), 335–344.

York, R., Rosa, E. A., & Dietz, T. (2003). Footprints on the earth: The environmental consequences of modernity. *American Sociological Review, 68*, 279–300.

York, R., Rosa, E.A., Dietz, T. (2004) http://scholar.google.com/citations?view_op=view_citation&hl=en&user=oIRDUjAAAAAJ&citation_for_view=oIRDUjAAAAAJ:zYLM7Y9cAGgC The ecological footprint intensity of national economies', *Journal of Industrial Ecology*, 8 (4): 139–154.

8

Transformation of Industrial Energy Systems

In this chapter, energy systems are described as components of interacting SES. At Seven analyses of modern energy systems from different disciplines are reviewed that help to develop an interdisciplinary theory of modern energy systems. Thereafter, problems of transforming the industrial energy system from fossil to renewable sources are discussed. Finally, practical experiences with the transformation of energy systems are described for two examples: development of wind power and bioenergy.

Schaeffer et al. (2011) discuss the consequences of climate change for energy systems and describe methodological problems of a new field of specialised research: the assessment of impacts of extreme weather events on the energy systems and possible consequences for energy planning and operation. Adaptation of the industrial energy system to climate change is not an alternative to its transformation from a system based on fossil energy sources to one based on renewable sources. Emissions of carbon dioxide from the use of fossil energy sources are a main factor of global climate change and the decarbonisation of economic systems is a main component of the transformation of modern society to a future sustainable society. The difficulties of changing capital- and technology-intensive energy systems are not shown with their climate change adaptation that happens simultaneously with the development of energy from renewable sources.

Systems Analyses of Modern Energy Systems: Different Perspectives

Environmental history, energy economics, and social ecology show the development of interdisciplinary forms of analyses of energy systems, where social-scientific and natural-scientific, empirical and theoretical knowledge is used to study energy systems in their social and ecological contexts. The three fields of research show the main factors influencing the development of modern energy systems.

1. *Energy analyses in environmental history:* A breakthrough in interdisciplinary environmental history is summarised by Barca as *incorporation of a social perspective and of the inequality issue in historical analyses of energy regimes.* Such historical studies show the combinations of technical, economic, territorial, and ecological system components in energy regimes and their connections with property regimes, social inequalities, and different lifestyles. In the study of the relationships between economy and ecology the forms of life in a society and the social structures of appropriation of energy sources are important (Barca 2011: 1311). Environmental history does not aim at developing an encompassing theory of society and nature, although various theories are used. These theories are reframed in three variants of historical energy analysis described by Barca.

    Development of energy systems in human history: Taking into account complexity theory in social, economic, and technology history, Debeir et al. (1986) carried out the first social-ecological analysis with a theoretical framework of interacting social and ecological regulation processes in the study of historical energy transitions. This approach is based on the bioeconomic research of Passet. Energy systems show the interdependencies between modes of production, social formations, and the biosphere as the important social and ecological contexts of human life and social action. Energy systems include ecological and technological aspects of energy, connecting to the social structures and different roles of social groups and classes in the appropriation and management of energy sources and conversion technologies. The dynamics of energy systems follow the logic of the social and economic reproduction processes of the society of which they are part. Historical analyses show the marks of the pharaoh's political religious rule, of the rule of the feudal landlord, or of the modern bourgeois state in energy systems. In the long historical perspective, factors driving the development of energy systems were the continually higher levels of technical specialisation and capital investment required for the use of mineral energy sources. With the industrial revolution, energy became a matter for investors, scientists, and engineers and an independent sector in the new economy. The social and economic influences on modern energy systems are summarised by Barca (2011: 1311ff) as follows:

Analysis of physical impacts of the human economy and its social and industrial metabolism in historical perspective: Industrialisation appears in the analyses (by Fischer-Kowalski et al.) as changing the flows of energy and matter in SES. The industrial metabolism develops through a stepwise decoupling of the supply of energy from land-related biomass and from human labour on the land. The energy regimes of societies shifted finally from tapping into flows of renewable energy towards the exploitation of finite stocks of fossil energy. These may have appeared at the beginning as very large and sufficient for human use for an indefinite time. With the transition from agricultural to industrial societies, the sustainability problems changed from problems of resource inputs and overexploitation of resources to output-related environmental impacts, habitat loss, and social inequality. The limits of resource input appear when the industrial energy resources are approaching their global limits. Sustainability problems of the fossil energy regime, especially energy scarcity and global pollution, include problems of distribution, which reinforce global inequalities. It seems that there are not enough resources for the total human population to become rich through industrialisation. The disasters related to climate change and toxic waste disposal will disproportionately affect the poor. To change this trend requires, according to Fischer-Kowalski et al., a new social-ecological regime with lower material and energy turnover per capita and a lower share of non-renewable resources (Barca 2011: 1313).

Main ideas of modern economic growth narratives: These ideas include the myths about modern growth-based economy—that economic growth can be perpetuated through increase of fossil energy consumption; that environmental and social costs are negligible; and that natural resources need to become private property for their productive use. Such ideas continue to inform international development policies and late attempts of industrialisation in the system structures of modern capitalism. Industrialisation does not happen in environment-friendly forms, bringing more negative consequences for the environment, health, and social justice wherever it starts. As analyses of oil policies have shown, the environmental, health, and social costs of increased oil flows are largely absent from government policy deliberations. The unequal distribution of burdens and benefits of increased oil production among countries, communities, and individuals are almost completely ignored and not discussed publicly (Barca 2011: 1314).

Barca analyses different constructions or narratives of the industrial energy system. The competing discourses show a similar dilemma as in the construction of the climate discourse: what does the construction of different energy narratives provide in terms of knowledge for the transformation of the industrial energy regime to a sustainable system? The social-ecological debate of energy systems, showing that sustainability problems of energy systems differ in the history of human societies, does not provide arguments that can be used in the present transformation debate (Krausman and Fischer-Kowalski 2010). The transformation of industrial energy regimes through energy from renewable sources does not mean to go back to older forms of wind or water for energy use. The use of renewable sources is reinvented in new technological forms, with technologies like electricity generation that were not used before in human history. For the introduction of renewable energy sources, more is required than historical analyses: theoretically underbuilt governance strategies for energy system transformation. Such strategies require knowledge from social-ecological or political-economic systems analyses of the modern economic world system. Empirical knowledge and normative reasoning with ecological rationality is not sufficient to guide the transformation.

2. *Energy analyses in economics:* The modern industrial energy regime has been analysed critically in the perspective of biophysical economy by Hall and Klitgaard, who discard in their conclusion neoliberal economics with the arguments that this thinking is not supported through empirical knowledge, violates the basic laws of physics, and has no consistent assumptions (Hall and Klitgaard 2012: 203). The authors avoid a more critical normative reasoning and political debate of the ideology of neoliberalism. The deregulation of national economies during globalisation has, for most countries, not resulted in efficiency increases: all economic growth was connected with increasing rates of exploitation of energy and other resources. But what can be the alternative ideas in economics that is so strongly directed by efficiency thinking? Obviously, the answer is difficult and cannot come from economics alone. The partial answer of the authors from biophysical economics is not a simple one; it is based on an economic model saying that development and increase in wealth occur only when the ratio energy resources/number of people increases. Their main argument is that wealth comes from nature and the exploitation of nature, much less from markets or their manipulation (Hall and Klitgaard 2012: 204). Although not precise in the formulation of the interaction of nature, human labour, and technology at this point, the argument approaches the reasoning of critical political economy (Altvater 1991) and the more recent ecological economics (Martinez-Alier et al. 2010) and social-ecological transformation analyses (Brand 2015a, b). The economic analyses of energy, as the historical analyses discussed above, needs to use ecological, physical, and technical knowledge in interdisciplinary reasoning.

In spite of their critical review of most economic theories applied for energy economics, the theoretical and interdisciplinary broadening of economics by Hall and Klitgaard (2012: 352ff) is limited for the purpose of formulation strategies of transformative governance. They do not adopt the perspectives of environmental and ecological economics, but they use as knowledge base the ecological and physical knowledge on energy consumption in ecosystems. They argue that humans do not differ from other species in energy requirements and nutrition. The guiding idea of ecological economics is summarised thus: economists could learn from ecologists about the many ways that nature has learned to live within limits (Hall and Klitgaard 2012: 280). The argument seems inexact and to neglect the point made in human ecology by Rees and others that human resource consumption is not biologically fixed but varies significantly with the forms of culture, organisation, and development of human societies. Hall and Klitgaard describe the socio-cultural evolution of humans which resulted in increasing exploitation of energy and material resources and the growth of natural resource use beyond the biological minimum and the level of population growth in affluent societies (Hall and Klitgaard 2012: 245ff). But their basic argument that economics can learn from ecology remains inexact and does not contribute much to the analysis of transformation of modern energy systems. It is similar to the older message of environmental movements of "learning from nature and following nature's lead". Such messages from ecology are insufficient for understanding the interactions between ecosystems, social systems, and societies. Ideas of the "good society" as a copy of the wisdom of nature remain socially naïve. The complexity of ecosystems and social systems needs to be shown in their interaction and the different forms of coupling between them (see Chap. 5). The question insufficiently answered by the authors, what economists can learn from ecology, connects to the more important one: does ecological learning provide sufficient knowledge for the transformation of modern energy systems? Often the learning is limited to the form that Hall and Klitgaard describe as, "There are alternatives to the forms of resource use in modern society and economy." At this point, ecological studies of energy flows in ecological systems do not help further; the social constraints and problems of resource use and technological development need to be analysed additionally to understand the dynamics of coupled social and ecological systems.

The high levels of energy consumption in modern human society require answers to the questions: What are the social consequences of reducing the consumption? To which levels can or should the consumption be reduced? The energy consumption of species in connection with food consumption is a "natural standard" for consuming, that is, necessary to survive, a biological minimum, varying between species. This minimum can be calculated for humans as species. With that calculation, one does not know more than that this level is exceeded considerably in all forms of human societies, throughout history, although in earlier phases, at much lower levels than in modern industrial society. Studying energy flows in ecosystems and calculations of the loss of 80–90% of energy as heat in each transfer of energy from one trophic level to another (Hall and Klitgaard 2012: 281) gives basic information about the thermodynamics of energy use for maintaining the biological metabolism of the species living in an ecosystem. This information does not help to find ecologically and socially acceptable levels of energy consumption in the societal metabolism of modern society.

Less difficult to calculate and practically important in the ecological debate of energy consumption is the "energy return on investment" (EROI), the energy obtained from an activity compared to the energy it took to generate it (Hall and Klitgaard 2012: 325). For the history of oil as fossil energy source in the industrial energy regime, it can be shown how the EROI was dropping during the last century and is very low today. The costs of oil production after the peak become ever larger for exploiting new sources, for example, in the deep sea. The global oil production is not easily calculated in EROI changes, but data from different countries seem to tell the message the costs of extracting and converting oil into energy are much higher than the value of the energy gained for use. The EROI becomes important in the comparison of different energy sources used in human society and for the development of "green" energy forms. Regarding modern agricultural systems, their energy problem is that they have become main consumers of energy in different forms of water, oil, and other fuels for use of machines and production of synthetic fertilisers that show negative ratios of energy input and output. These agricultural systems should now become, beside food producers, also producers of bioenergy that help to transform modern energy systems—in

extremely contradicting and controversial forms as the discussion below shows.
3. *Energy analyses in social ecology:* To develop a more systematic, theory-based analysis of transitions to sustainability implies intensive discussion of methods and knowledge practices to bring the sustainability debates away from simple normative and vision-based political ideas. Social-ecological research can help to formulate new strategies for energy transition as Fischer-Kowalski et al. (2012) show with the following arguments.

The socio-metabolic approach to transition argues that the appropriate unit of analysis to investigate social-ecological transitions in society is the socio-metabolic system or regime that connects social and ecological factors in natural resource use (see description in Chap. 4). Social-ecological transitions are transitions between socio-metabolic regimes that are rooted in the energy system of a society, including the sources and dominant conversion technologies of energy. Depending on the reasons and causes for and the speed of an energy transition, parts of the system may at a certain point in time be under different energy regimes: urban industrialised centres, for instance, may coexist with traditional agricultural communities, or industrialised countries with agrarian colonies. Such asynchronicity influences the course of transitions in the many countries of the modern world system. The socio-metabolic approach shares with complex systems theory the notion of emergence: one state cannot be deliberately transformed into another and the process cannot be fully controlled. The complexity of SES with self-organising dynamics such as metabolic regimes limits the possibilities of managerial governance by social or political institutions. Transitions may last for generations and it is difficult to identify the driving factors of socio-metabolic regime transitions over so long time; these factors may be changing. Difficult is also the assessment of the significance of conscious decisions by social actors and their collective efforts; the actor constellations change in unforeseeable ways in the long run of the transition processes. What can be analysed mainly are processes of structural change of coupled social and ecological systems with many variables. The socio-metabolic approach uses a relatively narrow set of these variables describing the society–nature interface for which reliable quantitative measurements can be obtained in different contexts. The approach can show empirically the interconnectedness of socio-economic changes and changes in ecological systems (including, e.g. population growth, diets, land use, and species extinction) and generate models for important biophysical requirements for the perpetuation of SES. When an energy regime changes, society and its metabolism change as well, as the ecological systems it interacts with changes. In such complex changes, it is at least possible to differentiate, with regard to social action and actors, between unintended consequences (such as resource exhaustion or pollution) and intentional change induced by society such as land use (Fischer-Kowalski et al. 2012: 24f).

With the question, what drives socio-metabolic regime transitions, the analysis of transformation to sustainability is redirected from policy and governance analysis to broader approaches and long-term change with different and varying constellations of multi-scale and multi-actor processes. The decisive point is that transitions include many interconnected processes and self-organising dynamics that cannot be fully controlled, where governance does not work. It seems that this point requires further elaboration in social ecological theory, beyond the reasoning with complex systems theory and emergence that are of formal kind and descriptive. What needs to be elaborated in a theory of social-ecological transformation is the social and ecological complexity of long transformation processes that require interdisciplinary knowledge syntheses, with explanatory components from systems analyses in sociology, economy, political economy, ecology. Structural change is one way to describe the broader transformation process beyond governance, but there are further processes of long-term change and other theoretical concepts to be used in complex processes of societal transformation: modes of production, world systems, complex social-ecological dynamics including Promethean revolutions, ruptures of path-dependence, social-ecological regulation of transformation of SES, and self-regulation of ecosystems.

Comparing the energy analyses in environmental history, energy economics, and social ecology reviewed above, two approaches are theoretically advanced and specified for analyses of energy transitions in the history of human societies: that of Debeir et al. (1986), and the social ecological approach by Fischer-Kowalski et al. (2012), both discussed in the comparative analysis of Barca. In both approaches, the analysis of modes of production is important to understand the transformation of energy systems and societies. The analysis of modes of production connects social ecology to political-economic analyses that are transferred by Fischer-Kowalski et al. in the concepts of societal metabolism and socio-metabolic regimes. The social-ecological theory recognises the significance of energy systems in social-ecological transformation and shows connections between global change, transformation of energy systems, and societal transformation to sustainability.

In the following sections, more detailed analyses are carried out to identify possibilities and pathways of transforming the industrial energy system through the use of new energy sources of wind power and bioenergy. The analyses show that the transformation process is not continuously progressing but is full of disputes, conflicts, contradictions, and interruptions and is redirected with new knowledge. In the processes of transforming energy systems struggle two logics of development with each other:

- That of maintaining the industrial energy system through a recombination and diversification of the energy mix, for which the use of renewable energy source seems a technically feasible and economically profitable solution
- The logic of a transformation of the energy system to a new, non-industrial, and ecologically sustainable system

Problems of Transformation of Energy Systems

Social-ecological analyses of the industrial metabolism show how economic growth and high levels of energy use together are driving economic development. The decisive difference to older energy regimes in human history is visible in

- the high level of technology required, and
- the high levels of per-capita consumption of material and energy resources in modern society in comparison to all earlier ones (see Fischer-Kowalski et al. 1997).

From this high per-capita input of resources, the largest part is energy. Given these facts, it is evident that modern energy regimes need to solve one main problem: how to provide and maintain continuous access to cheap energy for large numbers of consumers. This is not a consequence of modern energy conversion technologies alone and cannot be achieved through technical innovations; it becomes possible through a combination of various technical, economic, and social factors in the organisation of accumulation regimes in modern industrial society. The reserves of fossil energy resources seemed at the beginning of industrialisation practically unlimited and never exhausted through human consumption. Only few scientists could imagine that fossil energy sources ever could be used up completely.

Rapid and exponential global population growth, growth of private consumption (mass consumption), economic growth, accelerated urbanisation, and technical modernisation of lifestyles could not be foreseen at the beginning of industrialisation in European countries that industrialised first. Throughout the nineteenth century, industrialisation was experienced as generating misery and poverty, hunger, and unhealthy conditions of life in the industrial cities. High levels of mass consumption became possible relatively late in industrial countries, in the twentieth century, in the short period of the modern welfare state, of the Fordist accumulation regime, and of Keynesian economic policies. Cheap energy was and is primarily a problem of the combination of technology, capital, and cheap labour in the accumulations regimes. The Fordist accumulation regime with the assembly line and mass production of cars was the historical paradigm of such a solution. To repeat its success under worsening economic and ecological conditions, only with new technologies, is unlikely. The approaching of physical limits to growth in the second half of the twentieth century was one of the reasons for the crisis and end of Fordism. New accumulation regimes need to realise under worse conditions, with less energy resources available and deteriorating environmental quality, the access to cheap energy. This happens in a vicious circle of development maintained through growth, in which most hope is—unrealistically—set in technological innovations as the so-called fourth industrial revolution.

Krausman and Fischer-Kowalski (2010: 24) follow a social-ecological knowledge perspective in analysing the transformation of the present industrial energy regime and the national energy systems in different countries to identify new forms of sustainable energy systems. The energy futures are not given as possibilities to choose freely between combinations of energy sources, the combinations of energy sources in a socio-metabolic regime are determined by the mode of production, and only within the boundaries of this systemic structuring through additional factors as knowledge and technologies of energy conversion. Reviewing knowledge about energy systems and their transformation in human history and society, the authors have come to the following conclusion: a historical perspective and historical knowledge does not give clear answers in the sense of giving technical examples for sustainable energy systems that can be repeated in different modes of production or societies. But comparison of energy systems from different historical epochs provides important insights about the dependence of energy regimes from the socio-metabolic profile of a society that includes further factors. The global sustainability problems of the modern economic world system are to be taken into account in developing transformation strategies that reach across different scales (see Chap. 4 and Fischer-Kowalski 2007). For the modern industrial energy system, the high demand of materials and energy is caused by the complex systemic structure of industrial societies with large technical systems and the capitalist growth-based mode of production: this system driven by growth of energy consumption and economic growth prevents ecologically rational strategies of energy saving. Energy savings decided individually by consumers are not sufficient to reduce the high level of energy use in the economic system. Efficiency gains and low prices of energy in the modern market economy stimulate, as non-intended consequence, more use of energy and economic growth, the phenomenon known as "rebound effect". The following discussion shows energy systems as part of broader interactions between social and ecological systems, with tensions between three components of energy systems: the technical, social, and ecological components and their complicated interaction in the combined system.

1. *Energy problems as technical problems—energy technologies and technical converters:* In most scientific and political debates, also in the ecological discourse, energy use and problems appear as that of technologies, for the consumer energy is the use of a technology, very often electricity. The environmental problems of energy are mainly such of the enterprises that "produce" energy (economically seen) and their technologies of conversion in power plants and grids of distribution. The technologies, sources, and forms of energy differ strongly between historical societies so that it is difficult to compare the energy technologies of different historical societies. Also the scarcity of energy and natural resources is not absolute and the same for all societies, which is veiled through the use of the abstract term "scarcity". The practical forms, the social and economic consequences of scarcity differ strongly for each society. One important difference is

    – whether energy is connected to the use of material physical resources by the user without economic transformation through markets and as priced goods (in forms of subsistence production, dominant in older, pre-capitalistic societies), or
    – whether energy and other natural resources are economically valorised and mainly distributed through markets as priced goods (in modern society).

    The energy scarcity problem in medieval feudal-agricultural society in Europe was the physical scarcity of an important material good that was used for many purposes in everyday life: scarcity of wood generated through deforestation of large areas. It was not only energy scarcity but also the scarcity of a material resource used for many other purposes, as building material, and so on. The solution of the scarcity problem was twofold: reforestation, developing to an important technology only after the middle ages, and the transformation of the energy regime through the use of other energy sources, a technical innovation that was part of developing a new mode of production through industrialisation. In the transition to the modern industrial society, the forest problem was solved. But the use of fossil energy sources of coal and oil and the conversion technologies for these brought new problems and increasing risks; they generated the environmental problems of today. The risk spiral of technological innovation connected with the phenomena of exponential growth of economy, of resource use, and of population in modern society requires unprecedented efforts to change the growth-supporting mechanisms of the modern economic world system. The intensification and growth of resource use became historically possible through Promethean revolutions in human history, the Neolithic revolution that brought agriculture, and the industrial revolution, that brought modern society and its economic world system. For the first time in human history, at the end of industrial society, a "Promethean revolution" in the mode of production is to convert the long-term trend towards growing and intensified resource use into a socio-metabolic regime with less use and throughput of natural resources. How de-intensification of resource use and degrowth of the global economy can be initiated in a new "great transformation" is on the agenda with the discourse of social-ecological transformation. De-coupling of growth and human quality of life seems the direction of further development for which the transformation of the industrial energy system is the first decisive step. Decoupling of economic growth and human welfare is not possible through technological innovations and improved efficiency of resource use (dematerialisation) alone; it requires transformation of the interconnected mechanisms that keep modern society on its development path, and this will meet resistance, especially by political and economic elites in the old industrial countries and the newly industrialising countries. It seems better to initiate and guide the transition to sustainability in politically and economically controlled ways as "peaceful revolution" than leaving it to chaotic transformation enforced by the disastrous consequences of exponential growth: a polluted earth full with people, lack of resources, and an "overheated atmosphere".

2. *Energy problems as social problems—the economy of production and consumption:* Energy systems and limits of energy sources are problems in all historical societies, although the problems seem now more serious—ecologically seen. Socially seen energy problems are problems of access

to energy, and not always seen as energy problems in the meaning of the modern physical term of energy. This term received its present scientific meaning late in modern physics, and it is still theoretically discussed, differentiated, and classified in various forms and sources of energy (chemical, electrical, thermodynamic, etc.) and different possibilities to construct and calculate energy with terms as total energy, exergy (available energy), anergy (destroyed energy), and different modes of calculating energy flows (see further: Hall and Klitgaard 2012). Furthermore, energy is, in ecology and economics, part of the overstretched concept of resources (Lawrence 1993; Freese 1997). Energy has become an abstract, complex, controversially discussed and theory-dependent scientific concept (for a summary of the physical energy discussion, see Hall and Klitgaard 2012: 223ff). How to connect physical theories of energy as that of thermodynamics with social theories turned out to be continuously controversial, with disputed interpretations and misinterpretations, for example, in the attempt of Georgescu-Roegen to develop ecological economics on the basis of thermodynamic laws (for further discussion, see Bruckmeier 2013). Controversies about environmentally sound, viable, or preferable combinations of energy sources in socio-technical energy systems continue, also in ecological economics. Comparative analyses of the great transformations of energy systems in human history show the social structuring of energy consumption through the modes of production and the dominant forms of economic production. Such analyses are carried out in social ecology (Fischer-Kowalski et al. 1997) and other interdisciplinary analyses, including the path-breaking study of Debeir et al. (1986). With these analyses, the social and economic structuring of energy systems, connected with their technical and physical components in networks, is described in forms of energy regimes.

3. *Ecological scarcity—the ultimate energy problems:* The specific forms of nature–society interdependence in modern society are not sufficiently described as naturally given scarcity of natural resources, the guiding idea of neoclassical economics that the life on this earth suffers continually under the cold star of scarcity (Schneider 1967: 15). This is a social construction for specific purposes of an economic theory that was important in economic history, but always disputed, among other

reasons for its simplifications. The term of scarcity is used here in an unclear sense, not separating clearly

- scarcity as natural phenomenon analysed in ecology in the sense of the discussion in ecology (of the physical limits of natural resources on earth available for consumption through humans and other species), and
- scarcity as a socially generated mechanism in the modern market economy that connects scarcity to the assumption of unlimited human wants (a doubtful anthropological construction unlimited wants and limited means of satisfaction of wants), monetary valuation, competition, to all resources that can be exchanged and traded on markets.

With the economic reinterpretations of scarcity, the term loses its ecological meaning and reference to natural resources. But this does not mean that in modern society, the "natural" scarcity problems and physical and biological limits to growth vanish; they are socially transformed. Energy systems, as all other resource use systems, are in modern society connected to variables as access to, property and control and distribution of resources, which connects resource use with other social forms of inequality, problems, and conflicts. Conflicts about resource use in modern society imply the blending of natural and social forms, of physical and social limits, of multi-causal and multi-scalar forms of resource use. Problems of natural resource use in modern society are understood as interconnecting natural and social factors in resource use in social ecology, ecological economics (Martinez-Alier et al.), political ecology (Escobar, Watts, Peluso et al.), environmental history (Debeir, Déleage, Hémery, et al.), critical theory, and radical geography (Swyngedouw, Harvey, Brenner et al.). Such critical approaches develop from interdisciplinary analyses of energy problems and the development of energetic theories of society. Early energy research in environmental sociology built on several theories that deal with the relationship between energy and human activity (Rosa and Machlis 1983: 152), but not yet broad interdisciplinary theories. Today, and with important new interdisciplinary theories as that by Debeir et al. (1986), the development of interdisciplinary theories to analyse problems of transformation of energy systems is advancing in reconstructing the interrelations between social, technical, and ecological components of energy systems.

Industrial energy regimes develop not only in a systemic logic of modernisation and technological development, from the earlier coal regime to the present oil regime, but also in manifold tensions and conflicts between

- their *ecological qualities* (fossil and renewable sources and different environmentally friendly and environmentally destructive or polluting impacts through the development and use of these sources);
- their *social qualities* (economically cheap or expensive sources, forms, and technologies; private or public enterprises or cooperatives; small enterprises or multinational corporations; consumer practices of energy use; market dependence of energy consumption, energy prices, and market effects, e.g. rebound effect); and
- their technical qualities (dependence on large technical systems as power plants and distribution grids, with different forms of risks and vulnerability, e.g. in nuclear power plants).

In the following section, the problems inherent in the modern industrial energy system and its transformation are analysed with regard to these ecological, social, and technical problems of energy use.

Energy from Fossil and Renewable Sources

The present global climate change is strongly connected with the global industrial energy system as source of CO_2 emission. The ways out of the industrial system with its non-renewable, fossil sources of coal, gas, and oil are more complicated than the search for technical solutions or new conversion technologies. Integrated analyses of the social, political, economic, and environmental processes become necessary to develop transformation strategies and sustainable energy regimes. *Energy problems appear no longer as problems of specific energy technologies that can be technically optimised, but as problems through negative social and environmental consequences of energy systems.* The energy system is locked in the industrial socio-metabolic regime. The search and development of new energy forms continues since several decades in a situation where the globalised market economy struggles with the limits of fossil resources as well as with the negative environmental impacts of industrial energy systems.

1. *Unexpected consequences of developing "green energy":* The development of energy from renewable sources seemed necessary and justified through the search for environmentally sound forms of energy conversion and consumption. With this development appeared, more unexpectedly than foreseen, new difficulties and problems of rebuilding the modern industrial energy system that is trapped in large technical systems of energy conversion and infrastructures that are dependent on high investment and maintenance costs, financial capital, and natural capital. Competition and conflicts between different energy sources and strategies to develop them are part of the complicated social and economic dynamics of transformation. Experience with the development of wind energy, solar energy, and bioenergy, the important renewable energy sources developed so far, brought non-anticipated negative consequences that would not have been expected from the forms of "green energy". These consequences and effects are not such of the renewable sources in their ecological quality; they are social consequences and risks of the conversion technologies and economic forms of "production" of energy in enterprises (assuming that enterprises produce in an economic sense, although ecologically seen, energy cannot be produced). The global industrial energy system includes as the main components energy from coal, oil and gas, nuclear energy, energy from water, and, to a limited degree, of the renewable energy sources of wind energy, solar energy, and bioenergy produced on arable land. Energy from water and nuclear energy are the forms that developed rapidly during the twentieth century and both showed negative social and environmental consequences of the technology. Both forms developed in coherence with the technological and economic logics of the industrial energy system. *The energy from renewable sources develops as part of the established economic system where energy is dependent on the large industrial energy system that converts*

energy into one dominant form for consumption, electrical energy. The global energy system, although framed and legally managed at national levels, is locked in capital- and technology-intensive marked-based structures of economic development that have, with the neoliberal economic globalisation become "iron structures". Also renewable sources and green energy technologies show unexpected conflicts and non-intended consequences that cannot be explained from the energy sources themselves but from the contexts of economic, political, and societal systems in which they develop. With their practical application and spreading, the renewable forms of energy become part of large-scale and capital-dependent systems.

2. *The dilemma of nuclear energy:* The basic ideas for the renewal of industrial energy systems are simple, replacing environmentally risky and finite energy sources (non-renewable, fossil sources) through energy from renewable resources, accompanied by strategies to reduce energy consumption. These general considerations may be sufficient to justify sustainable energy systems, but they do not show the practical difficulties of changing modern energy systems. Nuclear energy was considered by the protagonists as new, unlimited, safe, and sustainable form of energy—as long as it did not show its unexpected and disastrous consequences in reactor accidents and unsolved problems with nuclear waste. It was not sufficient to become aware of the technical and environmental risks of certain forms of energy, as the social and political conflicts in Europe about the use of nuclear energy demonstrate. The last technical innovation in the industrial energy regime with nuclear energy started in the second half of the twentieth century, nearly from the beginning against the resistance from anti-nuclear movements. And throughout the development and use of nuclear energy, two contrasting constructions of risk with this energy form clashed:

   – A conventional probabilistic risk construction in technical and engineering sciences, where risk is calculated in quantitative terms and reduced to minor risks that could be neglected.
   – A more critical construction of risk in the environmental sciences and by environmental movements, where risks are not quantified, but the dangers of radiation and technical hazards and how to deal with them are assessed.

Environmental catastrophes as that in Chernobyl and more recently in Fukushima did not accelerate the termination of nuclear energy, although some European governments decided since the end of the twentieth century to phase out nuclear energy in their countries. Nowhere this is achieved until today, although the support for nuclear energy as a safe and environmentally sound technology is vanishing, not necessarily implying active resistance. The situation shows rather the deadlock of the industrial energy system. The process of transforming this system through renewable energy sources started in Europe with the pioneering use of wind power, especially in Denmark, since the 1960s. Discussed as an exemplary case of innovation below, wind power shows how difficult and slow changes of energy systems are, with many conflicts not foreseen at the beginning of the process.

3. *Transforming the industrial energy system:* The industrial energy system is a complex system with interacting social, economic, political, and technical components. The scientific explanation of energy in physics, with the laws of thermodynamics, is not all what needs to be known for the restructuring of the energy systems. The social process of transformation is full of contradictions, conflicts, and power fights. As in exemplary form experienced in the introduction of wind power, the conversion technologies become part of the existing modern "high-tech" systems for energy conversion and distribution. The transition from industrial to sustainable energy systems is a complicated process with changing combinations of energy sources, part of the broader strategies of sustainable development. *In the perspective of global sustainability the renewal of energy systems in modern society can be described in five main problems as follows:*

(a) *The industrial energy system is in a contradicting process of change:* This energy system can be described in its historical specificity as a system that decoupled energy use from the limits of the historically prevailing form of energy in pre-industrial agricultural societies. Solar energy and energy from physical labour of humans and animals, important sources in agricultural societies, vanished from the industrial system, also the older forms of producing energy from renewable sources of wood, water, and wind in local energy systems. The new system is constructed mainly for industrial and urban energy use, although this is not consequently possible: human society and economy cannot become independent from solar energy and energy in the form of human and animal labour, important in earlier forms of agriculture. Human labour has changed its forms, became labour using, managing, and controlling high-tech systems, machines, and laboratories of industrialised production. Animals have, as farm animals, become bioconverters of energy in the production of human food. In the industrial mode of production, the farm animal is a colonised and domesticated, constructed and genetically modified, and economically optimised producer of food products for humans.

The main sources of coal, oil, gas, and nuclear energy, varying in their composition from country to country, are combined with new, renewable energy sources, especially wind power, solar energy, and bioenergy from arable land. Different forms of technical innovation, economic restructuring, and transformation to environmentally sustainable energy forms appear in contradicting forms of change of the industrial energy system. Political and economic options for specific energy sources differ between political actors and governments. The dis-simultaneous development of energy systems, where old and new sources coexist and no consensus about future energy systems is found, results in country-specific forms of transformation, with interim solutions and changing strategies. The out-phasing use of coal, oil, and nuclear energy are temporarily favoured again in some countries, because of unexpected difficulties to rebuild market-dependent energy systems.

(b) *Wind power, solar and bioenergy brought new problems:* The new "green" energy forms are not always new in the sense that they are used for the first time in human history. Wind power is an old energy source that has lost significance during industrialisation. It is now socially reinvented again, although not in its old forms of local, autonomous, and small-scale conversion systems of windmills. In some countries, the use of renewable energy sources is more successful than in others, which can be explained through different political and economic strategies and interests, through the availability of renewable sources in a country, and partly also through efforts to reduce energy import that is possible in the networked energy systems of European countries.

The social context and the economic processes, not the goal of sustainability, cause problems in developing, introducing, and operating new energy forms. "Green" strategies, for example, of production of bioenergy from plants on arable land, become part of the contradicting and incoherent structures of modern society and globalised economy they aim to transform: they bring many conflicts, non-intended environmental social, economic, and environmental consequences, for example, socially unwanted land use or competing land use for food and energy production. In the present national energy systems, the tapping of renewable sources does not replace fossil fuel but add more forms and more energy output to the mix of sources (Hall and Klitgaard 2012: 219). Instead of transformation of the industrial energy regime, there seems to have started a conflict-prone experimenting with different forms of technical and economic optimisation and combination of various sources in the modernisation of the industrial energy system.

(c) *Dealing with technical illusions about renewal of energy systems:* Limiting the discussion of energy system changes to technical forms of energy conversion and technical fixes generates illusions about the change of energy systems solely through engineering, without transforming the social practices of energy consumption. Such changes need to deal with the problems of the economically trapped consumer in modern society. Energy consumers are dependent on the modern economic, social, and technical systems and urbanised lifestyles; they need to buy energy. The idea of the autonomous consumer deciding freely about the satisfaction of his needs (consumer sovereignty) is an ideological construction of the consumer situation in the modern market economy. The consumer revolution that is discussed in the ecological discourse touches the core problem but does not yet find the ways and means to develop a new consumption culture—this would require critical analyses of the economic constraints of consumption and the systemic nature of the societal metabolism.

(d) *Difficulties to realise effective forms of change through energy saving:* Ideas of energy saving and local energy systems as favoured by environmental movements turned out to be insufficient as long as they remained ideas for alternative lifestyles. These are practised outside the mainstream economy, in small groups and local experiments, as "ecological islands" in form of utopian lifestyles that did not touch the established industrial energy regime and the forms of consumption for the majority of the population. The necessity to change the energy system because of scarcity and finiteness of coal, oil, and gas as industrial energy resources causes such forms of lifestyle experiments as well as the organised efforts and policy reforms to develop renewable energy sources on countrywide availability, and both variants of change struggle with efforts of prolongation the offer of cheap energy in the industrial energy system through search of new oil reserves, deep sea drilling, or risky and polluting technologies as hydraulic fracking.

(e) *Difficulties to understand the social consequences of climate change that require transformation of energy systems:* Climate change is, in the perspective of the modern economic world system, not a problem in nature but an economic development problem and one of the transformation of the economic system. Similarly, transformation of industrial energy systems is a complex social and economic problem, not mainly an ecological one. Energy systems are interwoven with the conflicts in global and national, political and economic systems. The energy system transformation happens in times of increasing scarcity of natural resources, of economic and financial crises, overshadowed through global climate change. The ecological debate about "ways into postcapitalism" started more than ten years ago (Woltron 2004) with the critique of the inefficiency of neoliberal economic reforms to deal with the problems of global environmental change. So far, the debate has not generated sufficient results and ideas how to initiate a transformation of the societal system towards sustainability.

These five problems with the transformation of the industrial energy system show that the achievement of sustainability is a far more complicated process that cannot be discussed in terms of normative ideas of an ecologically sustainable society. Intra- and intergenerational solidarity in resource use as normative principles guiding sustainability transformation cannot replace a systems analysis of modern society and its interaction with nature. The normative ideas that guided the ecological discourse in science, politics, and environmental movements during the twentieth century, and still do, although their significance is decreasing with the progress of the sustainability discourse, have been summarised by Eckersley (1992: 45f) in four variants:

– Resource conservation (wise use of natural resources)
– Human welfare ecology (achieving environmental quality in the modern economic system)
– Preservationism (appreciating wilderness or protection of nature for its intrinsic value)
– Ecocentrism (more ecologically informed variant of preservationism to protect threatened populations, species, habitats, and ecosystems)

The normative views shaped the ecological discourse in Western countries, although with significant differences between countries. What is required beyond the normative views are *strategies with combined normative, empirical, and theoretical arguments for global transformation* and environmental governance based on systems analyses of interacting social and ecological systems. Climate change governance and transformation of the industrial energy system indicate the development of broadened strategies, based on interdisciplinary knowledge synthesis as efforts to achieve sustainability.

In European countries, the introduction of the new energy technologies in the market system showed many unforeseen problems. The energy system transformation became a process of repeated efforts, modified strategies, competing attempts of different actors, and trial-and-error-based innovation. This can be shown for each of the new forms of solar energy, wind power, and bioenergy in their introduction to markets. The established structures of the industrial energy system, its combinations of different energy forms and conversion technologies, and the use of large technical infrastructure systems and grids for energy delivery make the transformation a long and difficult process. The system transforms the largest part of energy in few forms of energy delivered to the users, such as electricity or gas. The new energy forms need to use these distribution nets (electricity produced through wind power), or develop new forms of local or individual use of the technology. The socio-economic organisation of modern industrial systems, with large corporations, market-dependent and capital-intensive processes, with private property as dominant legal form of ownership, enforces adaptation of the new energy forms. The policies of public support of development and market introduction differ in Europe from country to country and change in unforeseen ways. The introduction of new energy forms is strongly supported in some countries, in others lacking acceptance and deficits of implementation can be found. For wind power, the adaptation to the established energy system is visible in the commercial sector, where large enterprises own and manage the technology (in large windparks on- and offshore) and the energy delivery. Alternative and non-commercial forms developed only exceptionally, temporarily, and in some countries, for example, in Denmark (less in Great Britain: see Elliott 1997: 170f). In Denmark, where wind power technology is used for more than half a century, an alternative sector developed in cooperative forms, owned by the users, more in accordance with the ideas of environmental movements, of social and environmental justice, and solidarity of resource use. In general, the future of wind power development seems to be that of conventional commercial and corporate development.

No fast transformation of modern industrial energy systems can be expected. Thinking in terms of technology, development implies doubtful reductions of the complexity of energy systems until the problems appear as technical ones that require conventional forms of research and development and engineering approaches. Analyses of the social contexts of energy systems can help to understand the difficulties of transformative governance.

Approaching the Transformation of Energy Systems

Advances in the debate of transformation of industrial energy systems were achieved through changes of perspectives and ideas. Different perspectives of energy systems development came with the ideas of "dematerialisation", "degrowth", and "social-ecological transitions".

1. *An early debate about the change of the industrial energy system was part of the broader debate about dematerialisation of economic production* as a way of saving materials and energy that has been summarised elsewhere (Bruckmeier 2013). The industrial production processes that cause growing use of material resources cause also growing use of energy resources. Efficiency gains in terms of reducing the throughput of energy and materials in production systems are, however, annihilated through continued economic growth and increasing use of natural resources. The industrial system does not unfold by itself ecologically rational strategies of reducing material and energy flows; its economic mechanisms keep the industrial economy on the path of continued growth and drive growth and intensification of natural resource use. A transformation of the energy system cannot be consequently realised within the limits and constraints of the system, although the transformation starts within the industrial systems and limited saving effects can be achieved in the short run. Often technical change is understood as directing social change and enabling ecologically rational resource use, which fosters illusions about the solving of environmental and resource use problems through new technologies. Ways beyond technical fixes are shown in several studies and scenarios that show that nearly complete supply of energy from reliable sources and of strong reductions carbon dioxide emissions is theoretically possible until the end of the twenty-first century, although the transformation process may take other forms [Elliott (1997: 182ff); less clear are the predictions and constructions of energy futures by Hall and Klitgaard (2012: 397ff)].
2. *Dematerialisation was translated from technical improvements into social strategies for sustainable development with the ideas of decoupling growth*

and human welfare. With the ideas of decoupling growth and welfare, degrowth and social-ecological transformation, the debate approaches the ideas of a stationary state economy formulated in ecological economics. Technological innovations and improvements are not sufficient to achieve sustainability; the system mechanism of economic growth needs to be transformed. The possibilities to change the maladaptive coupling of the modern economic world system with ecological systems and with natural material cycles need to be shown in more detailed analyses of the functioning of SES, their multi-causal and multi-scalar processes. As discussed by Fischer-Kowalski et al. (2012), it is possible to demonstrate the interconnectedness of socio-economic changes and changes in natural systems in some empirically measurable processes and interactions: these can help to generate models for system perpetuation. Possibilities of social-ecological transformation can be clarified further with the elaboration of a theory of nature and society. The discussion of transformation, approached in the discourses of social and political ecology (Dietz and Vogelpohl 2005; Brunnengräber et al. 2008; Brunnengräber 2009; Leopold and Dietz 2012; Brand 2015a, b), requires more in-depth analyses of energy systems.

3. *An integrated framework for the analysis of social-ecological transitions between different energy regimes (outlined by Fischer-Kowalski et al. 2012), connects the analysis of transformation of energy systems and of societal transformation to sustainability.* In Europe, the historical transition into the fossil-fuel-based industrial regime is completed in an energetic and material stabilisation phase at high levels, whereas the new transition to renewable energy sources has just begun. Presently, a number of large countries are in transition into a fossil fuel energy regime through late industrialisation. The new transition strategies in European countries with renewable energy sources as the core happen in asynchronous societal changes, where, in other countries and parts of the world system, industrialisation just started. The changing role of human labour in large-scale and long-term processes of societal transformation is the second important component besides energy system changes that requires further analysis. Fischer-Kowalski et al.

analyse in which way the historical transition to industrial society has transformed human labour, and what can be learned from this, under changed conditions, for the role of labour in the new social-ecological transition to global sustainability. Six global megatrends, three originating from natural and three originating from societal drivers that affect the transformation of European economies either in more socially friendly form or with more crises, conflicts, and disasters are analysed. Knowledge about the social-ecological transition is used in three scenario variants: no policy change, ecological modernisation, and sustainability transformation.

The social-ecological analysis by Fischer-Kowalski et al. (2012) shows how far a transformation theory has developed from an integration of different knowledge sources. The analysis of long-term perspectives of transforming energy regimes as parts of interacting social and ecological systems requires combinations of several forms of analysis:

- Empirical case studies show the transition processes as complicated and contingent, also because of the desynchronised process of industrialisation, but do not show all blocking factors and the traps of energy transitions.
- Connecting analyses of resilience and sustainability, of adaptation and transformation of energy systems, and working with several methods of knowledge generation and synthesis develop more complex forms of interdisciplinary analyses, which require finally a theory of social-ecological transformation.
- The connected approaches include historical studies of transformations of mode of production and energy systems, empirical case studies of present transition processes, theoretical analyses of interacting societal and ecological systems, analyses of global social and environmental change, and scenarios of long-term systems transformation in the perspective of transformation to global sustainability.

It seems that at this point, the scientific and political discussion of transformation of energy systems ends—or is stuck in dealing with the new knowledge problems that come up in transformation analyses or in the practices of transformative governance. The example of introducing bioenergy production on arable land (see below) shows the dilemmas in the long process of transforming the societal metabolism of industrial society. The question "how to learn ways out of the manifold social traps and problems that appear in transitions" cannot be answered through research but only by way of combining different knowledge practices, where futures are theoretically reflected, constructed, and envisioned in a variety of methods (see Chap. 7). *The knowledge dilemma of energy system transformation and, more broadly, of societal transformation to sustainability, is of the kind "how can one know in advance what is necessary to do or not in a process that stretches in the long-term future, beyond the temporal horizons of social action and planning?"* The knowledge practices to deal with the future—the near future in forms of planning, the distant future in forms of prognoses, scenarios, envisioning, or in phantasm—cannot generate "knowledge in advance", and it seems unlikely that future studies, also with the help of new information technologies, create new forms and methods of looking in the distant future. Assuming that the future is unknown, that it is only cognisable in form of guesses, visions, and utopias, is, however, not the final conclusion. Other mental models and (re)combinations of knowledge forms and knowledge practices can be developed that help to find better ways into the future and better forms of transformative governance. The ecological discussion and research on resilience, sustainability, adaptive governance, and transformative agency has shown some ideas how to model the future, seeing it as generated through processes of change in complex adaptive systems. The most important ideas seem to imagine the future not only in the form of goals to achieve or to avoid but also as a process of continuous change. Governance processes can be modelled in continuous forms of monitoring and assessing processes of social and environmental change, learning and experimenting, simulation and anticipation, knowledge syntheses, and theory-based construction of potential development paths—these can be translated into processes of adaptation and transformation, where the future is negotiated. Such—limited—possibilities to improve strategies for societal transformation develop in social ecology as one of several scientific practices of interdisciplinary research, combination and recombination of knowledge. In the practices of environmental action and governance ideas as that of adaptive and transformative governance show the beginning of new combinations of different knowledge practices and the search for new knowledge on the way to sustainability. There are more alternatives to imagine the future than the two extremes of determined by systemic nature of societal or ecological systems or open and undetermined.

Transformation of SES to sustainability appears in the perspective of social-ecological theory as a long-term and complex process including managed and non-managed change. The practical experiences during the first steps of transformation of energy systems that show the challenges of societal transformation more concretely are discussed below with regard to the problems appearing with the introduction of wind power and bioenergy.

Renewable Energy Sources: Wind power and Bioenergy

In the long process of phasing out of fossil energy sources and nuclear energy, new conflicts within and between countries can be expected. Some of the new energy sources of the future are discussed and developed already, especially wind power, solar energy, and bioenergy; other forms are discussed less, for example, energy from water waves or thermal energy from the interior of the earth. Also research and development of energy created from nuclear fusion continues. In spite of intensive research funded by the EU, it is assumed that decades pass before electricity can be generated from fusion reactors—scenarios from the energy project of the International Thermonuclear Experimental Reactor envision the commercial use of fusion technology in the second half of this century (Lee and Saw 2010). In the long transition of the modern energy system also, risky bridging technologies and interim forms of energy technologies such as hydraulic fracking will be used.

Although non-renewable industrial energy resources may already be characterised as out-phasing, the transformation of energy systems with their different components is a slow process of many decades. It is not a process controlled by local, regional, or national decisions, although formally and legally seen, energy systems are national systems—even in the EU, where more and more regulation happens at supra-national level, for example, with regard to strategic decisions about sustainable development. In the final analysis, all energy systems are part of the global energy regime that is until today dominated by industrial, non-renewable energy sources, and economic and political interests organised to support the use of these conventional energy sources. In European countries, supported by the EU, policies to transform energy systems in the overarching process of transitions towards sustainability are on the way. It cannot be said, from the presently limited experiences, with the introduction of renewable energy sources, which of the green energy sources is the key for the transformation to a post-industrial energy regime—wind power, bioenergy, solar, thermal, and water energy are the forms presently discussed (the projections in studies by 'Greenpeace' see larger parts of energy to come from wind power and solar energy, less from bioenergy and hydro- or geothermal power: Elliott (1997: 182); further discussion for bioenergy: Haberl et al. (2010), see below). Every country has different national energy systems, "energy mixes", and strategies for energy system transformation; and with every energy source come specific problems, limits, and contradictory experiences as the examples of wind power and bioenergy production on arable land show.

Development and Introduction of Wind power Systems in Europe: Empirical Research

Wind power is relatively well investigated and the establishment of wind power systems has advanced in European countries in the past decade. As all components of future energy regimes, introduction of wind power is seen as part of the transformation of energy systems to sustainability. It can be assumed that the mitigation of many conflicts at different policy levels is necessary to achieve that goal. Presently, the quantity of renewable energy sources in national energy systems in European countries is still low, but national governments and economic actors redefined in recent years their interests for development of renewable energy sources, including wind power systems.

For long time, wind power development was suggested by a limited number of environmental actors, with arguments for the unlimited availability of this energy source and that it is not polluting the environment. With the advances in establishing wind power, the environmental and the social impacts and the interests involved are understood better in their conflicting dynamics. Wind power is no longer an exceptional technology or a future option, but a significant component of national energy systems; it is, with its integration in the national energy system, assessed in terms of economic value, of capital and investments, and maintenance and management costs as market-based technology, in addition to its ecological components. Such economic interests and components were not visible as long as wind power was in the phase of development and testing, of improving the construction of turbines. With the introduction of wind power in European energy markets, the process came under control of the established energy "producers", big private companies in the energy sector, although new firms appeared. It did not become an alternative economic subsector with small and local enterprises or cooperatives. The adaptation of wind power to the market conditions does not reduce its quality of a "clean" and renewable energy source. But the development and the total effects of wind power for transforming the industrial energy system are not easily predicted. In the energy policy of the EU, wind power is planned as a limited component in the energy mix that requires further on other sources. Technically seen, wind power has become a modern technology, differing from the historical forms of windmills that drove machines directly. The new forms of turbines, their building material, and the conversion of wind power to electrical energy that is distributed through large electricity nets to households and end users make the simple technology of windmills to a high-tech form of energy use.

The technology of wind power systems developed and differentiated since the beginning of the development of the energy source in Europe in the 1970s. The effects of wind power facilities on the environment are known better. The location of windparks has been optimised in complicated and conflicting decisions, and the limitations of wind power for the development of national energy systems have been experienced. For these limitations, especially the temporary unavailability of wind, solutions are sought more intensively, new and improved forms of energy storage for which presently only few and not optimal technologies are available (batteries, pumped-storage, compressed air energy storage, hydrogen storage).

What makes wind power a special case in difference to the other renewable energy sources, especially bioenergy production on arable land, is its different form of land use. It is not occupying large surfaces for food or energy production, can also be located on unproductive land and offshore. Also the environmental impacts of wind power are specific. Life cycle analysis and impact studies of bioenergy production revealed a number of—non-intended—negative environmental consequences of bioenergy which may not be as "green and clean" as it was seen when the ideas developed. Wind power, with all environmental impacts known so far, seems to have less negative environmental and social impacts than bioenergy production. A dominant conflict about its introduction was the location close to settlements, a paradigmatic form of NIMBY ("not in my backyard"). Many people did not want to have wind turbines in their neighbourhood, they were perceived, also because of their big size, as disturbing and ugly.

The discussion about adequate or non-adequate location of wind power systems is complicated and the values and interests related to wind power change. The scientific and political location discussion was intensive and is no longer left to casual local solutions, and public policies are now supporting wind power development. In Sweden (Bruckmeier and Böhler 2012), the discussion about location was summarised by the Swedish Nature Protection Agency as follows: As adequate locations are seen already exploited or used land areas (industrial areas, harbours, etc.) and the normal landscape that includes agricultural landscapes and non-protected forest landscapes—that is, the largest part of Swedish territory, as in many other countries. National parks, most nature reserves and "Natura 2000" areas with a nature protection status, and marine protected areas are assessed as non-adequate locations of wind power. Potential areas for location are certain nature reserves and protected areas and, in general, areas where the national interest for wind power location is more important than nature protection (NVV 2008). The decisive point in the location discussion is that systematic and comparative assessments of suitable locations are available. These assessments follow the general reasoning that nature protection is not always suitable for wind parks, whereas areas with resource use and different forms of land use are. This may evoke further debates why wind parks are not compatible with protected areas, taking into account that they are also a form of nature protection and sustainable resource use. The preliminary result of the Swedish location debate is as follows: wind power is seen as an energy technology and as land use that should be located there, where other forms of technologies and human land use also happen.

For the decision about offshore location of wind power systems, assuming that in Europe, large windparks will be built offshore, other arguments beyond suitable locations and possible effects on the marine environment (see Petersen and Malm 2006: 75) are coming to the foreground: the direct and indirect costs of wind power systems. The decisions between onshore and offshore location can be assumed to become complicated reasoning and interest negotiation, also when they are reduced to costs, excluding many other social factors. A literature review shows that optimal location for the future wind power development is a function of many variables, some of these relating to capital costs of wind power development. Studies in the UK show the on-land installation costs as about half of the size of offshore location. If offshore wind farms are located on deep water and at large distances from the coasts, the capital costs might be even larger. These direct costs indicate that on-land development from capital costs view might be most economically attractive, even accounting for better wind conditions offshore. However, the potential differences in external costs might change this relation (Ladenburg 2009: 179f).

In Europe, the development of wind power systems took different courses; in some countries, for example, Denmark, Spain, and Germany, it played an important role and the development started already long time ago, whereas in other countries, advances are rather slow. Denmark played a pioneering role in the development of wind power technology and in developing its economic organisation in two forms: in private company based and in cooperative forms, where local citizens own the technical system. The different sizes of wind power plants attract economic actors selectively: large windparks in offshore location may be of special interest for large and powerful energy companies, whereas small plants and single wind turbines may attract interest of local enterprises, actors, and inhabitants in local forms of development and use of energy.

For the analysis of wind power-related conflicts as in the Swedish research mentioned above, another perspective is more relevant than costs: that of wind power as part of long-term strategies to transform energy systems in the complicated transition to sustainability. Mitigation of manifold conflicts is required on the way to sustainability and conflict mitigation becomes a main task in transition management. For that purpose, it seems important to study conflicts and their resolution in detail, developing adequate forms of conflict mitigation instead of bypassing or suppressing conflicts with legal, political, or scientific support through powerful actors and institutions.

An ecological discussion of environmental advantages of wind power, characteristic in early phases of the discourse, when the environment-friendly nature of the energy source was the main theme, is no longer sufficient. The problems that come with the rebuilding of energy systems enforce the discussion of economic aspects of wind power. *For the long-term development of wind power systems, the following points show the issues to be discussed and clarified:*

1. *Location—the basic question is that of how to use space on land and in the sea.* This question comes back in the debate of energy generation. In the debate about wind power systems, offshore location could be seen as a way to avoid conflicts and the difficulties of matching interests of many stakeholders that are required for location on land. However, to find simple solutions without resistance of some stakeholders is unrealistic. Wind power cannot only be developed offshore, and solutions for location offshore and on land need to be sought through processes of mitigating local interest conflicts. As the debates about wind power and bioenergy show, energy production requires land and the use of land needs to be negotiated, taking into account ownership rights. Also the use of fossil sources requires land use decisions. It was not a realistic view of the use of fossil, subterranean energy sources that drove industrialisation, that they can be used without

land use decisions, seeing them as point sources of extraction that do not touch other and private land use interests. The mining of coal and the extraction of oil and gas have significant effects on land and land use, beyond local problems at the extraction points: land use for pipelines, transport, storage in tanks or deposits, energy conversion in power plants, distribution nets, and waste deposits. Additionally, the pollution of soils, water, and air need to be accounted for in location decisions.

2. *Technology—the technical research about wind power systems for land and offshore location is now advanced and the technology can be used in different locations.* The implementation problems do not arise from technological aspects only, but more from the divergent interests of actors and further social factors that are still insufficiently investigated. These include social acceptance of windparks, environmental awareness and reasoning of local populations, economic interests of different social groups, and the governmental and political decisions about wind power. To take into account the complex and varying interest constellations, including the resolution of conflicts in the processes of development, location decisions and use, differentiated and flexible strategies are required than the ones in earlier phases of wind power development where conflicts were often ignored, the "pro or contra" discussion mainly done with ecological arguments, and technology development happened without strong governmental support.

3. *The energy mix in national systems—wind power is planned in European countries as limited part of national energy systems that continue with the use of other sources, sometimes under premises of continuing high levels of energy consumption.* How much of the electricity consumed now and in future can be produced by wind power? In the EU, energy policy up to 20% are planned in the longer run, indicating the estimated market potential of wind power in the energy mix. For the long-term development of energy systems, the question is not that of maintaining high levels of energy consumption. If transformation of energy systems should happen, ways of reducing energy consumption and developing alternatives to large-scale grids (for energy distribution under control of few oligopolistic enterprises) need to be sought.

4. *The framing policies of sustainable development—wind power is an important component of sustainable development which implies multi-scale approaches, matching of interests, and conflict resolution.* As sustainable development is influenced and structured through national and international strategies, it seems important to take into and strengthen local contexts of and activities in such strategies. The conflicts emerging with wind power location are local conflicts and the interests of local inhabitants need to be discussed, negotiated, and compromises found as in all other conflicts about natural resource use. In the further development of strategies for sustainable development, the debate of transformation of societal and economic systems will evoke significant and lasting controversies, conflicts, and necessities to negotiate heterogeneous interests of actors.

Wind power is here discussed as an exemplary case to show the complicated interaction of social and ecological factors in the development of energy systems from renewable sources. These interaction processes and conditions of development differ for every form of energy, from fossil or renewable sources. The introduction of the new energy forms and their combination in energy systems will evoke new conflicts, also about the policy instruments, as the EU energy policy shows with the controversy about tradeable green certificates (Jacobsson et al. 2009). The problems and conflicts need to be discussed in the practice of introduction of "green" energy forms and for all forms of energy from renewable sources—solar energy, thermal energy, water energy, and bioenergy or agrofuels from arable land, and still less discussed forms as the use of energy from waves in coastal and ocean waters. Water energy is a specific case; it is already used since long time, in many countries, with large technical systems with dams, artificial lakes, and water turbines. The building of water power plants has evoked and evokes further strong resistance and conflicts, often from local inhabitants. The environmental effects of water power need to be discussed more critically: it is part of the interruption and significant disturbance of the natural water cycle that happened in the twentieth century, with a doubling of the water withdrawal since 1960, as the Millennium Ecosystem Assessment (2005) showed. The use of wood as renewable energy source and part of bioenergy production is in complicated forms interwoven with social, economic, and environmental problems and conflicts regarding deforestation.

The following discussion shows bioenergy production as a recently intensifying and conflicting process.

Development of Bioenergy: Empirical Research

The production of bioenergy in agriculture and on arable land creates many local conflicts about land use for energy production that was not originally designated for energy but for food production and is still needed for that purpose. The conflicts are less in European countries, where there is no lack of land for food production. But also in Europe, the change of agriculture form a producer of food to a producer of bioenergy is not socially negotiated and does not find consensus. Bioenergy production developed more as a side effect of the search for alternative forms of productive use of land that is no longer needed for food production. The policies of modernisation of agriculture, especially the Common Agricultural Policy of the EU, brought the trend that more and more farmland is temporarily taken out of production to reduce the overproduction of food for European agricultural markets. Although this was not a solution of agricultural and of land use problems, rather an interim solution delaying final decisions, it has, intended or not, supported the introduction of bioenergy production on agricultural land in Europe and also provided for farmers opportunities to develop and maintain their enterprises. In countries of the Global South, the bioenergy conflict is socially explosive: malnutrition and hunger of large parts of the population is a continuing problem. Not only use of fertile land for bioenergy production is under such conditions a source of conflicts but also deforestation for creating new land for commercial farming, and the more recent phenomena of buying of large areas of agricultural land by foreign governments and private companies from other countries that evokes conflicts about such land grabbing. Disputes and conflicts of interest emerge between land use for producing food for growing populations, for developing small-scale agriculture of poor agricultural producers that are still numerous, for the commercial and export-oriented forms of land use of cash-crop farming, and for commercial forms of bioethanol production (an exemplary case study: Martinez-Alier et al. 2010). Globally seen, it seems that expectations to replace larger parts of energy in industrial energy system through bioenergy are unrealistic. *Analyses of bioenergy and agrofuels show the following problems with the development of such forms of land use:*

1. *Social-ecological analysis of bioenergy as a conflicting development:* Bioenergy as energy produced from organic non-fossil material of biological origin is promoted as a substitute for fossil energy sources to reduce greenhouse gas emissions and dependency on energy imports. Haberl et al. (2010) discuss the potential of development of bioenergy that amounts presently to about 10% of humanity's primary energy supply. The authors review recent literature on the potential of global bioenergy supply until 2050, taking into account technology development, food demand, and environmental targets. The estimations vary extremely (from 30 to over 1000 Exajoule/EJ per year). The high estimation is seen as implausible because of overestimation of the area available for bioenergy crops and of too high yield expectations (resulting from extrapolation of plot-based studies to less productive areas). According to the authors, the global technical primary bioenergy potential in 2050 is in the range of 160–270 EJ per year, considering sustainability criteria. The potential of bioenergy crops is at the lower end of previous estimations. Residues from food production and forestry could provide significant amounts of energy based on an integrated optimisation (cascade utilisation) of biomass flows (Haberl et al. 2010: 394).

   The authors conclude that no scientific study is available to clarify satisfactorily the many scientific issues related to future bioenergy potentials in the next decades. Factors of uncertainty include availability and suitability of land for energy crops; development and potential of yield increases; future area demand for food, conservation, and other purposes; trade-offs with other environmental goals; water availability; and climate impacts. Also when human behavioural practices are intimately related to relatively stable cultural and other socio-economic factors, future habits of food consumption can hardly be predicted. All studies reviewed showed shortcomings (Haberl et al. 2010: 401). This implies that quantitative assessments of the potential

of bioenergy development, as in this analysis, need to be complemented through social studies of the processes of bioenergy production that take into account the socio-economic conditions and the policies of bioenergy production in the countries of the Global South, as studies of agrofuel production show.
2. *Political-ecological analysis of production of agrofuels:* Leopold and Dietz (2012) connect the commercial production of agrofuels (e.g. biodiesel and ethanol produced on agricultural land) to the broader problems of nature–society interaction and the multi-scalar and transnational connections of this process.

Nature–society interaction: Production and consumption of agrofuels shows interlinkages across and between policy fields, scales, and the spheres of nature and society. The relationship between nature and society is understood by the authors as dialectical and the spheres as being mutually constitutive. Social development is always dependent on the metabolism of nature, and the metabolism of nature is dependent on the social appropriation, use, and transformation of nature. The natural world is utilised in the global forms of production and consumption; it is changed irreversibly by humanity and humanity is changed irreversibly through its use of nature. Nature can be understood as being socially produced, for example, through anthropogenic climate change. The social practices of utilisation and acquisition of natural resources are dynamic in time and space and governed by processes of political-economic and political-institutional change, affected by power relations and cultural differences. The currently changing forms of land rights, practices, rules, ownership relations, and recent land use for agrofuel production in Africa, Latin America, or Asia are in many ways the result of colonial and post-colonial politics of resource distribution and political-economic transformations.

Multi-scalar and transnational constitution: The view of society–nature relations by the authors emphasises the multi-scalar nature and transnational constitution of society and nature. Changing local conditions of land use and land distribution are interlinked with political-institutional, political-economic, and discursive changes at national and global scales of social action and vice versa. Agrofuel policies show that access to, use of, and control over natural resources as land and water cannot be understood as purely local or national but unfolding on new levels of action, on different scales, and through new constellations of power relations. New constellations of transnational capital and transnational geopolitical space and power are emerging between and amongst developing and developed nations. Land with its productive capacity has become an economically scarce resource over which countries try to obtain control. Current efforts to cope with the challenges of climate change, development, and energy systems imply shifting social, political, and economic power relations (Leopold and Dietz 2012: 6f).

The two main arguments from this political ecological analysis are that of

- studying bioenergy production in the framework of nature–society interaction, requiring the elaboration of the theory that has been discussed here, and
- analysing the multi-scalar constitution of the governance processes that structure bioenergy production.

Both requirements coincide, for the analysis of energy system transformation, in the analysis of social-ecological regulation of the relations between society and nature. This analysis is, as long as the theory of nature–society interaction is not systematically elaborated, possible in the preliminary forms of combining several methods, empirical and theoretical ones, as formulated above for the social-ecological discourse, referring to the framework developed by Fischer-Kowalski et al. (2012).

The conflicting and contradicting forms of bioenergy production according to these interdisciplinary studies include the following:

- Bioenergy production is influenced by the processes of economic globalisation and connecting with the industrialisation of agricultural production. The globalisation of agrofuel production is assessed by critical ecological researchers as consolidating the unequal power relations between industrialised and non-industrialised countries, between the Global North and South, and as resulting in a partial and unequal transformation of these power structures in the countries of the South (Danker et al. 2013: 3, 31ff).
- The EU policy for supporting renewable energy sources in Europe evokes conflicts between the countries of the Global North and

South. The planned use of biofuel in the EU exceeds the possibilities of production in Europe and stimulates import of commercially produced agrofuels from countries in the Southern Hemisphere and through the neglect of social criteria in bioenergy development (Leopold and Dietz 2012: 13ff).
- The gender relations, gender-specific access to resources, and division of labour create or enforce inequality, are woven into social relations of production and land use, and are reproduced in agrofuel production in naturalised forms (Wasser et al. 2012: 18).

Bioenergy in various forms through plant production on fertile land and wind power are two important sources of renewable energy, showing different social and environmental consequences. Both technologies and their commercialisation illustrate that green energy technologies are not as environmentally sound and socially sustainable as they were supposed to be when the ideas developed in the ecological discourse. Many non-anticipated consequences and conflicts appear with the development and market introduction of these technologies, showing the real complexity and problems with the use of renewable energy sources. More or less unexpected conflicts include the following:

- In wind power development, conflicts are more within the countries of the Global North where wind power is in rapid development. Conflicts are about location of windparks and problems connected to the appropriation of wind power by big energy companies that act in international energy markets and make the green energy to a commercial component of large-scale industrial energy systems.
- In the development of bioenergy and agrofuels, strong conflicts emerge at several levels, within countries in the Global South and North and between countries in the North and South. These conflicts are caused through the transnational relocation of bioenergy production and its concentration in countries of the Global South where the production clashes with social and environmental goals of national and global food production.

Conclusion: Future Energy Systems

The specificity of modern industrial energy systems is that their ecosystem components have been more and more neglected with the decoupling of energy from land use that happened with the development of fossil energy sources. Modern energy systems are part of large-scale socio-technical and economic systems that imply risky decisions, high investments, and long time-frames of development and planning. The capital-intensive energy systems do not easily open for decentralisation, local and autonomous development that appear as ecologically rational solutions. The transformation of modern society and its societal metabolism is in all phases, now at the beginning and probably later on with advancing transformation, a process that can only, to a limited degree and partly, be planned, managed, and technically and economically rationalised. Transformation of energy systems is a long process with unforeseeable consequences, risks, conflicts, and surprises. In the transformation of the global energy regime of industrial society, that is, on the political agendas, one of the first lessons that had to be learned is, social consensus about nature protection and sustainability (found in many European countries) does not make the transformation process easier and more peaceful. It is a process with continuous social, political, economic conflicts and unexpected side effects. At the levels of countries and national economies, it is a competitive race for energy resources where contradicting trends and doubtful technical ideas or bridging strategies generate interim solutions that disturb and distort ecologically rational strategies of developing energy systems with renewable sources.

The arguments of environmental sustainability are mainly arguments of intergenerational solidarity and the conditions of life for future generations. In the transformation process, the environmental interests clash in extreme forms with the social and economic interests of present generations, especially the vested interests of powerful political and economic actors.

1. *The basic idea of developing sustainable energy systems through the use of renewable energy sources is clear, simple and convincing—but not*

sufficient for dealing with the transformation problems. Energy cannot be reused, is lost through dissipation from higher to lower quality, according to the thermodynamic laws of physics. Fossil energy resources are limited and create problems of environmental pollution, thus causing the necessity to develop energy systems with renewable sources of energy. With the development of "green energy", elements of sustainability should be built into industrial energy systems. The quality of wind power and bioenergy as energy from renewable sources is not sufficient to develop and introduce them in the markets. For both energy forms, their commercial introduction clashes with many other requirements in social and economic development. The process of energy transition turns out to be one where energy technologies are commercialised and instrumentalised for doubtful strategies of "green economy" that block the transformation of industrial systems.

2. *To overcome the "roadblocks" to sustainable energy systems of the future requires the use of knowledge from analyses of interacting social and ecological systems, in the final analysis knowledge from a theory of nature–society interaction.* This seems to introduce large quantities of knowledge in the analysis of energy transition—knowledge that cannot be seen as necessary when energy systems are reduced to sociotechnical systems. The processes of global social and environmental change in which energy systems are embedded do not leave many alternatives than to use knowledge available from the analysis and assessment of social-ecological change in coupled SES. To transfer this knowledge from social-scientific research in the ecological discourse means: to test, verify, and correct it in continually improving social practices of knowledge application and in strategies of social-ecological regulation and transformative governance. The renewal of the sustainability process implies to use knowledge and experience developing on the way of transformation to deal with the solution of problems and conflicts that emerge in the process.

3. *The discussion of global governance and environmental governance does not take up systematically ideas and knowledge from the social-ecological and political-ecological analyses of energy transition.* These ideas remain contested, but neglecting them and reducing the governance of transformation to policy perspectives and frameworks for organising and

coordinating environmental, climate, and energy policies nationally and internationally is insufficient. An alternative can be as follows: broadening the scope of governance analyses for the development of multi-scale governance strategies and to work with knowledge from interdisciplinary analyses of social-ecological systems. This results after shorter or longer time in approaching the complexity of transformative governance, showing possibilities to connect governance processes with autonomous forms of social-ecological change that cannot be managed and governed. Such learning of improvements of governance strategies seems necessary in the long process of societal transformation to sustainability.

References

Altvater, E. (1991). *Die Zukunft des Marktes. Ein Essay über die Regulation von Geld und Natur nach dem Scheitern des „real existierenden Sozialismus".* Münster: Westfälisches Dampfboot.
Barca, S. (2011) 'Energy, property, and the industrial revolution narrative', *Ecological Economics*, 70 (7): 1309–1315.
Brand, U. (2015a, June 10–12). *How to get out of the multiple crisis? Contours of a critical theory of social-ecological transformation.* Paper presented at the conference The Theory of Regulation in Times of Crises, Paris. Accessed November 10, 2015, from https://www.eiseverywhere.com/retrieveupload.php?
Brand, U. (2015b). Sozial-ökologische Transformation als Horizont praktischer Kritik: Befreiung in Zeiten sich vertiefender imperialer Lebensweise. In D. Martin, S. Martin, & J. Wissel (Eds.), *Perspektiven und Konstellationen kritischer Theorie.* Münster: Westfälisches Dampfboot.
Bruckmeier, K. (2013). *Natural resource use and global change: New interdisciplinary perspectives in social ecology.* Houndmills, Basingstoke: Palgrave Macmillan.
Bruckmeier, K., & Böhler, T. (2012). *Wind power and bioenergy development – Social dynamics of land use conflicts in Swedish rural and coastal areas.* Research report, SECOA-Project. School of Global Studies, Gothenburg University.
Brunnengräber, A. (2009). Die politische Ökonomie des Klimawandels Ergebnisse sozial-ökologischer Forschung, Band 11. München: Ökom-Verlag.

Brunnengräber, A., Dietz, K., Hirschl, B., Walk, H., & Weber, M. (2008). *Eine sozial-ökologische Perspektive auf die lokale, nationale und internationale Klimapolitik*. Münster: Westfälisches Dampfboot.

Danker, H.-C., Dietz, K., Jaeger, N., & Thomas, W. (2013). *Die Globalisierung der Agrarkraftstoffe – Produktion, Handel und Akteure*. Working Paper 7. Berlin: Fair Fuels?

Debeir, J.-C., Deléage, J.-P., & Hémery, D. (1986). *Les servitudes de la puissance: une histoire de l'energie*. Paris: Flammarion.

Dietz, K., & Vogelpohl, K. (2005). Raumtheoretische Überlegungen im Konfliktfeld Klima. Projekt „Global Governance und Klimawandel". Diskussionspapier 03/05. Freie Universität Berlin, Fachbereich Politik und Sozialwissenschaften.

Eckersley, R. (1992). *Environmentalism and political theory: Toward an ecocentric approach*. Albany, NY: State University of New York Press.

Elliott, D. (1997). *Energy, society and environment: Technology for a sustainable future*. London: Routledge.

Fischer-Kowalski, M. (2007, September). *Socioecological transitions in human history and present, and their impact upon biodiversity*. Presentation to the Second ALTER-Net Summerschool, Peyresq, Alpes de Haute-Provence. Accessed August 25, 2015, from https://www.pik-potsdam.de/news/public-events/archiv/alter-net/former-ss/2007/11-09.2007/fischer-kowalski/presentation_fischer-kowalski.pdf

Fischer-Kowalski, M., Haas, W., Widenhofer, D., Weisz, U., Pallua, I., Possanner, N., et al. (2012). *Socio-ecological transitions: Definitions, dynamics and related global* scenarios. Project NEUJOBS, European Union. Accessed August 25, 2015, from http://www.neujobs.eu

Fischer-Kowalski, M., Haberl, H., Hüttler, W., Payer, H., Schandl, H., Winiwarter, V., et al. (1997). *Gesellschaftlicher Stoffwechsel und Kolonisierung von Natur: Ein Versuch in Sozialer Ökologie*. Amsterdam: G+B Verlag Fakultas.

Freese, L. (1997). Environmental connections. *Advances in Human Ecology*, Supplement 1, Part B.

Haberl, H., Beringer, T., Bhattacharya, S. C., Erb, K.-H., & Hoogwijk, M. (2010). The global technical potential of bio-energy in 2050 considering sustainability constraints. *Current Opinion in Environmental Sustainability*, 2, 394–403.

Hall, C. A. S., & Klitgaard, K. A. (2012). *Energy and the wealth of nations: Understanding the biophysical economy*. New York: Springer.

Jacobsson, S., Bergek, A., Finon, D., Lauber, V., Mitchell, C., Toke, D., et al. (2009). EU renewable energy support policy: Faith or facts? *Energy Policy*, 37(6), 2143–2146.

Krausman, F., & Fischer-Kowalski, M. (2010). *Gesellschaftliche Naturverhältnisse: Energiequellen und die globale Transformation des gesellschaftlichen Stoffwechsels.* Social Ecology Working Paper 117. Vienna: Institute of Social Ecology.

Ladenburg, J. (2009). Stated public preferences for on-land and offshore wind power generation – A review. *Wind Energy, 12,* 171–181.

Lawrence, J. R. (1993). Can human ecology provide and integrative framework? The contribution of structuration theory to the contemporary debate. In D. Steiner & M. Nauser (Eds.), *Human ecology: Fragments of anti-fragmentary views of the world.* London: Routledge.

Lee, S., & Saw, S. H. (2010, July 4–7). *Nuclear fusion energy – Mankind's giant step forward.* Paper presented on the 2nd International Conference on Nuclear and Renewable Energy Sources, Ankara, Turkey. Accessed November 28, 2015.

Leopold, A., & Dietz, K. (2012). *Transnational contradictions and effects of Europe's bioenergy policy: Evidence from Sub-Saharan Africa.* Working Paper 4. Berlin: Fair Fuels?

Martinez-Alier, J., Pascual, U., Vivien, F. D., & Zaccai, E. (2010). Sustainable de-growth: Mapping the context, criticisms and future prospects of an emergent paradigm. *Ecological Economics, 69*(9), 1741–1747.

Millennium Ecosystem Assessment. (2005). *Current state and trends.* Global Assessment Reports, Vol. 1. www.millenniumassessment.org

NVV (Naturvårdsverket). (2008). *Naturvårdsverket och vindkraft i Sverige. Vindkraftseminar Stickelstad.* Presentation by Alexandra Noren.

Petersen, K. J., & Malm, T. (2006). Offshore windmill farms: Threats or possibilities to the marine environment. *Ambio, 35*(2), 29–34.

Rosa, E. A., & Machlis, G. E. (1983). Energetic theories of society: An evaluative review. *Sociological Inquiry, 53*(2-3), 152–178.

Schaeffer, R., Szklo, A. S., André Frossard Pereira de Lucena, A., Soares Moreira Cesar Borba, B., Pinheiro Pupo Nogueira, L., Pereira Fleming, F., et al. (2011). Energy sector vulnerability to climate change: A review. *Energy, 38*(1), 1–12.

Schneider, E. (1967). *Einführung in die Wirtschaftstheorie, 1. Teil: Theorie des Wirtschaftskreislaufs.*13. Auflage. Tübingen: Mohr.

Wasser, N., Backhouse, M., & Dietz, K. (2012). *Zur Bedeutung von Geschlecht in der Agrarkraftstoffproduktion.* Working Paper 5. Berlin: Fair Fuels?

Woltron, K. (2004). Wege in den Postkapitalismus. *GAIA, 13*(1), 11–18.

9

Conclusion: The Coming Crisis of Global Environmental Governance

The report of the Worldwatch Institute in 2013—Is sustainability still possible?—and the annual reports since then indicate the necessity to deal with threats to sustainability and growing doubts about the possibility of transformation to sustainability. The intensifying search for natural resources in remote locations and fragile ecosystems as the arctic or the deep sea since the beginning of the twenty-first century shows the efforts to maintain the metabolic regime of industrial society. Large parts of the global population living until today in non-industrialised countries should be integrated in the economy of growth. The aims are less to reduce poverty and hunger and to create decent conditions of life for the poor in the Global South. In practice, such improvements are only side effects of the maintenance of capital accumulation and economic growth through further industrialisation. Non-growing capitalism is a contradiction in itself. The lack of natural and social resources for global industrialisation and the global boundaries of natural resource use make further unlimited growth impossible. What seems impossible in the modern capitalist world system, to end industrialisation and economic growth, is now on the agendas of environmental research and global governance: as sustainable development, rethought in social ecology with the term of social-ecological transformation. To address this "impossible transformation" more consequently requires, according to the theory discussed in this book, other perspectives and knowledge practices in environmental research, policy, and governance.

Social-ecological transformation and *transformative governance* are not yet effective although the ideas are discussed in science and policy, indicating the search of new ways to deal with global environmental problems. The notion of social-ecological transformation spreads quickly in the political debates, which does not mean that social-ecological knowledge and research is used. Also the concept of *social-ecological agency* is not always interpreted with the knowledge synthesised in social ecology. The new terms are used with different interpretations; sometimes old ideas are reformulated with the new terminology. Transformative agency is described by Westley et al. (2013) in terms of resilience and the adaptive cycle. Their guiding idea of safe space is developed from the sociological theory of communicative action by Habermas. In social-ecological research and theory, the terms of social-ecological agency and transformation are used in other forms.

The environmental discourse of global change uses an alarming rhetoric: this century is to become one of global "overshoot and collapse" should there be no success of earth system governance. The time window for organised, coordinated, and mediated transformation of industrial society to sustainability is closing, as many events, especially global climate change and the increasing difficulties of environmental governance indicate. Little time seems left to deal with the imbalances and the system contradictions of the globalised economy. Time is too short for learning, adaptation, and transformation, as many environmental researchers say, arguing in the tradition of the "limits-to-growth" studies. In the climate discourse, views prevail that only few decades, until the middle of the twenty-first century, are left to attain solutions for climate change regulation that can help to avoid environmental catastrophes. More voices are heard that it is already too late for non-catastrophic transformation. Simultaneously uncertainty increases; future changes of society and the ways of transforming society to global sustainability appear as unknown. Ideas differ strongly: how to build "ecological rationality" (Mol), to "re-embed the markets" (Polanyi), to "regulate nature-society interaction" (social ecology), to build a new socially and environmentally just "world order" through global environmental governance?

To deal with the knowledge problems in environmental science and policy requires a series of changes in the forms of knowledge production and application. One possible way, suggested here is, *to clarify the conditions, the forms and the implications of dealing with global social and environmental change theoretically and empirically, through empirical and theoretical knowledge and different forms of knowledge generation in interdisciplinary knowledge syntheses.* This clarification is attempted with the social-ecological theory that connects societal and ecological theories to understand the interaction of modern society and nature. The interaction happens partly through global systems and system mechanisms, and partly through particular, culturally specific systems and mechanisms. With interdisciplinary knowledge syntheses, the research on global change can be integrated and the forms of global environmental governance and of transformative governance, agency, literacy, and learning can be understood and described better.

A first step of rebuilding of environmental governance would be *to learn from the failure of prior environmental research and collective action since the take-off of global environmentalism in the 1970s*, with the "limits-to-growth" debate. Taking the example of climate research and policy discussed in this book, it could be said: more could be learned from controversies about constructing world systems forty years ago than that what is visible in the neoliberal climate change narrative. The controversy between the "limits-to-growth" model of Meadows and Meadows and the more differentiated model of Pestel and Mesarovic showed: global processes of social and environmental change cannot be constructed in undifferentiated global models of growth and development; necessary are methodologically differentiated forms of translating scientific knowledge and results of research and modelling in knowledge practices of collective action, transformative governance, and social-ecological regulation. Geographically, ecologically, and socially (culturally, politically, and economically) differentiated systems and processes in the economic world system and the ecological earth system need to be analysed. Huge amounts of data and high numbers of mathematical equations, necessary for mathematical modelling of complex systems, cannot replace theoretical knowledge about modern society and its interaction with nature in many subsystems and processes. When the future of global societal development is mathematically modelled and calculated, it is still open and unclear, which governance strategies are required to catch the varying and differentiated social consequences of global change in different countries and for different social groups. In spite of addressing some of the problems in a concomitant analysis of climate change, the social consequences of climate change have been neglected by the dominant climate research and policy.

In the further scientific and political discourses about global environmental governance should develop new forms of collective learning and reflexivity, as discussed in Chaps. 7 and 8. The examples of climate change and world modelling are limited examples to discuss problems of scientific learning; the reflection is limited to methods and data for modelling. Mathematical modelling as data-dependent and theory-averse method cannot replace theoretical analyses of interacting SES, requires such theories as structuring knowledge to formulate models. Expectations of the kind "data can replace theory with the advances of world models based on new information technology" were premature and misleading. More useful would be the integration of scientific knowledge in more complex forms of "epistemic triangulation" of empirical research, mathematical modelling, and interdisciplinary theories. Thus, the methodological deficits of each of these modes of knowledge generation could be compensated.

Transformation to sustainability is discussed in social ecology, in the political discourse, and in scenarios for sustainable development. In the transfer of knowledge from research to policy, many ideas and much knowledge are lost and filtered away—the process is not always transparent and the criteria of knowledge use unclear. In policy debates and scenario formulation, the term of transformation remains a metaphor for a complicated process. The visions and worldviews count more than theoretical and interdisciplinary knowledge about modern society and SES. In the global governance discourse, the neoliberal vision of a "green economy" supports illusions of an easy and quick transition to sustainability, without changes of societal and economic systems, through simple policy reforms and with market-based policy instruments, as paradigmatically discussed in payments for ecosystem services (Chap. 5). Although these approaches and methods are more and more criticised in science and in environmental politics because of their dysfunctionality, no changes of paradigms of knowledge use, research practices, and political action happened. The idea of "resource capitalism" (Urry 2011) in environmental sociology is more critical than older ideas of "ecological modernisation" and development of "ecological rationality" in environmental politics (Mol), but creates more uncertainty than clarification of possibilities of social-ecological transformation. Urry himself formulates some doubt whether sustainability is possible within capitalism and the present economic accumulation regime, but does not achieve more consequent reasoning. He was criticised (Gustafson 2012) for the technical fix concept of resource capitalism that does not offer solutions to climate problems and a confusing conceptualisation of politics of scale. A broadening of the sociological perspective and an interdisciplinary opening for further ideas and knowledge develops in the social-ecological research about transformation and transformative governance. This research can create alternative and further ideas for future knowledge and policy practices. Ideas to renew and change the environmental discourse and the policy practices should be formulated as possibilities, not as recipes for improving natural resource management. Recipes in form of imperatives for resource management are often formulated in applied ecological research and devalued the success of adaptive management—in spite of its necessity and the important insights it brought through a reflexive use of ecological knowledge. The simplifications in applied ecological research in forms of recommendations and instructions for resource managers do not improve managerial practices, but create sometimes more uncertainty. The problems and possibilities of changing and transforming SES (for adaptive management, see Chap. 5; for resilience analyses, see Chap. 3) are not systematically, critically, and theoretically reflected. Knowledge from science and research cannot be directly infused in the practices of natural resource management, as the idea of scientific management used in the adaptive management debate seems to say. The knowledge comes from one (scientific) discourse and needs to be translated for use in another (political) discourse. To carry out new empirical research and apply it directly in managerial practice is an epistemological shortcut.

In the scientific and political knowledge practices, not just the knowledge applied needs to be changed, but the ways of knowledge generation, transfer, and application. This is inexactly discussed as reflexive knowledge practices, more exactly described social ecology and in terms of second- and third-order environmental governance.

To advance in the change of knowledge practices in environmental science and governance, the arguments developed in the chapters of this book can be systematised in the following suggestions:

1. *The main conclusion from the discussion of a social-ecological theory of nature and society is as follows: A theory of transformation to sustainability needs to compile, evaluate, and synthesise the knowledge about the interaction of social and ecological systems in modern society in several forms.*
   - In an action-theoretical *analysis of multi-scale governance processes and regulation of the interface of society and nature* that helps to identify changes in terms of collective action, resource management, policy, and governance. This implies to be aware of limits and constraints of policies as means of societal transformation.
   - In a more complex system, theoretical analysis needs to be dealt with the *systemic processes of society*: societal reproduction including symbolic and material reproduction of the modern societal systems, multi-scale processes and changing forms of reproduction of the modern economic world system.
   - Parallel to the analysis of systemic processes in society is that of *ecosystem development and change* to be dealt with in the transformation of social and ecological systems. The knowledge about forms of global environmental change that enable or restrict certain options of transformation to social and ecological sustainability is mainly from ecological research.
   - The analysis of *intersystemic processes and networks of interacting SES* is the core of the theory of interaction of society and nature in terms of historically specified societal and ecological systems analysis.
2. *The knowledge from interdisciplinary SES analyses shows ways to criticise and develop present forms of environmental governance. Theoretical knowledge can be included in knowledge transfer and application*, in the

translation of scientific knowledge in strategies of social-ecological regulation, of building of institutions, and of transformative governance (described in exemplary forms in Chaps. 7 and 8). Although these strategies may turn out to be inadequate in future, it can be *learned by experience how to improve continually strategies for societal transformation*, with bounded ecological rationality, in situations of insufficient knowledge, insecurity, and an unknown future.

These two suggestions from the social-ecological theory can be translated in a series of further, more concrete ideas to change knowledge practices in environmental research and governance, in attempts to deal with the cognitive crises in conventional environmental research and governance:

3. *What is the "unit of analysis" and the "unit of governance" that connects research and governance practices?* To improve environmental governance is not primarily a question of the disciplines that can or should provide the knowledge. A more important question is: which knowledge is necessary to understand the systemic interaction of modern society and "modern" ecological systems? Interaction processes in coupled SES are the core theme of environmental research and governance, which implies neither research nor governance can be done without knowledge about social and ecological system components. This knowledge cannot be provided in disciplinary and specialised research—no discipline has the exclusive knowledge about the forms of interaction and coupling between both systems components.

Inter- and transdisciplinary knowledge syntheses and practices improve the chances to find knowledge about social-ecological transformation, but syntheses can also fail and need to be reflected critically. Interdisciplinary knowledge production can always be criticised because it transgresses the safe space of canon- and rule-guided disciplinary traditions. Examples discussed in all chapters show insufficient forms of interdisciplinarity: constructions of shortcut hybrid terms, for example, "socio-natures" that misdirect analyses of nature-society interaction; mainly politically motivated and world-view-dependent forms of knowledge syntheses (e.g. in parts of the transdisciplinarity debate: Nicolescu 1994); interdisciplinarity and scientific knowledge practices that construct new concepts, terminologies narratives, but do not advance to create transformative knowledge and agency (the example of actor-network theory discussed in Chap. 7); "experimental interdisciplinarity" in forms of combinations of knowledge understood as partial, provisional, and incomplete (Whatmore 2002); interdisciplinarity that is guided by a disciplinary core of knowledge that is set as standard and seen as not negotiable (exemplified in forms of ecological research and its selective integration of social-scientific knowledge). Social-ecological interdisciplinarity is not protected against inadequate and doubtful interpretations and conclusions. The critical discussion and reflections of the knowledge used for the construction of a nature–society theory should help to validate the knowledge. In this way, the theory that guides knowledge syntheses attempts to control the ecological, the construct, and the content validity of the knowledge generated in a historically specific analysis and theory of nature–society interaction.

4. *How can the analyses of societal systems and SES be done in theoretically backed and practically applicable forms?* The complex systems analyses from political economic and other theories of modes of production in human societies are restructured through a simplified reconstruction of the core mechanisms of societal reproduction that maintains the main theoretical arguments. These core mechanisms are described as societal metabolism. The metabolism is reconstructed in social ecology theoretically and studied in empirical research, a main part of the research done in this field. Critical systems analyses of the industrial metabolic regimes can, in the practices of social-ecological research and knowledge use, generate a series of more specific and concrete suggestions to change knowledge practices in environmental research *and* in environmental governance. The following suggestions seem the most important ones.

5. *The identification of the social subject(s) acting in processes of social-ecological transformation* is a main condition for improving strategies of transformative governance. In the complex process of transformation act many heterogeneous groups and actors. The process of change has no class subject in the sociological sense, but subjects in form of

transformative action groups and networked management of natural resources. The question, which actors are part of the many subjects of change cannot be formally answered (through legal rights and formal authorisation of actors), not on the basis of formal power of actors and power relations, and only insufficiently through the intuitive methods of stakeholder analysis that are widespread in practice, but hardly substantiated through social-scientific knowledge. At least the following questions should be asked to *identify the specific social subjects at different levels of governance and the potential interests that influence the possibilities of a transformation to sustainability:*

- The political interests of political actors and citizens
- The economic, not always politically articulated interests of resource users and other economic actors
- The social interests that are only partly articulated in political discourses, to a large degree in independent social processes (specific interests of social classes and social groups, of social and cultural movements)
- The interests that can be called new forms of general will or interests in this phase of development of modern society (especially the interests to maintain the ecological processes and life-supporting systems for the present and the future generations).

It seems evident that the identification of the collective subjects and actors in transformative governance requires empirical research and theoretical analyses, otherwise the question is continuously unanswered who should participate in collective action for social-ecological transformation. First concrete forms are transformative action groups at local and community levels. More complicated is the identification of actors and forms of action at supra-local levels.

6. *Analyses of systemic, institutional, actor-related factors (e.g. vested interests) and the "system contradictions" in modern society that block transformation to sustainability or maintain path-dependent development:* For an interdisciplinary theory of social and ecological systems and for transformative governance practices, it is essential to be aware of the conditions of success and the "roadblocks of transformation", which is the main reason to develop a theory. Even in advanced approaches of

ecological research and in strategies as adaptive governance, the blocking factors are insufficiently analysed, ignoring large parts of social-scientific knowledge, and formulating inexact reasons, for example, "institutional inertia" or "lack of leadership". The question is to some degree one of power asymmetries between actors, and to some degree one of knowledge of potential hinders, but more than that one of systemic structures and functions of societal and ecological systems. Among the strong hindrances are the ones that have been called in prior chapters autonomous processes in SES that cannot be managed or only indirectly. Blocking factors overlap with the autonomous social and systemic processes that make the transformation difficult: economic growth, population growth, competition between national economies and countries about access to resources and chances of development, modernisation, and industrialisation, and the ecosystem functions and processes that need to be maintained in the long run. With regard to such factors, modern society appears as "blocked society", blocked in much more complicated forms than the blocked development through bureaucratisation that has traditionally been described in the sociology of organisations, in salient form by Crozier (1971). Modern society is locked into the impasse of an industrial society and the social-ecological transformation is a transformation to a new mode of production. If that is not theoretically analysed, transformative governance will remain weak, as examples of the present global scenarios show and visions of a new industrial revolution that ignore relevant knowledge, for example, that by Hawken et al. (2000) on "Natural Capitalism", that of "resource capitalism" (Urry 2011), or of the "second machine age" envisioned by Brynjolfsson and McAfee (2014) that should bring abundance where hitherto scarcity or economic stagnation and crisis were seen as the problems to deal with.

7. *The pathways and the duration of sustainable development are inexact and weak points in environmental research and governance. With the theoretical analysis of SES dynamics and roadblocks of transformation, the projections can be improved and become more realistic.* About the trajectories and the duration of social-ecological transformation, little is said in the advanced discussion of global transformation scenarios to sustainability, paradigmatically by Gerst et al. (2014) in the three scenario

variants of "conventional development", "policy reform", and "great transition". The conclusion of the authors, similar to that of many other scenario analyses, is as follows: There are plausible scenarios that remain within safe bio-physical operating space or the planetary boundaries of resource use and achieve various development targets. However, dramatic social and technological changes are required to combat the risks of conventional development and growth (Gerst et al. 2014: 123). About the forms, practices, and phases of these changes, one does not read much in the global scenarios; they argue: it is enough to know the planetary boundaries of resource use to make the changes possible. In the scenarios, no theoretical knowledge from societal systems analyses is used to improve scenario variants. Global governance models and global scenarios alone do not show how to deal with the many problems that need to be dealt with in the transformation process. To think "beyond scenarios", in terms of processes that are not or cannot be reflected in scenarios, is a step further that remains to be done. At this point of scientific ignorance about the future, the discussion of sustainability seems to dissolve in speculations and formulation of utopias, different in form but not in content from simpler visions about the good society in the environmental discourse. To seek for alternatives to speculation, discussing and combining different forms of knowledge use and knowledge practices (as exemplified in the climate change debate in Chap. 7) can be methodological ways to deal with the "veils of ignorance", some of which are theoretically constructed, others may indicate final limits of scientific knowledge.

8. More detailed and exact descriptions of the *processes and phases in the long social-ecological transformation* are the final demands from environmental and social-ecological research. The process of sustainable development becomes more transparent when it is reconstructed as

   – *temporally structured* in a sequence of phases of change;
   – *spatially differentiated* in transformation strategies for different countries, such in the centre and in the periphery of the modern world system;
   – *socially differentiated* according to the components of socio-metabolic regimes and the reorganisation of processes of natural resource use and management. The main forms of transformative agency can be

described in terms of governance, regulation, and other forms of collective action. To initiate transformation and maintain it through different phases, the forms of collective action discussed so far are important: newly built transformative action groups and networked forms of natural resource management. Further forms of action need to be developed in the process. The constraints of societal and ecological systems, their systemic dynamics and requirements of reproduction, should not be seen as factors preventing social-ecological transformation, but as difficulties to be aware of. Theoretical analyses of the complex internal dynamics of modern society and "modernised nature" are needed to guide the governance practices.

As a critical theory, the social-ecological theory of nature and society should not only help to develop new and improved forms of transformative governance which is the most important theory–practice link of this theory, but the theory should also be able to give arguments why certain theories of society or ecological theories are excluded from its construction and remain competing theories (see Chaps. 2 and 3). Furthermore, the theory should identify its knowledge limits, the main limit given with the impossibility to synthesise all knowledge that would be required for the analysis of nature–society interaction. A widespread knowledge practice in environmental research is to give primacy to empirical research and to reduce theoretical analysis to concepts, models, and frameworks to structure empirical research. The social-ecological theory discussed here works with other assumptions: the limits of knowledge are not given with empirical research; empirical research and theoretical analyses of coupled SES are interacting in more complicated forms; the limits of knowledge are finally dependent on the theoretical construction and validation of the interdisciplinary knowledge web.

The social-ecological theory provides knowledge about system transformation which is only to a limited degree generated in recent social-ecological research: it is more the result of connecting, integrating, and synthesising knowledge from different theories and fields of research, natural- and social-scientific knowledge. The broad theme of nature–society interaction is studied in several disciplines and theories. Alternatives to the social-ecological theory that take up its core theme develop in other theories. Some of these theories are complementary to, others competing with the social-ecological theory, and further ones more specific and limited. Aspects of a theory of society or nature can be found in other theories not discussed here: the theory of kaleidoscopic dialectics that builds from the discourse on post-colonial development (Rehbein); the feminist theories that analyses gender relations and inequality in a globalising society (Walby); and theories from the human-ecological (Schnaiberg) and political-ecological discourse (Lipietz). Different interpretations and explanations remain, and controversies about societal transformation continue—in improved forms. Beyond the single theories, an intertheoretical discourse creates new spaces of knowledge that can create further possibilities to develop the interdisciplinary theory of society and nature. Social-constructivist, post-modern and some variants of feminist and post-colonial theories reproduce the critical arguments of ethnocentrism, false constructions of globality, and neglect of non-Western cultural and cultural-anthropological perspectives in sociological or economic theories of modern society. With that the controversy between social theories becomes one about cultural relativism and universalism. Theoretical approaches that can communicate and mediate knowledge exchange between contrasting culturally relativistic and universalistic theories are rare. One important example is the attempt of Wolf (1982) to combine the macroscopic theoretical reasoning of political economy of modern capitalism as world system with culturally and locally specific, microscopic anthropological research. This analysis shows the hybrid nature of modern capitalism developing throughout its history in two incompatible forms, as cultural diffusion, colonialism, and globalisation of its economic systems described in world system theory, and as internalisation of elements of non-Western cultures and societies in the modern forms economy and modes of production, where it was not possible or not necessary to change these. Another, more recent example, the civilisational analysis of Arnason, was discussed in the construction of the social-ecological theory (Chap. 2).

References

Brynjolfsson, E., & McAfee, A. (2014). *The second machine age: Work, progress, and prosperity in a time of brilliant technologies.* New York: W.W. Norton.
Crozier, M. (1971). *La Société bloquée.* Paris: Le Seuil.

Gerst, M. D., Raskin, P. D., & Rockström, J. (2014). Contours of a resilient global future. *Sustainability, 6*(1), 123–135. doi:10.3390/su6010123.

Gustafson, S. (2012, August 28). Review of: Urry, John 2011 Climate change and society. *Environment and Planning D Society and Space.*

Hawken, P., Lovins, A., & Lovins, L. H. (2000). *Natural capitalism: Creating the next industrial revolution.* Boston: Little, Brown, and Company.

Nicolescu, B. (1994). Charter of transdisciplinarity. In: *Manifesto of trandisciplinarity* (K. C. Voss, Trans.). Albany, NY: State University of New York Press.

Urry, J. (2011). *Climate change and society.* Cambridge: Polity Press.

Westley, F. R., Tjornbo, O., Schultz, L., Olsson, P., Folke, C., Crona, B., et al. (2013). A theory of transformative agency in linked social-ecological systems. *Ecology and Society, 18*(3), 27. doi:10.5751/ES-05072-180327.

Whatmore, S. (2002). *Hybrid geographies: Natures cultures spaces.* London: Sage.

Wolf, E. (1982). *Europe and the people without history.* Berkeley, CA: University of California Press.

Index

A

actor network theory, 16, 26, 40, 296–8, 300, 301, 392
adaptive governance, 9, 223, 246, 257–67, 270, 274, 275, 313, 394
adaptive management, 11, 147, 173, 235–81, 289–90, 313, 325, 329, 330
  adaptive co-management, 239, 246, 255, 258, 261
Adger, Neil, 99, 113, 293
Agency, 39, 95, 114, 160–3, 277, 308, 318, 319, 321, 392
  societal agency, 386
agriculture, 149, 176, 200, 350, 374
agrofuel, 373, 376–8
Allen, Craig, 247–9, 252, 255, 263, 264
Amstutz, Marc, 17, 20, 55
Anthropocene, 41, 48, 73, 74, 148

Arnason, Johann Pall, 18–20, 34, 397

B

Barca, Stefania, 338–40, 346
Bauman, Zygmunt, 14, 24, 36
Beck, Ulrich, 17, 18, 20, 24, 58–62, 318
Bell, Daniel, 16, 55
Biermann, Frank, 35, 158, 267, 271, 274, 314
bioenergy, 343, 354, 357–8, 361, 366–7, 369, 373–8, 380
biological metabolism, 343
Blok, Anders, 3, 17, 292, 296–9, 302
Brand, Ulrich, 4, 35, 128, 145, 158, 166, 319, 321, 326, 341, 363
bridging concept, 91, 96, 139, 281

© The Editor(s) (if applicable) and The Author(s) 2016
K. Bruckmeier, *Social-Ecological Transformation*,
DOI 10.1057/978-1-137-43828-7

399

400 Index

Brunnengräber, Achim, 35, 275, 286, 292, 304, 319, 321, 363

C

carrying capacity, 77, 132, 137, 190, 329
Castells, Manuel, 16, 22, 41, 47, 190
civilisational-cultural analysis, 34
climate change
    adaptation to climate change, 173, 250, 287–324
    mitigation of climate change, 292–4, 306
climate policy, 109, 154, 155, 160, 291–293, 296, 297, 301, 307, 314
coastal areas, 285–330
coastal development, 285, 310, 312–3, 319, 324, 327
colonisation of nature, 41, 87, 187, 202, 208, 209
commodification of nature, 27, 49, 189, 227
common pool resources, 222
complex adaptive systems, 76, 84, 210, 365
conflicts, 9, 20, 53, 54, 160, 162–4, 172, 246, 262, 292, 293, 299, 308, 314, 316, 318, 322, 324, 327, 329, 330, 352–6, 359, 364, 366, 367, 371–4, 377–80
constructivism, 281
coupled social-ecological systems (SES), 10, 201

coupling of social and ecological systems, 70, 71, 78, 85, 87, 185, 199, 201, 203, 210
critical theory, 11, 16–21, 25, 31–3, 41, 48, 51–6, 61, 110, 128, 147, 207, 217, 396
cultural anthropology, 24, 33
cultural ecology, 24, 162, 213

D

Daly, Herman, 9, 81, 131–3, 159
Degrowth, 9, 57, 81, 133, 159, 162, 165–6, 362, 363
Dematerialisation, 160, 350, 362–3
Dietz, Kristina, 197, 363, 376–8

E

ecological distribution conflicts, 33, 163, 164, 316
ecological economics, 9, 75, 81, 85, 90, 133, 159, 267, 342, 351
ecological footprint, 116, 159, 175, 190
ecological Marxism, 17, 26, 38, 302
ecological modernisation, 16, 26, 29, 58, 115, 132, 141, 168, 302, 303, 389
ecological theories, 69, 75, 83, 86, 89, 195, 396
ecology, 1, 2, 5, 10, 12, 15, 17, 28, 41, 45, 47, 69–119, 127, 128, 135, 138, 145, 150, 155, 159, 166, 185, 187, 197–199, 202, 203, 205, 206, 208, 209, 230, 235, 252, 267, 303, 304, 306,

322, 326, 338, 342, 344, 346, 351, 352, 360, 365, 388, 390, 392
ecosystem ecology, 69, 216
ecosystems, 6, 11, 18, 42, 47, 69–73, 75–81, 83–9, 93–5, 98, 99, 102–6, 110, 112, 113, 129, 138, 151, 154, 155, 166, 169, 173, 184, 186–8, 190–2, 196, 198–201, 210–2, 215–219, 223–9, 232, 238, 246, 250, 254, 259, 263, 275, 280, 286, 288, 312, 322, 329, 342, 343, 385
ecosystem services (ESSs), 11, 28, 79, 83–5, 89, 173, 183–232
energy regime, 11, 40, 129, 148, 162, 167, 338–41, 345, 347, 348, 353, 355, 359, 363, 364, 366, 379
energy sources, 11, 12, 202, 289, 337, 343, 346, 348, 350, 351, 354–9, 366–9, 379
energy system, 11, 160, 173, 289, 337–381
environmental flows, 16, 62
environmental governance, 39, 44, 104, 146, 159, 161, 162, 174, 193, 235–281, 289, 298, 307, 311, 316–18, 320, 321, 380, 385–97
environmental policy, 8, 37, 86, 89, 96, 115, 137, 219, 229, 241, 269, 289, 318
environmental research, 1, 4, 10, 11, 29–31, 43, 84, 94, 164, 230, 238, 242, 267, 278, 297, 308, 329, 386, 391, 396
environmental sociology, 17, 28, 33, 197, 292, 302–4, 389
epistemology, 7, 47, 206

F

Farley, Joshua, 79, 220, 223, 224
Fischer-Kowalski, Marina, 41, 42, 44, 109, 128, 145, 151, 152, 154, 157, 166, 167, 199, 319, 340, 344–8, 363, 377
Fischer-Lescano, Andreas, 17, 20, 55
Folke, Carl, 79, 89, 113, 114, 194, 246, 247, 250, 255, 258–60
food production, 149, 173, 325, 374, 375
Frankfurt School, 19, 32, 52
Freese, Lee, 6, 17, 351

G

Georgescu-Roegen, Nicholas, 81, 131, 133, 351
Gerst, Michael, 101, 102, 192, 394, 395
Giddens, Anthony, 16, 20, 24, 25, 58–62
global environmental change, 1, 3, 23, 48, 77, 84, 104, 106, 148, 162, 165, 195, 200, 208, 287, 292, 309, 320, 325, 390
global governance, 9, 23, 45, 102, 157, 268, 271, 274, 317, 318, 380, 385, 395

globalisation, 23, 26, 29, 34, 40, 56, 57, 59, 60, 62, 148, 150, 272, 317, 341, 355, 377, 397
global South, 23, 55, 56, 58, 116, 137, 142, 151, 167, 271, 296, 327, 374, 376, 378, 385
global sustainability, 101, 109, 137, 138, 155, 158, 164, 168, 172, 176, 319, 320, 326, 348, 356, 386
great transformation, 43, 58, 138, 141, 149, 161, 168, 272
growth
  economic growth, 62, 130, 136, 142, 144, 151, 156, 160, 162, 163, 190, 224, 254, 263, 312, 315, 327, 340, 347, 348, 350, 362, 363, 394
  exponential growth, 9, 92, 102, 162, 350
  population growth, 102, 129, 142, 161, 347, 394
Gunderson, Lance, 58, 248, 252

H

Haberl, Helmut, 33, 128, 152, 158, 166, 199, 326, 367, 375
Habermas, Jürgen, 16, 18, 19, 21, 45, 51–6, 386
Hall, Charles, 9, 72, 324, 341–3, 351, 358, 362
historical-geographical materialism, 41, 44, 195
Holling, Crawford Stanley, 88, 247

Honneth, Joas, 19, 26, 54
human appropriation of net primary production (HANPP), 190
human ecology, 17, 28, 41, 131, 166, 208, 342
human nature, 3, 52

I

industrial ecology, 82, 85, 303, 306
industrial energy systems, 9, 11, 337–81
integrated coastal zone management (ICZM), 312, 327–30
interaction of society and nature, 5, 15–62, 127, 137, 190, 194, 205, 215, 390
interdisciplinarity, 70, 90, 267, 391, 392
  interdisciplinary knowledge, 23, 28, 31, 46, 70, 83, 87, 88, 107, 108, 172, 186
  interdisciplinary research, 8, 83, 86, 89, 193, 203, 239, 278, 326, 365

J

Jamison, Andrew, 35, 86, 172, 192, 300–2

K

Klitgaard, Kent, 9, 341–343, 351, 358, 362
knowledge integration, 11, 29, 87, 105, 110–19, 172, 176,

183, 209, 231, 237–43, 281, 297, 308
knowledge synthesis, 2, 8, 10, 23, 28, 46, 74, 83, 87, 90, 91, 107, 108, 158, 160, 176–7, 191–211, 237–41, 278, 280, 281, 317, 386
knowledge transfer, 11, 39, 90, 109, 235–281, 285

L

land grabbing, 374
land use, 12, 41, 88, 129, 148, 153, 221, 260, 294, 358, 370–2, 374, 375, 379
Latour, Bruno, 17, 36, 297, 318
Lele, Sharachchandra, 79, 220–4, 227
Leopold, Aaron, 196, 363, 376–8
limits to growth, 9, 77, 101, 116, 132, 194, 347, 386, 387
local knowledge
    local ecological knowledge, 47, 242, 298
Luhmann, Niklas, 16–18, 20, 22, 25, 45, 46, 49–51, 53, 55, 58, 184, 207

M

marine governance, 201, 313–14
Marx, Karl, 16–19, 25, 32, 41, 51, 60, 303
material and energy flow accounting (MEFA), 33, 175, 190
metabolic rift, 303

modes of production, 18, 22, 26, 27, 29, 30, 32, 38, 42, 61, 73, 105, 146, 165, 192, 194, 203, 261, 345, 346, 348, 351, 392, 397
Mol, Arthur, 16, 58, 61, 62, 115, 132, 190, 217, 302, 303, 386, 389
Moscovici, Serge, 24, 27, 31
multi-scale governance, 104, 236, 263, 266–277, 322–3, 326

N

natural resource management, 11, 96, 111, 137, 190, 235, 238–239, 241–5, 248, 250, 257, 261, 266, 269, 319, 389
natural resource use, 29, 43, 47, 57, 83, 84, 87, 106, 118, 125, 136, 149, 151, 164, 190–2, 200, 202–4, 231, 239, 252, 310, 342, 344, 352, 385
nature-society interaction (society-nature interaction), 4, 5, 12, 16–18, 21, 24–30, 32, 36, 37, 39, 40, 45, 54, 56, 69, 71, 73–92, 96, 98, 105, 107, 108, 166, 169, 191, 193, 202, 204, 206–9, 211, 218, 231, 232, 235, 242, 259, 271, 297, 327, 375–7, 391, 396
new ecological paradigm (NEP), 41, 131

O

O'Connor, James, 32, 59
Offe, Claus, 56, 257, 268, 269, 318
Ostrom, Elinor, 88, 164, 171, 174, 175, 192, 194, 223, 239–41, 246, 259, 270, 316
overuse of resources, 136, 329

P

philosophical anthropology, 24
physical economy, 33, 341
physical resources, 129, 349
planetary boundaries, 40, 43, 77, 84, 110, 132, 166, 190, 395
Polanyi, Karl, 49, 149, 151, 386
political ecology, 5, 28, 33, 69, 83, 84, 162, 326, 363
political economy, 26, 28, 32, 45, 60, 166, 200, 203, 341, 397
post-industrial society, 16, 21, 55–8, 61
postnormal science, 240, 243, 246, 278
post-political environmental consensus, 298, 299, 301
Promethean revolution, 138, 345, 350

R

Raskin, Paul, 126, 127, 158, 326
rebound effect, 348, 353
reflexive modernisation, 16, 20, 58–62, 172, 318
regulation, 8, 35, 44, 110, 117, 128, 146, 157, 162, 265, 266, 268, 275–81, 305, 317, 321, 325, 338, 345, 377, 386, 390
renewable energy sources, 11, 306, 340, 354, 356, 358, 359, 363, 366–368, 377, 379
reproduction, 8, 19, 22, 27, 29, 30, 32, 37, 38, 50–4, 58, 74, 75, 83–5, 91, 105, 109, 153, 192, 198, 200, 205, 210, 288, 309, 339, 390, 392, 396
resilience
  ecological resilience, 97–9, 113, 246, 308, 319
  social-ecological resilience, 46, 97–9, 104, 113, 118
  social resilience, 97, 104, 113, 114
resilience research, 97, 99, 100, 113, 165, 192, 194, 220, 257, 262, 308, 317, 319
resource flows
  global resource flows, 33, 37, 135, 146, 161, 164, 175, 236, 303
Rice, James, 33, 154, 156, 164, 166, 175, 197
risk society, 16, 18, 20, 29, 58–62, 318
Rist, Lucy, 255–257
Ritsert, Jürgen, 5, 16
rural development, 325
rural sociology, 200

S

scarcity, 117, 133, 152, 171, 173, 326, 349, 351–2, 359, 394

scenario, 100–3, 147, 154, 158, 167–9, 249, 295, 326, 362, 364, 388, 394, 395
Schägner, Jan Philipp, 225
Schnaiberg, Allan, 17, 32, 166, 197, 302, 397
social ecology, 2, 3, 10, 70, 78, 128, 158, 161, 165, 171, 183–232, 380, social–ecological systems (SES)
   social-ecological theory, 1, 2, 5, 10, 15, 28–31, 33–5, 37–46, 69–73, 83–6, 89–93, 103–10, 125, 138, 150, 152, 157, 165–70, 183, 193, 194, 200, 202, 207, 209, 210, 215, 217, 220, 228–232, 236, 245, 261, 267, 269, 271, 273, 277, 279, 285, 309, 322, 324, 325, 327, 346, 366, 387, 390, 391, 396, 397
   social-ecological transformation, 1–12, 15–62, 69–119, 125–77, 183–232, 235–81, 285–330, 337–81, 385–97
social movement theory, 292, 300, 301
societal metabolism, 5, 9, 32, 33, 37, 38, 44, 46, 71, 81, 84, 92, 100, 109, 118, 145, 151, 153–5, 161, 166, 192, 194, 202, 209, 217, 219, 304, 343, 346, 365, 376, 392
   industrial metabolism, 82, 339, 346
   urban metabolism, 44, 159, 175

societal relations to nature, 4, 22, 44, 48, 52, 85, 205
sociological theories, 15–30, 34, 37, 38, 40, 45, 49–62, 83, 386
sociology of flows, 190, 217
socio-metabolic regimes, 38, 44, 51, 58, 99, 109, 149, 151, 152, 166, 192–194, 203, 205, 344, 345, 350, 354, 395
space, 24, 25, 38, 43, 54, 60, 101, 170, 186, 210, 211, 286, 298, 311, 371, 376, 386, 391, 395, 397
steady state economy, 136
sustainability
   sustainable development, 10, 42, 80, 100, 103, 105, 115, 125–47, 154–65, 171, 176, 219, 223, 266, 270, 272, 273, 289, 294, 309–327, 362–3, 372, 373, 385, 394, 395
   sustainable resource management, 164, 236, 240, 243, 245, 253, 262, 279, 313, 317, 318
sustainability paradoxes, 319
sustainability science, 194, 195, 199, 220, 278
Swyngedouw, Eric, 44, 292, 298, 299, 302, 352
systems ecology, 75, 76, 216

T

techno-nature, 205
theory of communicative action, 16, 19, 21, 51–5, 386

theory of society
- theory of society and nature, 4, 5, 10, 17, 28, 31, 37, 40–4, 72, 205, 209, 338, 397

thermodynamics, 343, 351, 356

transdisciplinarity, 23, 242, 255, 391

transformation
- social-ecological transformation, 1–12, 15–62, 69–119, 125–77, 183–232, 235–81, 285–330, 337–81, 385–97

transformation of energy systems, 289, 337, 346–53, 362–7, 372, 379

transformative agency, 110, 160–2, 285, 308, 318, 319, 321, 322, 326, 365, 386, 395

transition
- transitions to sustainability, 8, 100, 101, 103, 108, 111, 128, 148, 149, 157, 160, 162, 169, 267, 303, 312, 315, 318, 321, 330, 344

U

unequal exchange (unequal ecological exchange), 33, 150, 151, 156, 164, 175, 190, 197, 306

Urbanisation, 1, 41, 57, 149, 161, 175, 302, 347

Urry, John, 21, 292, 304, 389, 394

V

vulnerability, 2, 10, 39, 46, 69–71, 84–6, 91–96, 99, 100, 103–8, 111–12, 116–19, 145, 147, 160–1, 170, 176, 198, 208, 224, 230, 236, 241, 274, 293, 315, 353

W

Wallerstein, Immanuel, 16, 26, 34, 153, 156, 166

Windpower, 337, 346, 356–8, 361, 366–73, 378, 380

Wolf, Eric, 26, 397

world ecology, 34, 81, 128, 150

world society, 18, 19, 49–51

world system theory, 16, 26, 34, 71, 150, 304, 397
- economic world system, 18, 24, 26, 50, 104, 149, 151, 163, 277, 311, 348, 350, 359, 363, 387, 390

Y

York, Richard, 197, 292, 302, 303, 306